Villages, Fields and Frontiers

Studies in European Rural Settlement in the Medieval and Early Modern Periods

Papers presented at the meeting of the Permanent European Conference for the Study of the Rural Landscape, held at Durham and Cambridge, England, 10—17 September 1981

edited by

B. K. Roberts and R. E. Glasscock

BAR International Series 185

1983

B.A.R.

122 Banbury Road, Oxford OX2 7BP, England

B.A.R.-S185, 1983 :'Villages, Fields and Frontiers'

ISBN 9780860542384 paperback
ISBN 9781407333656 e-book
DOI https://doi.org/10.30861/9780860542384
A catalogue record for this book is available from the British Library
This book is available at www.barpublishing.com

CONTENTS

(b) Economic

PERMANENT EUROPEAN CONFERENCE FOR THE STUDY OF THE RURAL LANDSCAPE

DURHAM-CAMBRIDGE MEETING, 10-17 SEPTEMBER 1981

INTRODUCTION

A sustained interest in the study of the rural landscapes of Europe and their historical antecedents has been evident in a series of meetings organised by a study group associated with the International Geographical Union which in 1976 adopted the name 'Permanent European Conference for the Study of Rural Landscape' (Conférence Européenne Permanente pour l'Etude du Paysage Rural). A central secretariat is based in Paris by courtesy of Professor X. de Planhol. The Durham-Cambridge symposium was in fact the eleventh, with earlier meetings at Nancy in 1957, Vadstena in 1960, Bangor-Leicester in 1964, Würzburg in 1966, Liège in 1969, Belfast in 1971, Perugia in 1973, Warsaw in 1975, Rennes in 1977 and Roskilde in 1979. Each meeting was built around paper sessions and field excursions, and with one exception the papers have subsequently been collected together as publications, which, while following a limited number of key themes, demonstrate the rich diversity of work in progress and the need for comparative studies at the international scale. The opportunity for discussion, for comparison, for agreement and disagreement, and not least for encouragement and stimulation, represents a vital ingredient of these meetings, and in 1981 the sessions were attended by some sixty scholars representing eleven countries.

The Durham-Cambridge meeting was consciously built around a series of field seminars whose content was designed to illustrate and complement three themes on which papers were initially offered, namely, regulated villages, frontier settlements and rural industry. Recognizing the substantial costs now involved in assembling such a group of scholars as much time as possible was invested in field study, and participants were invited to make only relatively short oral contributions. Our thanks are due to all of them for the skill and good humour with which they complied with these constraints.

Two field seminars were organized by B.K. Roberts (University of Durham) on The Landscapes of Cockfield Village and Fell and The Landscapes of Edlingham, illustrating work in progress on sites in County Durham and Northumberland, and the latter was extended by a visit to the town of Alnwick (the focus of work by M.R.G. Conzen) and the frontier landscapes of the Roman Wall. M.J. Sill (Newcastle Polytechnic) organised another field-seminar on The Mining Villages of East Durham, while the journey from Durham to Cambridge allowed a visit to the important deserted village site of Wharram Percy, Yorkshire and a demonstration of its features in some detail by J.G.Hurst. R.E.Glasscock (University of Cambridge) organized the second part of the meeting in Cambridge where C.C. Taylor (Royal Commission on Historical Monuments) concluded the field programme with a study of *The Settlement of the Cambridge Region.* As all those who participated will have excursion notes it has been decided not to reproduce an account here: some of the Cockfield material is included in Chapter 2 of this volume.

- - - -

For an account of the meeting published in French see:- F. Dussart, 'Les réunions de la "Conférence européenne permanente pour l'étude du Paysage rural" tenues à Durham et à Cambridge en Septembre 1981' in *Bulletin de la Société géographique de Liège*, No. 18, 18e année, Sept. 1982, 97-100.

Editing a volume of papers in different languages presents some difficulty. Four papers are published in French, the remainder in English. We are especially grateful to continental colleagues who prepared versions of their papers in English. This was not always easy for them, nor has it been for us, the editors. As one contributor wrote ... "I have tried to offer my paper in English: I hope the result is not indecent". We have tried to ensure that it isn't! At the same time, and as long as the general meaning is clear, we have retained many words and phrases of English which are unmistakably those of contributors for whom English is a second language. They have their own charm.

- - - -

ACKNOWLEDGEMENTS

The editors wish to record their thanks to the many institutions who, through grants, supported the attendance of participants in this meeting. Also to the many individuals who assisted in the running of the meeting and in the preparation of this volume.

At Durham we are indebted to the Vice-Chancellor, to Professor W.B. Fisher, and to members of the technical and secretarial staff of the Department of Geography. The help given in various ways by M.J. Sill, Tim Unwin, John Clifford, Nikki Gregson, Michael Alexander and the Cockfield Field Reeves, is gratefully acknowledged as is, at Cambridge, the help of Michael Young, Ian Gulley, Bob Coe, Dennis Blackburn and Jean Lucas, all of the Geography department, and Christopher Taylor of the Royal Commission on Historical Monuments. Our thanks are due also to Shell and to Durham City Council for the provision of maps.

We are indebted to the British Council for its sponsorship of a visitor from Poland and to St. John's College, Cambridge, for its support of another visitor from eastern Europe and for a grant towards publication.

LIST OF PARTICIPANTS

BECKER, Prof. H.	University of Bamberg, G.F.R.
BRINK, Dr. S.	University of Uppsala, Sweden.
BUCHANAN, Dr & Mrs R.H.	Queen's University Belfast, N.Ireland.UK.
DUSSART, Prof. F. & Mme	University of Liège, Belgium.
EIGLER, Dr F.	Werner-von-Siemens Gymnasium, Weissenberg, G.F.R.
FLATRÈS, Prof. P.	University of Paris, Sorbonne, France.
FOUTREIN, Mlle C.	University of Lille, France.
FRANDSEN, Dr K-E.	University of Copenhagen, Denmark.
GIBSON, Mr A.J.	University of Lancaster, U.K.
GLASSCOCK, Dr R.E.	University of Cambridge, U.K.
HALL, Mr D.	Fenland Field Officer, Cambridge, U.K.
HANSEN, Prof. V.	University of Copenhagen, Denmark.
HANSEN, Mr N.	University of Copenhagen, Denmark.
HARTEN, Mr J.D.H.	University of Utrecht, Netherlands.
HILDEBRANDT, Prof. H.	University of Main, G.F.R.
HURST, Mr J.G.	Inspectorate of Ancient Monuments, London, U.K.
JONES, Prof. G.R.J.	University of Leeds, U.K.
KRENZLIN, Prof. A.	University of Frankfurt, G.F.R.
KRINGS, Prof. W.	University of Bamberg, G.F.R.
LEENDERS, Dr K.A.H.W.	Regional Authority, Rotterdam, Netherlands.
LEISTER, Prof. I.	University of Marburg, G.F.R.
LEWAN, Dr N.	University of Lund, Sweden.
LINKE, Dr M.	Martin Luther University, Halle-Wissenburg, G.D.R.
MATZAT, Prof. W.	University of Bonn, G.F.R.
MORGENSEN, Dr A.K.	University of Aarhus, Denmark.
NITZ, Prof. H.J.	University of Gottingen, G.F.R.
PELTRE Prof. and Mme J.	University of Nancy II, France.
PINARD, Prof. J.M.L.	University of Limoges, France.
RENES, Mr J.	Soil Survey Institute, Wageningen, Netherlands.
ROBERTS, Dr B.K.	University of Durham, U.K.
SCHUYF, Dr J.	University of Utrecht, Netherlands.
SZULC, Dr H.	Polish Academy of Sciences, Warsaw, Poland.
SERENO, Prof. Dr. P.	University of Torino, Italy.
SHEPPARD, Dr J.	University of London, U.K.
SIMMS, Dr A.	University College Dublin, Republic of Ireland.
SPORRONG, Prof. U.	University of Stockholm, Sweden.
STANLEY, Mr M.J.	University of Edinburgh, U.K.
TIGGESBAUMKER, Dr G.	University of Bamberg, G.F.R.
ULSIG, Prof. E.	University of Aarhus, Denmark.
UNWIN, Dr P.T.H.	University of Durham, U.K.
VAN DER VAART, Dr J.	Fryske Academy, Netherlands.
VERHOEVE, Mr A.	University of Ghent, Belgium.
VERVLOET, Dr J.A.J.	Soil Survey Institute, Netherlands.
WINDELHED, Mr B.	University of Stockholm, Sweden.

Durham only

ATKINS, Dr P.	University of Durham, U.K.
CLIFFORD, Mr J.	University of Durham, U.K.
GREGSON, Miss N.	University of Durham, U.K.
HODGSON, Mr R.I.	University of Manchester, U.K.
MULDERS, Miss P.J.W.M.	University of Utrecht, Netherlands.
SILL, Mr M.J.	Polytechnic of Newcastle-upon-Tyne, U.K.
VAN DER WIELEN, Mr K.E.	University of Utrecht, Netherlands.

Cambridge only

BAKER, Dr A.R.H.	University of Cambridge, U.K.
BUTLIN, Prof. R.A.	University of Loughborough, U.K.
TAYLOR, Mr C.C.	Royal Commission on Historical Monuments, U.K.

PUBLICATIONS OF THE PROCEEDINGS OF PREVIOUS P.E.C.S.R.L. MEETINGS

Géographie et Histoire rurales. Actes du colloque international organise par la Faculté des Lettres de l'Université de Nancy (Nancy, 2-7 Septembre 1957), in Annales de l'Est, Memoire no. 21, Nancy, 1959, 252 p.

Morphogenesis of the agrarian cultural landscape. Papers of the Vadstena symposium at the XIXth International Geographical Congress, August 14-20, 1960, in Geografiska Annaler, vol. XLIII, 1961, pp. 1-328.

Beiträge zur Genesis der Siedlungs-und Agrarlandschaft in Europa. Rundgespräche vom 4. Juli bis 6. Juli 1966 in Wurzburg veranstaltet von der Deutschen Forschungsgemeinschaft unter Leitung von Helmut Jäger, Anneliese Krenzlin und Harald Uhlig, in Erdkundliches Wissen, Heft 18. Geographische Zeitschrift, Beihefte, Wiesbaden, 1968, 211 p.

L'habitat et les paysages ruraux d'Europe. Comptes rendus du symposium tenu à Liège du 29 Juin au 5 Juillet 1969. Volume publié par les soins de F. Dussart dans Les Congrès et Colloques de l'Université de Liège, 58. Université de Liège, 1971, 472 p.

Fields, farms and settlement in Europe. Papers presented at a symposium, Belfast, July 12-15, 1971. Ed. by R.H. Buchanan, R.A. Butlin and D. McCourt. Published by the Ulster Folk and Transport Museum, Cultra Manor, Holywood, Co. Down, 1976, 161 p.

I paesaggi rurali Europei. Atti del convegno internazionale indetto a Perugia dal 7 al 12 Maggio 1973 dalla Conférence européenne permanente pour l'étude du paysage rural. Deputazione di storia patria per l'Umbria. Appendici al Bolletino, No.12. Perugia 1975, 613 p.

L'évolution de l'habitat et des paysages ruraux d'Europe. Actes de la Conférence tenue à Varsovie en Septembre 1975. Volume rédigé par Mme Maria Kietczewska-Zaleska avec la collaboration de M. Jerzy Grzeszezak et Mme Teresa Lijewska. Polish Academy of Sciences. Institute of Geography and Spatial Organization, in Geographia polonica, 38, Warszawa 1978, 304 p.

Paysages ruraux européens. Travaux de la Conférence européenne permanente pour l'étude du paysage rural. Rennes-Quimper 26-30 Septembre 1977. Travaux publiés par les soins de P. Flatrès, Université de Haute Bretagne, Rennes, 1979, 629 p.

Collected papers presented at the Permanent European Conference for the Study of the Rural Landscape held at

Roskilde, Denmark, 3-9 June 1979. Organizing committee Karl-Erik Frandsen, Sven Gissel, Viggo Hansen Ed. Copenhagen 1981, 226 p.

N.B. - A notre connaissance, les réunions de Bangor-Birmingham-Leicester n'ont fait l'objet que d'un bref compte rendu: G.R.J. Jones and H. Thorpe. - Symposium S 4a: *The rural landscape and its evolution.* July 11th to 20th, 1964, dans *20th International Geographical Congress. Congress Proceedings*, ed. by J.W. Watson, London, 1967, pp. 221-229. - Pour toutes les réunions, de nombreux comptes rendus ont paru dans des revues géographiques, notamment dans la *Revue géographique de l'Est*, Nancy, par X. de Planhol (voir tomes I-1961, VI-1966, IX-1969 et XV-1975) et dans le *Bulletin de la Société géographique de Liège*, par Jacqueline Claude et F. Dussart (voir nos. 3-1966, 5-1969, 7-1971, 9-1973, 12-1976 et 14-1978).

LIST OF FIGURES (titles abbreviated)

LIST OF PLATES

I : VILLAGES

Introduction

'Shape, size and character vary greatly. No village is quite like any other. And this, of course, is the glory of it, that every village is an individual place.' Thomas SHARPE, The Anatomy of the Village (1946).

The first paper of this volume, by HURST, reviews the development of the plan of a single European village, the deserted village of Wharram Percy, Yorkshire, England, and draws upon archaeological, documentary and topographical evidence accumulated in over a quarter of a century of work. The essential simplicity of this analytical description belies both the complexity it reveals, the far-reaching repercussions of the methods it demonstrates, and the importance of the conclusions it reaches: Wharram offers both a vista of what can be achieved and a warning of the pit-falls involved in all morphological studies - the analysis of settlement forms revealed by late map sources and the landscape itself. ROBERTS also examines a single village, Cockfield, Co. Durham, used as a focus for a field seminar during the 1981 Durham-Cambridge meeting, and seeks to provide a bridge between the detail of HURST'S perspective, his own wider views of the north English village, and the broader aspects of settlement study seen in European work, where Uhlig has demonstrated the need to work towards a measure of terminology standardisation. In their review paper HARTEN and SCHUYF catalogue and document the rich harvest of traditional plans found within the Netherlands, a country occupying a pivotal position in any appraisal of European rural settlements, influencing and being influenced by both maritime and continental forces. The importance of achieving an understanding of the structure of rural settlements in this densely populated region is demonstrated by VAN DER VAART who reminds us that each village is following a trajectory through time, from the past, through the pressures of the present, to a future whose shape may be uncertain : an awareness of origins is an essential requisite for a sensitive appraisal of the way inherited forms can be adopted to contemporary and future needs, and he illustrated this with reference to the peripheral greens, or brinken, found in many Dutch villages. TIGGESBÄUMKER'S contribution, on the foundation of regular planned settlements by German colonists in nineteenth century Transcaucasia, is a welcome reminder of the complex decisions underlying such ventures, and although the temporal context is late and the geographical location marginally European, this paper usefully counterbalances EIGLER'S discussion of the tenuous evidence for planned settlements established a thousand years earlier, in the forelands of the Franconian Jura, and brings together a classically formulated series of arguments for suggesting that the forms visible today originate at that remote time.

The limitations of medieval evidence for examining colonising movements are all too apparent, for only documents can truly indicate motives, hopes, driving forces and the ingredients underlying failure or success. HURST'S arguments illustrate the problems faced by all scholars using morphogenetic analysis : simplistic retrogressive projections are all too easy - and Hurst uses this derogatory description of his own early model of Wharram - but more evidence and more experience will add depth to interpretations. In other contexts ROBERTS has argued for the need to identify four separate chronological levels in settlement studies; an absolute chronology, probably unattainable in practice; a chronology of hierarchical importance, involving status within a settlement hierarchy, the character of their place-names and their appearance in the documentary record; a chronology of the renders, services and dues provided by each place, and finally, a chronology of the features visible in a given landscape, today or on earlier maps. Sometimes the final three chronologies can be in accord, but often they are in discord; in practice it is usually difficult to decide which, and the arguments linking a given morphology with a documented phase of settlement often appear very tenuous when subjected to vigorous testing.

The final two studies of this section look towards very specific aspects of village morphology : PELTRE examines the distinctive character of the structures associated with the building lines facing onto the street greens of Lorraine villages, where private and public space meet in a complex transition zone of defined rights, while FRANDSEN provides a vital link between this section on villages and that, to follow, on fields. The parson's farm, the glebe, has tended to represent a more stable unit of landownership than the ordinary farms of the peasantry and Frandsen uses this fact as a key for an examination of village-field relationships, recognising that, just as HURST is able to identify archaeological layers, it is possible, given the appropriate documentation, to identify in a single village the presence of several organisational levels, each constructed upon antecedents, which may be completely or only partially destroyed in the processes of adaptive change which take place. The Great Land Register of 1682 provides the foundation for a study of village glebe lands on the Danish island of Falster.

CHAPTER 1

THE TOPOGRAPHY OF WHARRAM PERCY VILLAGE

J.G.Hurst
(Department of the Environment, London)

The deserted medieval village of Wharram Percy, in the former East Riding of Yorkshire, is situated near the north-western scarp of the chalk Wolds half way between York and Scarborough. It was finally deserted about 1500, at the time when many villages in England were abandoned because of a change from arable to sheep farming, so that a single shepherd replaced the peasant families who had cultivated the land. The general historical background to deserted medieval villages is discussed in Beresford (1963) and Beresford (1971). When the earthworks, grass-grown and almost undisturbed for over four centuries, (Fig.1:1.) were first studied it was thought that they simply represented the buildings, roads, or boundaries of the late medieval village that were in use at the time of its depopulation; excavation, however, has shown that they do, in fact, illustrate the complex development of the site over the last 4,000 years.

Work at the site was first undertaken by Professor M.W. Beresford in 1950, and in 1952 it was chosen as the main research project of the Deserted Medieval Village Research Group. Full interim reports have been published each year in MVRG 1953-1981. Over the last 30 years many aspects of the village have been investigated : two peasant house sites (Areas 10 & 6), the 12th century manor house in toft 10 (ANDREWS & MILNE 1979); the parish church (HURST 1976), parsonages, watermills and fishpond, ironworking, the road system, boundaries and field systems (for latest interim reports see Beresford & Hurst 1976 and 1979), and, during recent years, the complete parishes of Wharram Percy and Wharram le Street to put the site in its context both in place and time (HURST 1980). The ecology and palaeoecology of the village have also been studied, together with the economic evidence afforded by the numerous small finds. Due to the generosity of the late Lord Middleton the site is now in the care of the Department of the Environment. It is intended to display all the above aspects of the medieval village to the public, as this is the one place in England where this can be achieved.

No detailed description or interpretation of the earthworks has previously been attempted. As recent survey, excavation and aerial photography have brought to light many new facts this seems an appropriate time to attempt a general summary account. The features surviving from the Prehistoric and

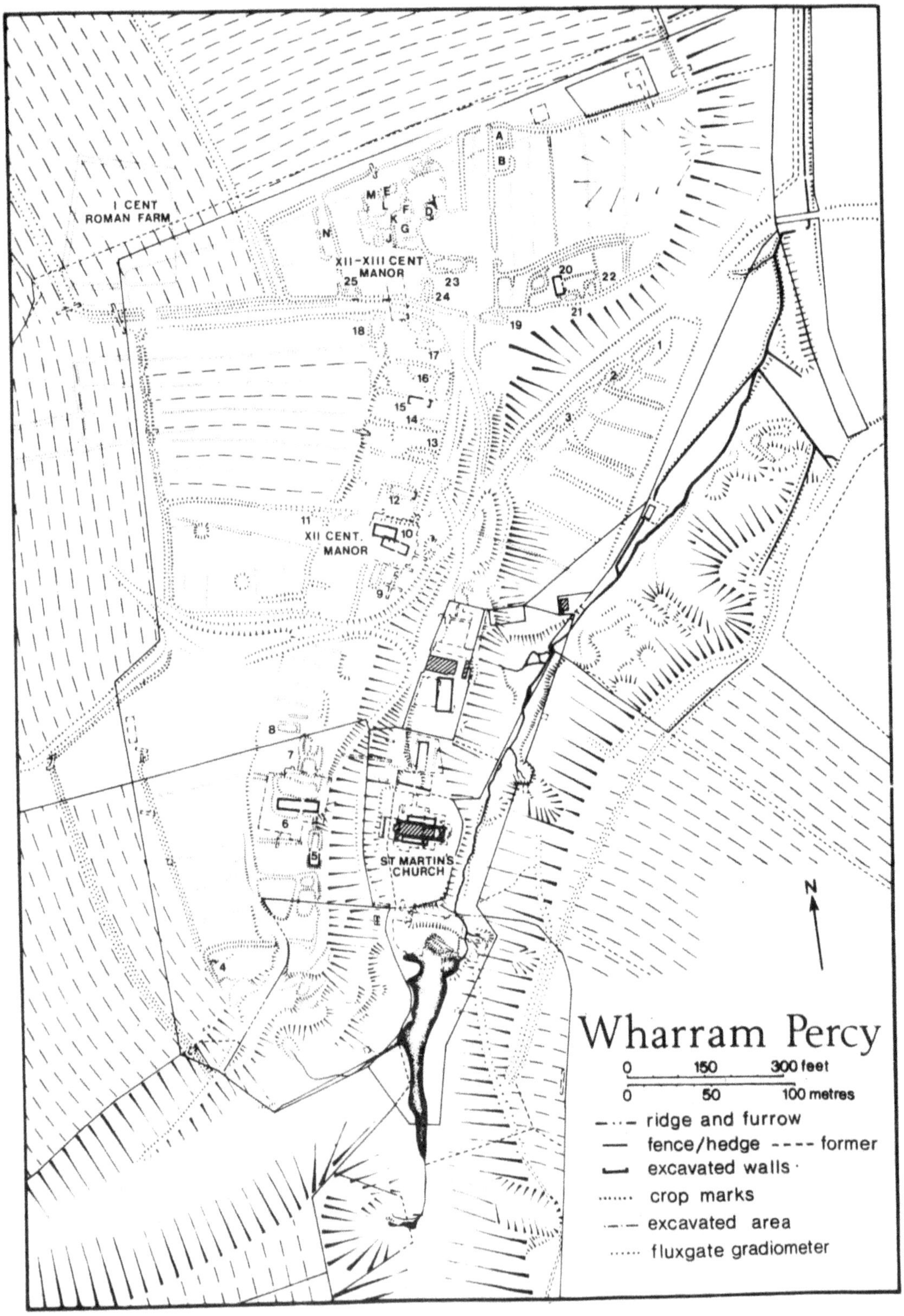

Fig. 1:1 Wharram Percy, Yorkshire: general plan of the deserted medieval village area.

Romano-British periods will first be described (Fig. 1:2), then the network of through roads and access tracks (Fig. 1:3), followed by an account of the east, west and north rows of enclosures which formed the main medieval village plan. Finally there will be an attempt at interpretation and dating, though this can only be a very preliminary account at this stage (Figs. 1:4 & 1:5), since the details still have to be worked out together with the full implications of recent discoveries.

THE PHYSICAL SETTING (Fig. 1:1)

The plateau of the chalk Wolds is dissected by numerous steep-sided winding valleys, mostly dry. The main earthworks of the village are situated on the plateau at about 160m above sea level. In the valley, where lies the only remaining medieval building (the former parish church of St Martin), are the springs which form the headwaters of the Settrington Beck; above this point, Deep Dale is dry. On the floor of the valley are heavy Upper Jurassic clays, but the church and modern cottages stand on a well-drained terrace about 5m above the valley floor and 10-14m below the lip of the plateau.

The generally smooth valley sides are broken by occasional embayments, usually associated with springs, while the gentle plateau slopes are traversed by several eastward-trending and extremely shallow 'valleys', which merely notch the edges of the plateau. One of these 'valleys' is indicated on the plan north-east of building 22; another is north of building 7 and debouches immediately above the cottages, and a third is followed by the fence south of building 5; they are thought to be of periglacial origin.

PREHISTORIC FEATURES (Fig. 1:2)

There are four major linear features on the western plateau which may be interpreted as prehistoric, dating back to the Iron Age and possibly the Bronze Age. The most substantial and striking of these is a lynchet which runs for nearly 400 m from the northern boundary of toft 4 near Drue Dale to the northern boundary of toft 17 where it is overlain by the medieval building 18. The lynchet ends at the main sunken road R1 (Fig. 1:3) which runs from the plateau down into the valley, passing between the north and west rows. R1 was certainly used in the Roman period and could have originated as a prehistoric droveway for bringing cattle up on to the plateau from the valley. Another linear feature runs west-east along the northern boundary of the north row. At its eastern end it forms a south-facing negative lynchet, enhanced by the periglacial 'valley' whose north edge it follows; further west it is still traceable through the complex of the northern manor, which was extended northwards over it to a new boundary bank in medieval times. Running north from this, and possibly extending southwards to the west-east road,

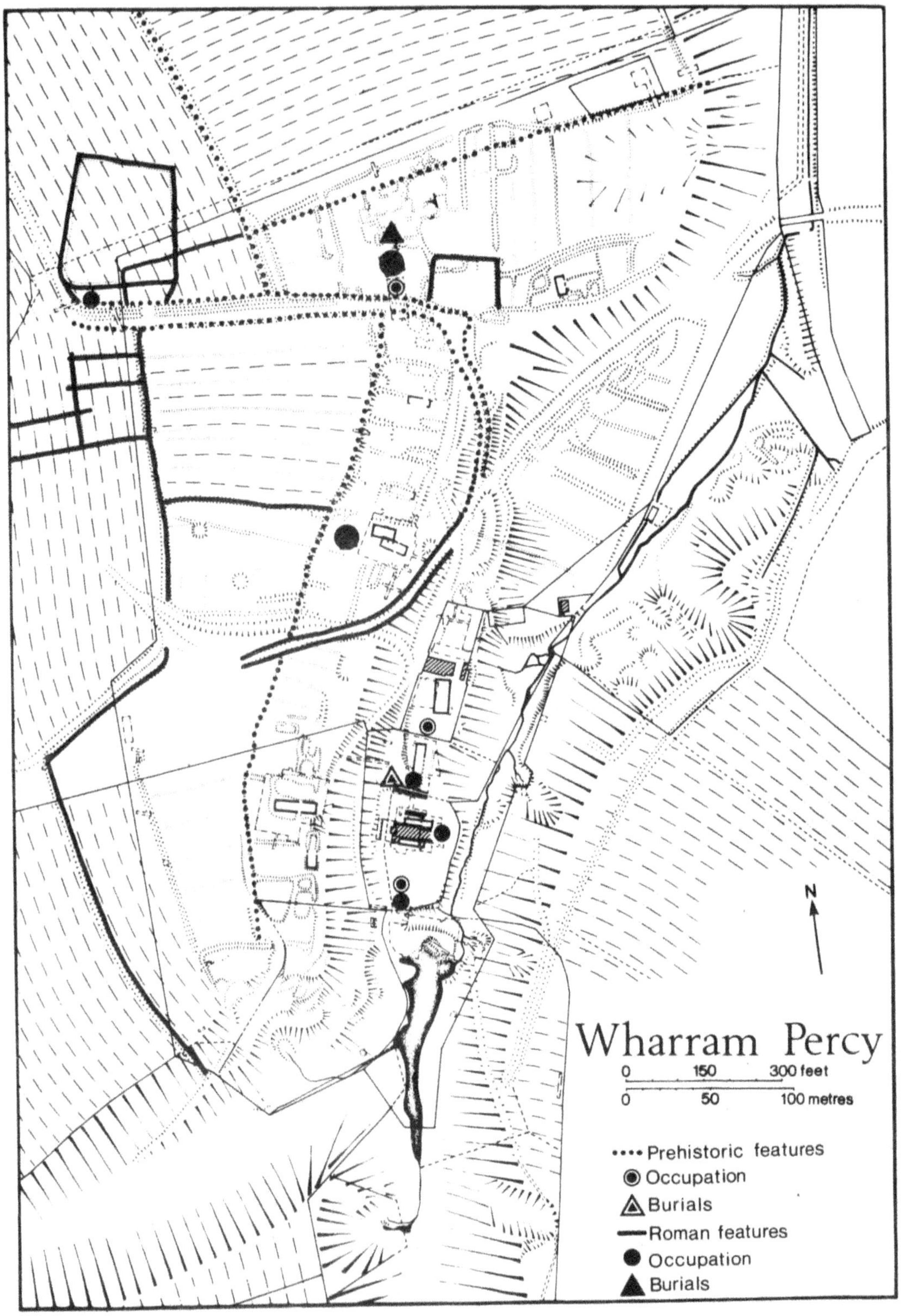

Fig. 1:2 Wharram Percy: prehistoric and Roman features which form the basic framework for the laying out of the medieval village.

is a major medieval headland which was so massive that it still shows as a major cropmark despite thirty years' ploughing. It is therefore very likely that this originated in another prehistoric linear earthwork.

These linear features, apparently laid out in the prehistoric period, thus form the framework within which the subdivisions of the Romano-British landscape were placed.

ROMANO-BRITISH FEATURES (Fig. 1:2)

The western boundary of the crofts of the west row appears to be laid out entirely along a Romano-British boundary, including the unusual south-westerly change of direction to the west of croft 8. This bend may be associated with track Tr1 (Fig. 1:3) up from the terrace between crofts 8 & 9, following the second periglacial 'valley', and Tr1 thus may also be Romano-British in origin. Tr5 cuts through the prehistoric lynchet, so it is not likely to be primary. The boundary between crofts 10 and 12 has also been shown to be Romano-British, adding support to the possibility that crofts 9 and 10 formed a Romano-British farm area which was later utilised for a medieval manor house.

The sequence of the Romano-British features to the north-west of the village is uncertain as there are possibly two periods. First there is the cropmark of a one-acre (0.4 ha) trapezoidal earthwork (1st century Roman farm on plan) which either was extended southwards to R1 or was made smaller; this farm may be associated with the series of small irregular enclosures which extend south of it to the west of the main west boundary of crofts 13-17. Secondly there is the rectangular area north of R1 and west of the medieval headland; this seems to be pre-medieval as it is overlain by ridge and furrow and squares off this corner of the village by prolonging the lines of the north lynchet and the western boundary of the crofts. It is only in this corner of the site that there seems to be a palimpsest of different periods on varying alignments. All other features fit into a basic rectangular framework.

The area 24-23-19 is deeply terraced into the hillside in a much more massive form than would be needed for the medieval manorial outbuildings. This also may therefore be the site of a major Roman building.

PREHISTORIC & ROMANO-BRITISH OCCUPATION (Fig. 1:2)

Evidence for prehistoric occupation is slight but is present in four places. Recent excavations in the north manor have produced a number of prehistoric sherds, flints and a stone axe but as yet no structures, so there may be occupation nearby. In the south of the cottage garden the digging of a modern cesspit produced evidence for an early Iron Age round

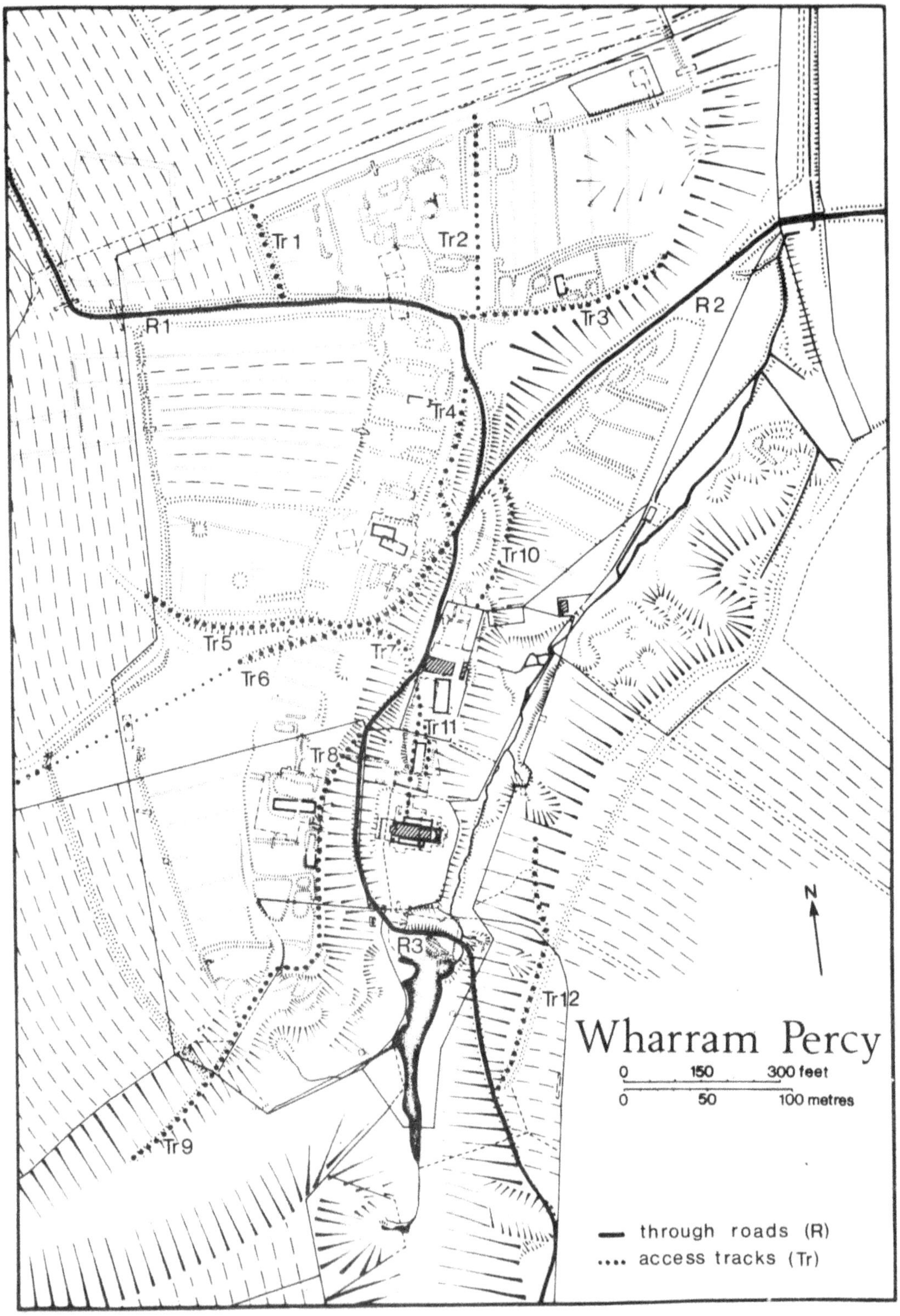

Fig. 1:3 Wharram Percy: the roads and access tracks.

hut while there were pits and ditches, in a filled glacial hollow south-west of the church, datable to the late Iron Age. To the north-west of the churchyard an Iron Age crouched burial was found with a Carbon 14 date of 80bc ± 70.

In the Romano-British period there was settlement in the north-west trapezoidal earthwork, in the north manor area and in three places on the terrace: north of, under, and south of the church. It is not clear if these latter form part of one farm or more. In toft 10 there is the likelihood of a settlement but its exact nature and position has yet to be determined. One Romano-British burial was found in the north manor. All the structures seem to be simple timber buildings, suggesting a series of farmsteads, although in the north manor current excavations are producing evidence for a major stone structure which may possibly be a villa associated with the terraced area 24-23-19. This evidence, and the three other farms, all quite close together (about 150m apart) is in contrast to the general settlement pattern of the area, which normally comprises Romano-British farms about every kilometre.

THE MEDIEVAL ROAD SYSTEM comprises three through roads and a series of access tracks (Fig. 1:3)

Through roads. A west-east road R1 which comes in from Malton, a market borough from the early 12th century, passes to the south of the north row and north of the west row, then makes a sharp turn to the south to descend the valley side to join R2, which runs north-east along the valley to the west of the northern end of the east row. At the northern end of the village it turns east across the beck then up on to the plateau, via a narrow but deep tributary valley, and so south-eastwards to Towthorpe.

A north-south road R3 starting from the Y junction formed by R1 and R2 continues south along the terrace to the west of the east row, making a detour halfway up the slope west of the parish church to avoid the late-medieval parsonage and churchyard; in earlier times it may have continued straight on past the west end of the church. South of the church it descends the hill again, turns east past the south of the churchyard, over the fish/millpond dam and then turns south-east to climb the eastern valley side and continue southwards along the lip of the plateau, with open fields on the east and Deep Dale on the west, towards Raisthorpe, Burdale and Thixendale.

There are numerous access tracks (Tr) mainly linking the village with the open fields. In the north row Tr1 runs from R1 northwards to the open fields, passing between the west boundary of the north manor and the north-south ridge and furrow to join the major north-south headland where the ridge and furrow changes direction. Tr2 runs northwards between the main manorial complex and the animal area to the east through what was earlier possibly a toft and croft. Tr3 runs along the southern boundary of the north manor extending the line of the west-east road but ending at the edge of the plateau above the steep valley side. Its exact link with the main system is uncertain as it seems to be closed by a bank at its west end, but this may be later, closing the track when it was no longer required to give access to the tofts.

In the west row Tr4 runs along the crest of the slope to the east of tofts 17-13, giving access to them, then descends the slope to join the main north-south road. From this junction the more deeply sunken Tr5 climbs the slope to run westwards between tofts 9 & 8 to give access to the western open fields. Tr6 is a bifurcation running south-westwards but its date is uncertain as it seems to overlie earlier tofts but at the same time to be on an easier line. Tr7 is a short hollow-way from this junction down to the terrace west of the cottages. Leaving the parsonage bypass halfway up the slope Tr8 is another terrace-way giving access to tofts 6 & 5. Tr9, running from the east of toft 4, then descends the northern side of Drue Dale to run westwards along the bottom of the dale and gives access from the village to the south-western part of the township, avoiding the wet valley bottom at the junction of Drue Dale and Deep Dale.

To the east of R3, Tr10 runs southwards, to the centre of the 18th century courtyard farm buildings. This track is likely to be post-medieval in date and overlies earlier tofts to the north of the present cottage garden. To the south of the present cottages Tr11 runs southwards along the terrace, giving access to the main building of the 18th century farm and to pass east of the post-medieval vicarage. In medieval times there was a similar track which passed the medieval vicarage and then became a footpath giving access to the churchyard through a gate in the north wall of the churchyard boundary. To the east of the valley Tr12 is a terraced track which runs northwards from near the crest of the slope: starting from R3, it runs along the boundary bank with the open field, then descends the east side of the valley just to the south of a spring. This presumably gave access to Deep Dale for animals grazing in the valley bottom east of the church, bypassing the mill and the later fishpond.

THE EARTHWORKS OF THE MEDIEVAL VILLAGE (Fig. 1:1)

The earthworks of the buildings, tofts and crofts form a regular two-row street green village with a head-row at the northern end. A row has been defined as a morphologically distinct unit, in which the homesteads, together with their associated outbuildings and curtilages, tofts and crofts, are arranged in a line, which may be regular or irregular in character (ROBERTS 1970, 235-6). (1) The east row on the terrace forms an arc from north to south following the curve of the valley; (2) the west row runs from north to south along the east edge of the plateau; (3) the head-row runs from west to east, closing the northern limits of the west and east rows, and with them delimiting a wedge-shaped green. Thus, the medieval plan is partly determined by the topography and partly by the features of earlier occupation. The east row, on the terrace, had to be long, narrow and curved, reflecting the terrain. The west row had more scope, lying on the open ground of the plateau, but the medieval western boundary seems to have been determined by a Romano-British line. The change of direction between crofts 8 and 9 seems to follow the edge of the second periglacial 'valley' or hollow. The head-row had as its northern limit, with only a small extension, a prehistoric linear earthwork.

The East Row contains at its northern end three clear house sites along the road (1-3) and several partially levelled earthworks which are likely to cover earlier buildings. There are at least five crofts, with the faint traces of a sixth. Recent air photographs have shown other parallel banks to the south. This would leave room for three or more tofts before the present cottage garden. To the west of Tr10 there are earthworks of houses which may have extended along here in medieval times. It is likely that these house sites were abandoned in the 14th and 15th centuries, leaving only the northern three. Sample excavations in the area to the west of the pumphouse, by the beck north-east of the cottages, have shown thick garden soil, confirming their interpretation as crofts, but with no finds after 1300. It therefore seems likely that at least the southernmost crofts were abandoned quite early and may have been included in the green, but because of later disturbances it is not yet clear what was underneath the present cottages and their gardens in the medieval period. There could have been more tofts or the area might have been glebe land, though it was not glebe from the 18th century onwards.

If the boundaries enclosing the parsonage, church, mill and pond are regarded as equivalent to further, specialist tofts, extending the sequence southwards, this creates a regular east row parallel with the west row for most of its length, but curving slightly eastwards to the north and thus forming a wedge-shaped expansion of the green at the north end of the village. This green, it may be noted, consisted almost wholly of the steep uncultivated valley side between the plateau and the terrace, so it would be the ideal location for communal grazing. Greens are by no means always on flat ground; for example several in County Durham such as Middlestone (ROBERTS 1972, 40, Fig. 21 & pl. VIIB) are in a similar situation.

The West Row may be divided into three parts. Tofts & crofts 9 & 10 in the centre seem to correspond with a Romano-British farm which was later the 12th-century manorial enclosure of the Chamberlains. After the abandonment of this

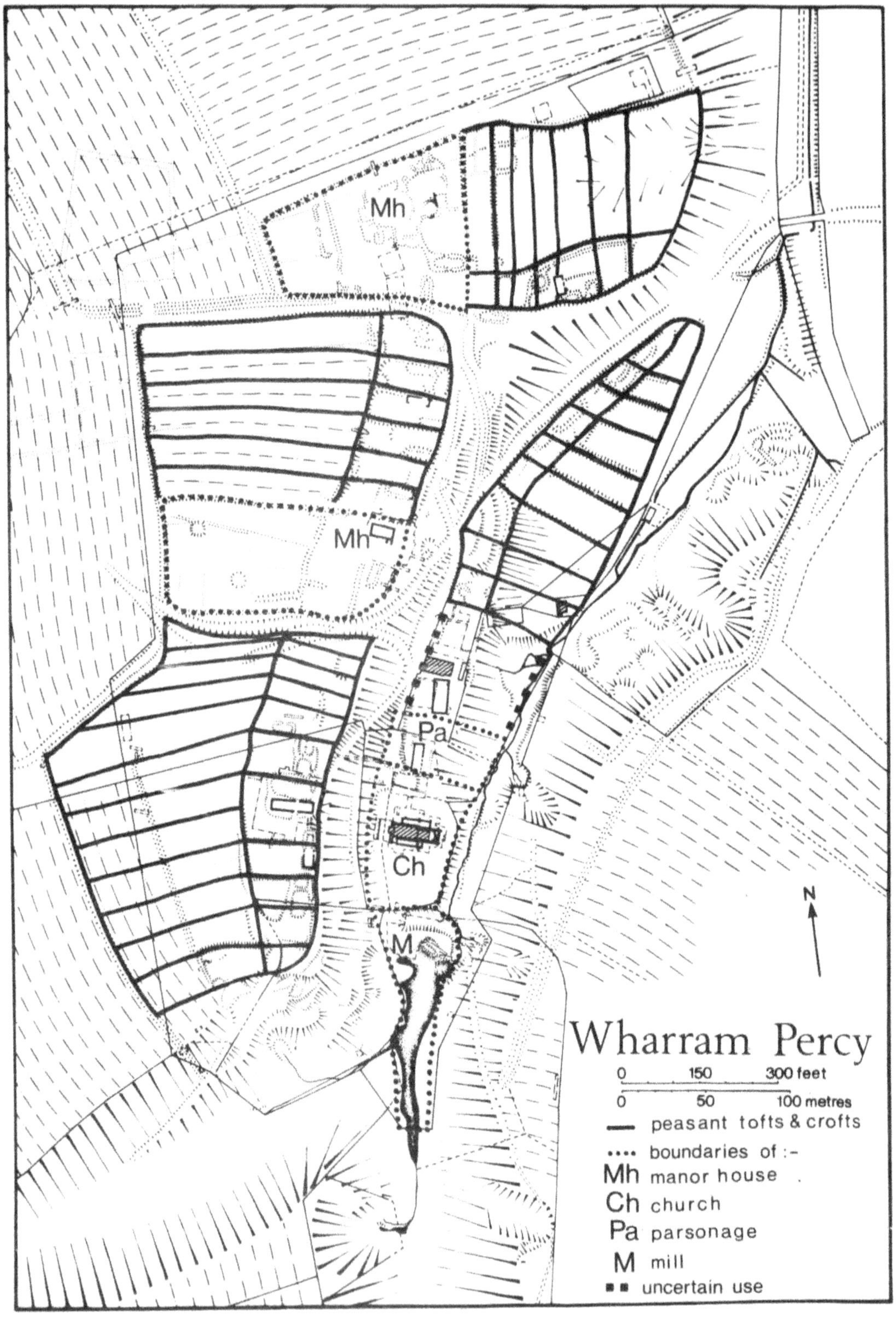

Fig. 1:4 Wharram Percy: schematic plan of a possible early medieval layout of three rows, comprising some 30 peasant tofts, two manor houses, church, parsonage and mill, with a central green.

manor in 1254, when the Percies acquired the whole village, the site was derelict and used for quarrying until the early 14th century, when it was laid out as peasant tofts and crofts forming a block with 12 to 17 (ANDREWS & MILNE 1979, 26-8).

The northern block of tofts (12-17) has a very regular lay out, but the houses placed with their gable end facing the street are likely to date only from the second half of the 15th century, as was shown by the excavations in Area 10 and by the many traces of the earthworks of earlier houses on varying alignments in tofts 12-17. The tofts were divided from their long crofts to the west by the prehistoric lynchet, which was variously cut back or revetted by the householders along its length. The six crofts have a regular pattern each being 60ft wide (18m) and covering the remains of two selions, each the average 30ft (9m) width common to most of the ridge and furrow at Wharram. But the tofts seem to have changed their boundaries e.g. toft 12. As the western boundary has been shown to be Romano-British in origin, and the selions lie west-east rather than north-south as in the main field system west of the village, the dates when this area was ploughed and when it was formed into crofts are uncertain. It was originally thought that houses were not built here, and the ridge and furrow divided into crofts, until the replanning of the village after 1254, but the apparently regular village plan, which is at latest 12th century, suggests that the change may have taken place earlier in late Saxon or early Norman times.

The southern block of the west row is separated from the 9 & 10 manorial enclosure by Tr5. There is a regular layout of tofts east of the prehistoric lynchet, which has been much reduced in this area, while to the west there seems to have been originally a series of long crofts stretching westward as far as the Roman boundary. At some time the western ends of the crofts appear to have been shortened and this western area incorporated into the regular north-south ridge and furrow. Because of the orientation of the western boundary the croft boundaries do not continue in straight lines westwards from the toft boundaries but are angled to run in a more southerly direction. In the early medieval period the tofts may have formed a solid block but it is not clear if the northern crofts between 8 and 9 were ever constructed, because of the problems caused by the southern bifurcation of the sunken road coming up from the valley. Tr6 cuts across what would, in a regular layout, have been the northern crofts, and seems to have run south-westwards parallel with the boundary and headland which run obliquely across the ends of the crofts. In late medieval times the 3 or 4 tofts north of toft 8 appear to have been abandoned, and the house sites levelled, to form a westward extension of the green. Tr6 is unlikely, however, to date from this period as it seems to follow the line of Romano-British track. Toft 6 at least was laid out by the 12th Century (ANDREWS & MILNE 1979, 42-4) and the whole block may date to the same period.

The North Head-row contains as its major feature a large complex in which are the main earthworks of the 12th and 13th century Percy manor house. In the centre is the hall (K) with living quarters to the west and the service rooms to the east. To the west is a separate enclosure containing the large barn (N), while to the south is a yard containing an excavated 12th century grain-drying kiln. The living quarters of the manor house, the grain storage, and the grain preparation were thus three separate components, themselves separated from the rest of the village by the main west-east road. The northern part of the complex was extended northwards beyond the prehistoric west-east lynchet, presumably to give more space.

To the east there is a large space open to the east with two structures (A & B), which are too wide to be roofed, in the north-west corner. These may be interpreted as animal enclosures and it is likely that the whole area was also for animals with its eastern side open to give access to grazing on the valley side. In its southern part is a 15th century courtyard farm (20-22) which may have been the demesne farm or the centre from which the estate was run after the main manorial buildings went out of use in the 14th century. This is the only enclosure in the village which contains a courtyard farm; all the others contain long-houses or smaller buildings. This therefore suggests higher status since courtyard farms do not become common on the Wolds until the 16th or 17th centuries, as can be seen at Duggleby (BERESFORD & HURST 1976, Pl. II,8) and Towthorpe, the latter not being deserted till about 1700.

Recent air photographs, however, show that in the earlier medieval period the situation in this large eastern area was different, as at least five parallel divisions are visible running north-south. These would form a series of five crofts, with the possibility of another which in the main manorial period seems to have been left as a wide access way (Tr2) from the village to the northern open fields. To the east there is room for certainly one and possibly two further crofts before the valley edge. The croft boundaries continue, to the south of an east-west bank, under the courtyard farm, which takes account of them in its formation. There is therefore little doubt that before the courtyard farm was built there were at least six and possibly seven or eight tofts and crofts laid out regularly to form a north row. The date of their abandonment, or perhaps their deliberate clearance, is at present uncertain. It could have been in the 12th century when the Percy Manor was fully established or, more probably, after 1254 when the two manors were amalgamated - possibly about 1300 when houses were laid out over tofts 9 & 10 where the Chamberlain manor house had been: it has always been a problem as to why new houses were built here in the 14th century, when there is likely to have been a decline rather than an expansion of population. But the need to rehouse the evicted tenants from the north row would provide an explanation. If it was as late as 1300 it is likely that the manor house was no longer used as there are no major groups of pottery after this date. Almost the whole northern enclosure, and not just its eastern half, could therefore have been turned over to animals and might be the park mentioned in 1320 (BERESFORD 1979, 20-1), the only features retained being the courtyard farm together with the western enclosure and its barn which might be that mentioned in an Inquisition Post Mortem of 1368 (BERESFORD 1979, 21), but the IPM could also refer to the building of the courtyard farm 20-22.

INTERPRETATION AND DATING

It is very difficult to establish the sequence and dating of the earthworks since, despite 30 seasons' excavation, only 5% of the site has been excavated and even large sample areas can be misleading. The classic case has been that eight seasons' excavation of an area about 30m square in toft 10 showed very slight traces of Roman or Anglo-Saxon occupation, while only a short distance to the west large numbers of sherds are now coming to light. This means that not only is it dangerous to make general statements about the development of the site from quite large excavated parts, but, for example, a sample square in tofts 12-17 would not conclusively determine when this area was laid out. Therefore, although in twenty seasons' excavation only a handful of Anglo-Saxon pottery has been found on the terrace north and south of the church it is unwise to be too dogmatic and state that there was a gap in occupation, as this hiatus might be filled at any time. The following interpretation therefore is very tentative, the more so as most of it is based on a study of the earthworks and air photographs without checking by excavation.

In the 1960s, after the discovery of the unexpected 12th century manor house in toft 10 and bearing in mind that area 6 did not appear to be laid out until a similar period (ANDREWS & MILNE 1979, 43-4), a very simplistic development from a single nucleus was worked out for the village (HURST 1972).

It was the development of the parish survey in the 1970s, together with the archaeological discovery on the village site that many of the 'medieval' boundaries were in fact Roman or prehistoric in origin, that led to doubts which were clinched by the remarkable discovery in 1975 of a middle Saxon sunken hut in the north-east part of the north manorial area, 440m north of the church. Study of the north manor showed that its complex plan, with buildings round several courtyards, which had naively been thought to be 14th or 15th century, was in fact 12th and 13th century and so contemporary with the toft 10 manor house: there were in fact two manors recorded in Domesday (BERESFORD 1979, 5-6). With the discovery that the whole parish was covered by Romano-British farms at approximately one kilometre intervals (HURST 1981) it was clear that there was a pattern of dispersed settlement which might not have become nucleated until the late Saxon period. It is remarkable that these local discoveries were all coming in thick and fast at a time when other places in England were producing evidence for polyfocal settlements and late nucleation (TAYLOR 1977 and FOARD 1978).

As a result a whole series of models may be suggested for a possible development from three or more centres to form the three rows, but recent discoveries, which seem to make the three rows much more regular than they previously appeared, now make it a high probability that the three rows were planned or replanned together at one time and that the present earthwork plan is the result of a gradual breakdown of this pattern. In particular it is possible to reconstruct a hypothetical regular plan which was laid out with tofts and crofts averaging 60ft (18m), the width of two selions, across. Some, however, were considerably narrower as at the west end of the north head-row. The actual alignment and lengths depend on the topography and the constraints of earlier boundaries. Figure 1:4 shows a schematic ideal plan of a regular two-row street green village, with a head-row at the northern end, as it might have been laid out in the 12th century with about 30 tofts, contrasted with Figure 1:5 which interprets the present earthworks showing how the regular plan was modified by the 14th and 15th centuries when only 16 tofts remained. This halving of the number of tofts equates very closely with the historical evidence (BERESFORD 1979, 13).

An alternative hypothesis is that this laying out took place in late Saxon or Scandinavian times at the period of nucleation, but several factors, and especially the heavy emphasis on a main period of activity in the 12th century with the construction of the two manor houses and the laying out of the southern part of the west row, make a post-Conquest date more likely in the present state of our knowledge. This possibility of early Norman planning immediately suggests that this was the time also that the regular laying out of the open fields in long strips took place. In this way an irregular furlong laid out west-east was taken in to form crofts 12-17 when the main field system was laid out north-south. This would fit with the suggestions for Holderness (HARVEY 1981) and in County Durham (ROBERTS 1972) and the general likelihood of major changes in the early twelfth century when the Percies and the Chamberlains seem to have come to Wharram and established themselves as resident lords.

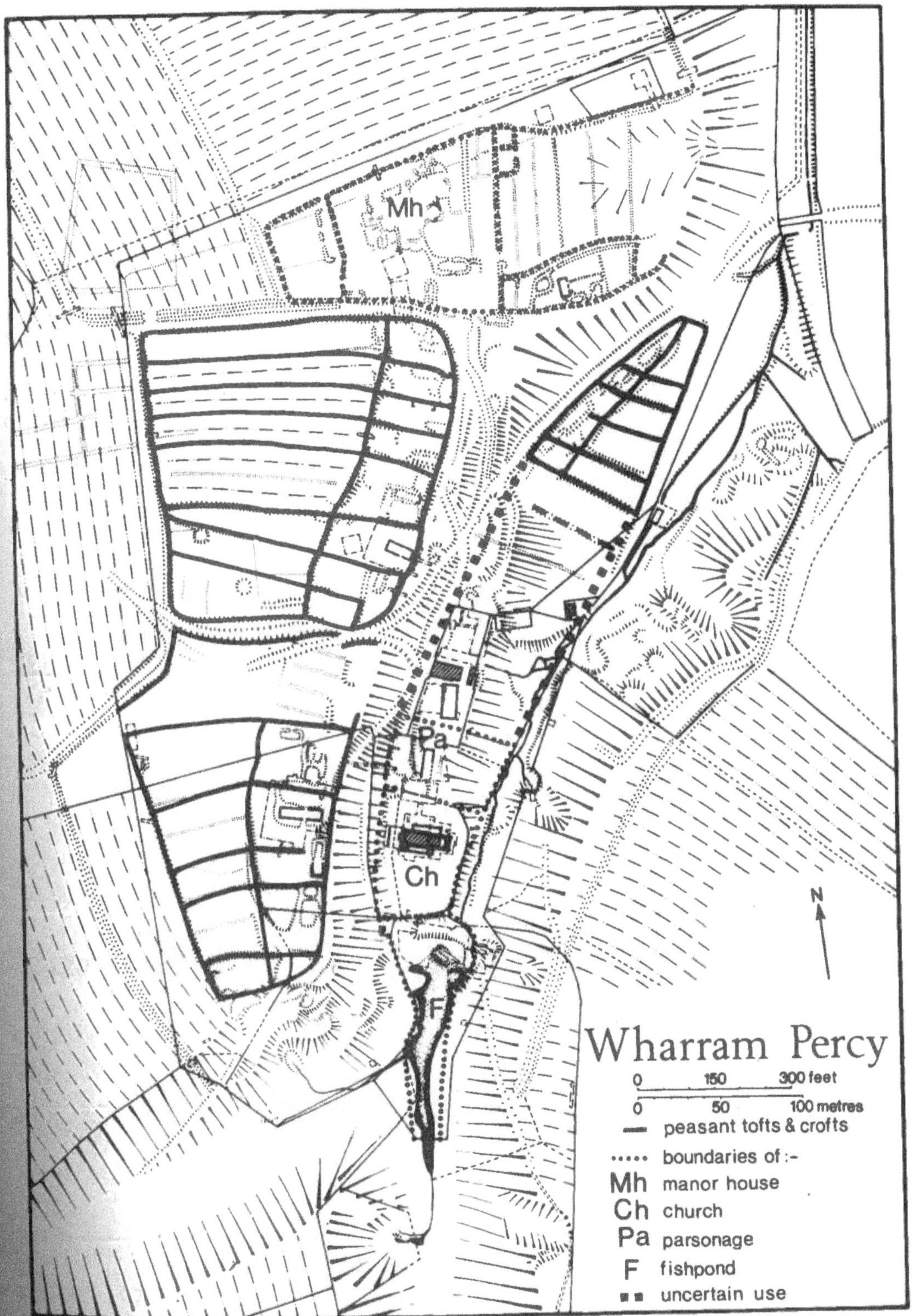

Fig. 1:5 Wharram Percy: late medieval plan showing the break-down of the regular rows, with about 16 peasant tofts, one manor house, church, parsonage and fishpond, and an enlarged green.

Acknowledgements - Over one hundred specialists in many disciplines are involved in the Wharram Research Project, so over the years I have benefited from discussions with many of them over the village plan. Based on an original 1:2500 survey by the Ordnance Survey in 1952, it was fair-drawn at 1:625 by G.L.Worsley in 1954 and redrawn with corrections in 1961 by R.P. Kirby. During the last twenty years R.T. Porter has steadily improved the plan, producing a major revision at 1:600 in 1969, based on ground surveys and air photo interpretation, and bringing it up to date each year since. In recent years W.J. Hopkins has been making detailed 1:120 or 1:100 surveys of certain areas of the village but this revision is not yet complete and the positions of features such as the croft boundaries should not be taken as precisely correct. The general interpretation presented in this paper, with new drawings by J. Thorn, is a statement which it is hoped will form the basis for further discussion. I am most grateful to R.T. Porter for his many suggestions for improving the text, though the final version must remain my responsibility.

REFERENCES

ANDREWS & MILNE 1979: D.D. Andrews & G. Milne eds., Domestic Settlement I: Areas 10 & 6. Vol. I of Wharram: A study of Settlement on the Yorkshire Wolds ed J.G. Hurst, Society Medieval Archaeology monograph, 8 (1979).

BERESFORD 1963: M. Beresford, The Lost Villages of England, (London, 1963 fourth impression).

BERESFORD 1971 M.W. Beresford, 'A Review of Historical Research (to 1968)' in Deserted Medieval Villages: Studies, edited by M. Beresford and J.G. Hurst (London,1971).

BERESFORD 1979: M.W. Beresford, 'Documentary Evidence for the History of Wharram Percy', in Andrews and Milne (1979). 5-25.

BERESFORD & HURST 1976 & 1979 M.W.Beresford & J.G. Hurst, 'Wharram Percy: a Case Study in Microtopography', in P.H. Sawyer ed., Medieval Settlement, (London, 1976), 114-144, reprinted in P.H. Sawyer ed., English Medieval Settlement, (London, 1979), 52-82 with supplement, giving the results of the Research Project up to 1978, on pp. 83-5.

FOARD 1978: G. Foard, 'Systematic Fieldwalking and the Investigation of Saxon Settlements in Northamptonshire', World Archaeology, IX (1978), 357-74.

HARVEY 1981: M. Harvey, 'The Origin of Planned Field Systems in Holderness, Yorkshire' in T. Rowley, ed., The Origins of Open-Field Agriculture, (London,1981), 184-201

HURST 1972: J.G. Hurst, 'The Changing Medieval Village in England', in P.J. Ucko et. al. eds. Man Settlement and Urbanism, (London,1972), 531-40.

HURST 1976: J.G. Hurst, 'Wharram Percy: St. Martin's Church', in P. Addyman & R. Morris eds., The Archaeological Study of Churches, Council British Archaeology Research Report, 12 (1976), 36-9.

HURST 1979: J.G. Hurst, 'The Wharram Research Project', in N.J. Higham ed., The Changing Past, (Manchester, 1979),

HURST 1980: J.G. Hurst, 'Archaologische Forschungen in der Siedlungskammer Wharram, England', Offa, XXXVII (1980), 301-8.

HURST 1981: J.G. Hurst, 'Wharram: Roman to Medieval', in Vera I. Evison ed., Angles, Saxons and Jutes; Essays Presented to J.N.L. Myres, (Oxford, 1981), 241-55.

MVRG 1953-1981: Annual Reports of the (Deserted) Medieval Village Research Group (1953) - (1981).

ROBERTS 1970: B.K. Roberts, 'Rural Settlement' in J.C. Dewdney ed., Durham County and City with Teeside, British Association for the Advancement of Science, (Durham, 1970), 235-50

ROBERTS 1972: B.K. Roberts, 'Village Plans in County Durham: A Preliminary Statement', Medieval Archaeology, XVI (1972), 33-56.

TAYLOR 1977: C.C. Taylor, 'Polyfocal Settlement and the English Village', Medieval Archaeology, XXI (1977), 189-93.

CHAPTER 2

OF LANDSCAPES AND WORDS

B.K. Roberts
(University of Durham)

Fieldwork involves tension: on one hand it is concerned with the particular features creating character and qualities of an individual location, but on the other hand a field situation presents a challenge, to generalise and to escape from the constraints of the individual case towards a broader view. Whatever attitudes and objectives are taken into the field, and whatsoever specific questions are asked, this tension, and the degree to which any work undertaken is inductive or deductive, finds a focus in the terminology used and the frameworks adopted to classify the observed features. The word-symbols used to characterise and describe the basic observations reflect very much the 'state of the discipline' and this in turn limits or extends the value of comparisons and contrasts. The application of precise and generally agreed terms represents a foundation for the integration of the results of each case-study into a wider framework, and such terminology must relate to associated systems of classification. As Harvey notes 'Classification is, perhaps, the basic precedure by which we impose some sort of order upon the vast inflow of information from the real world. By grouping sense-perception data into classes or "sets" we transform a mass of unwieldy information so that it may be more easily comprehended and more easily manipulated' (HARVEY, 1969, 326-49). There are, as he points out, dangers involved, for classifications can become inflexible to the point of actually inhibiting research, and we must always strive to separate our classificatory system from the objectives of our enquiry, the difference between means and ends. In this paper a brief re-examination of one of the field-study locations visited during the 1981 meeting will form the basis for a discussion of some of the questions of terminology and classification involved, particularly with reference to rural settlement forms, and the relationship between the terms being formulated and defined by the author and those of Professor Uhlig's corpus (1972).

COCKFIELD, CO. DURHAM : THE FIELD STUDY

Cockfield, Co Durham (NZ 1324) is a rural village which expanded during the nineteenth century as a result of the presence of extractive industries in the immediate vicinity, coal mining and the quarrying of dolerite ('whinstone'). These industries have now receded, leaving behind landscapes of great complexity which present many problems of interpretation. These are made more profound by the paucity of documentary material. Three particular problems may be

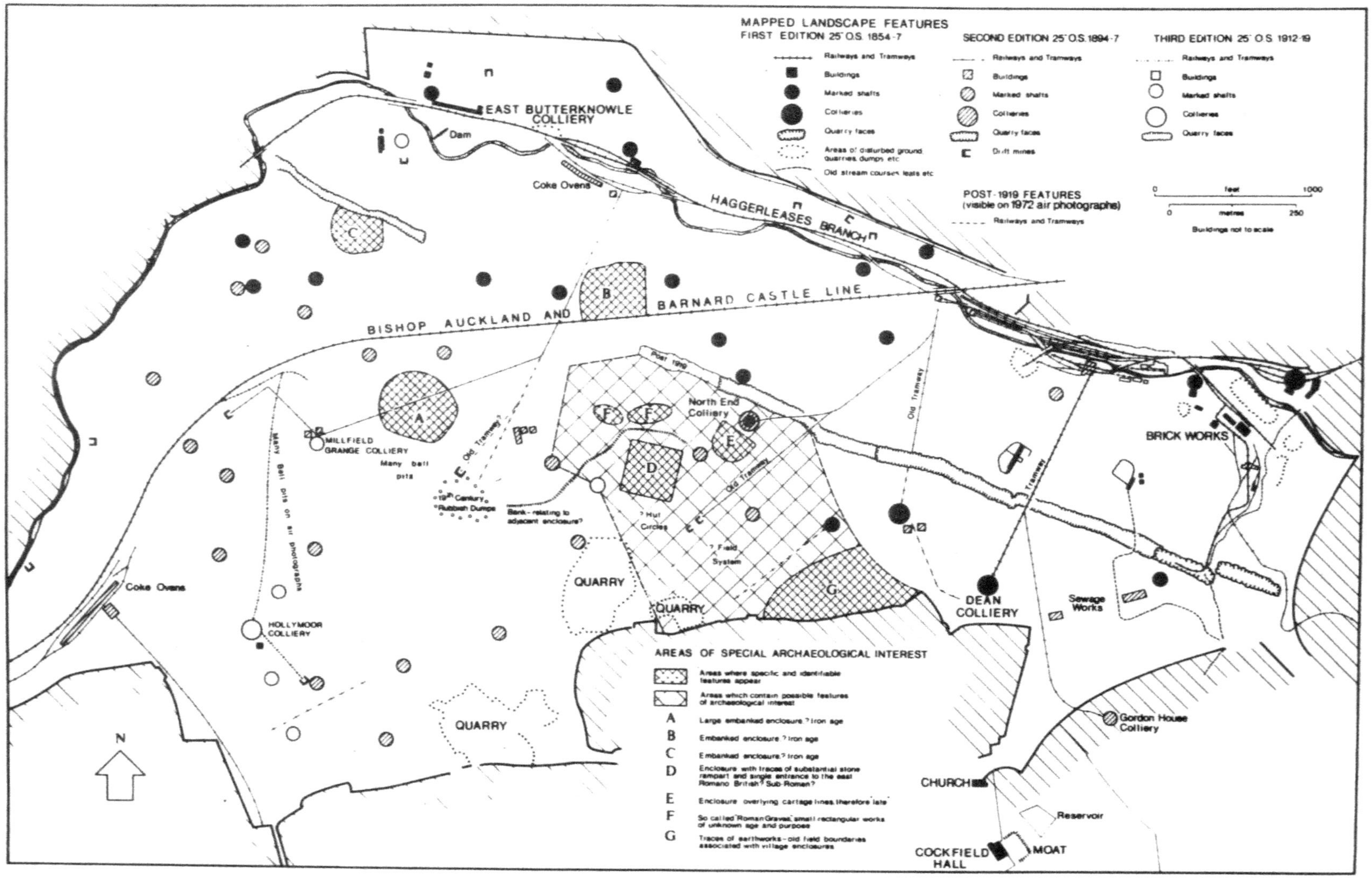

Fig. 2:1 Cockfield Fell: interpretational map.

isolated for brief comment and analysis:

(i) the first is concerned with disentangling the varied genetic layers which are present within the landscape;

(ii) the second is to attempt to structure what can be observed within an explanatory description;

(iii) the third relates to problems of management, planning and preservation generated by the occurrence of such landscapes within close proximity to a functioning settlement.

COCKFIELD: THE LANDSCAPES

Cockfield village and fell (Plates 2:1 and 2:2) lie on a low east to west ridge at a general height of 200-230m. To the north of the village lies the fell, an area of about 330ha of unenclosed stinted pasture, controlled to this day by a committee of Field Reeves. It is an exposed situation, and the village is, in the words of a mid-nineteenth century Sanitary Inspector, 'exposed to every wind that blows' (CFB, 30-40). The fell is rich in earthwork remains, and so far as is possible these have been indentified and commented upon in Figure 2.1, using dated Ordnance Survey maps to create approximate chronological horizons. In 1868 the rough grazings were legally converted into stinted pasture (CFB, 67 et seq.): on the 330 ha. some 1100 stints, or grazing units, were allocated, 238 to the Duke of Cleveland, 227 to John Bowes, and the remainder to 41 other smaller landowners. Each stint unit represents a single 'sheep stint' but other beasts could be grazed using the following conversion scale:

1 sheep stint	=	1 sheep or ewe with unweaned lamb
5 sheep stints	=	1 cow or beast 3 years old or upwards
$3\frac{1}{2}$ sheep stints	=	1 two years old cow or beast (bullock)
$2\frac{1}{3}$ sheep stints	=	1 year old beast
8 sheep stints	=	1 horse or mare
6 sheep stints	=	10 geese

The fell is still used for grazing, following essentially these rules, but pigeon-houses (crees) and garages are now permitted. It can only be admitted that the rough grass and bracken of this unenclosed landscape probably conceals large numbers of unidentified features, but the visible monuments are virtually unique, for example the great grooves, which seem to represent lines where carts have scored deeply into the ground. Their date is uncertain, but it seems likely that they antedate the collieries present by the mid-nineteenth century. Enclosure D (Fig 2.1) appears to be the remains of a substantial square stone-built enclosure, with traces of ramparts between 3 and 5m wide. This is clearly a very different category of site to Enclosure A, which is almost certainly prehistoric, and the closest parallel to D is a massive site in the hills to the north, Hamsterley Castles. Both may in fact be 'Dark Age' in date constructed in the centuries immediately following the collapse of Roman control (ROBERTS,

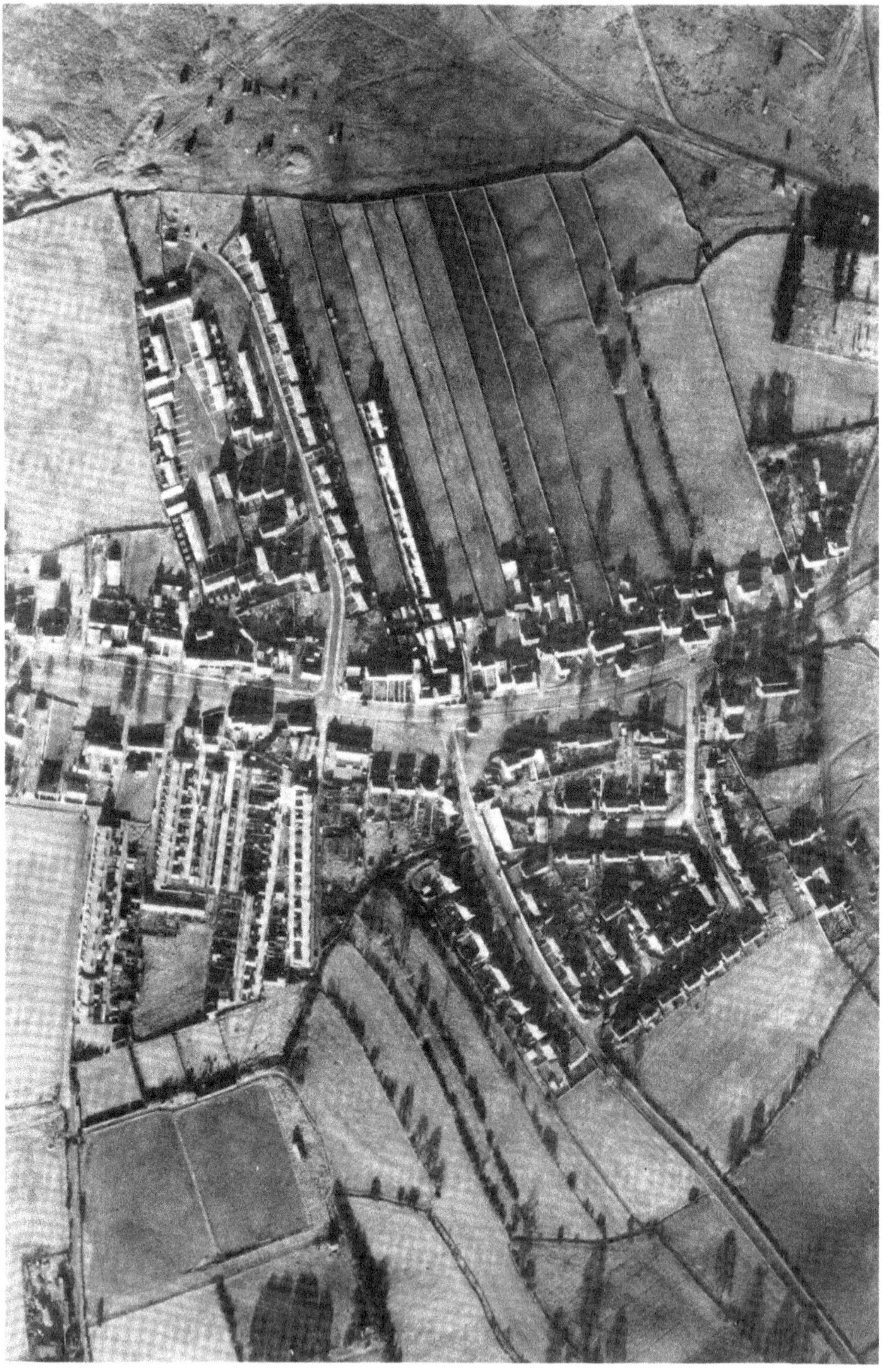

Plate 2:1 Cockfield village. Durham County Council and B.K.S. Surveys Ltd.

Plate 2:2 Cockfield Fell. Durham County Council and B.K.S. Surveys Ltd.

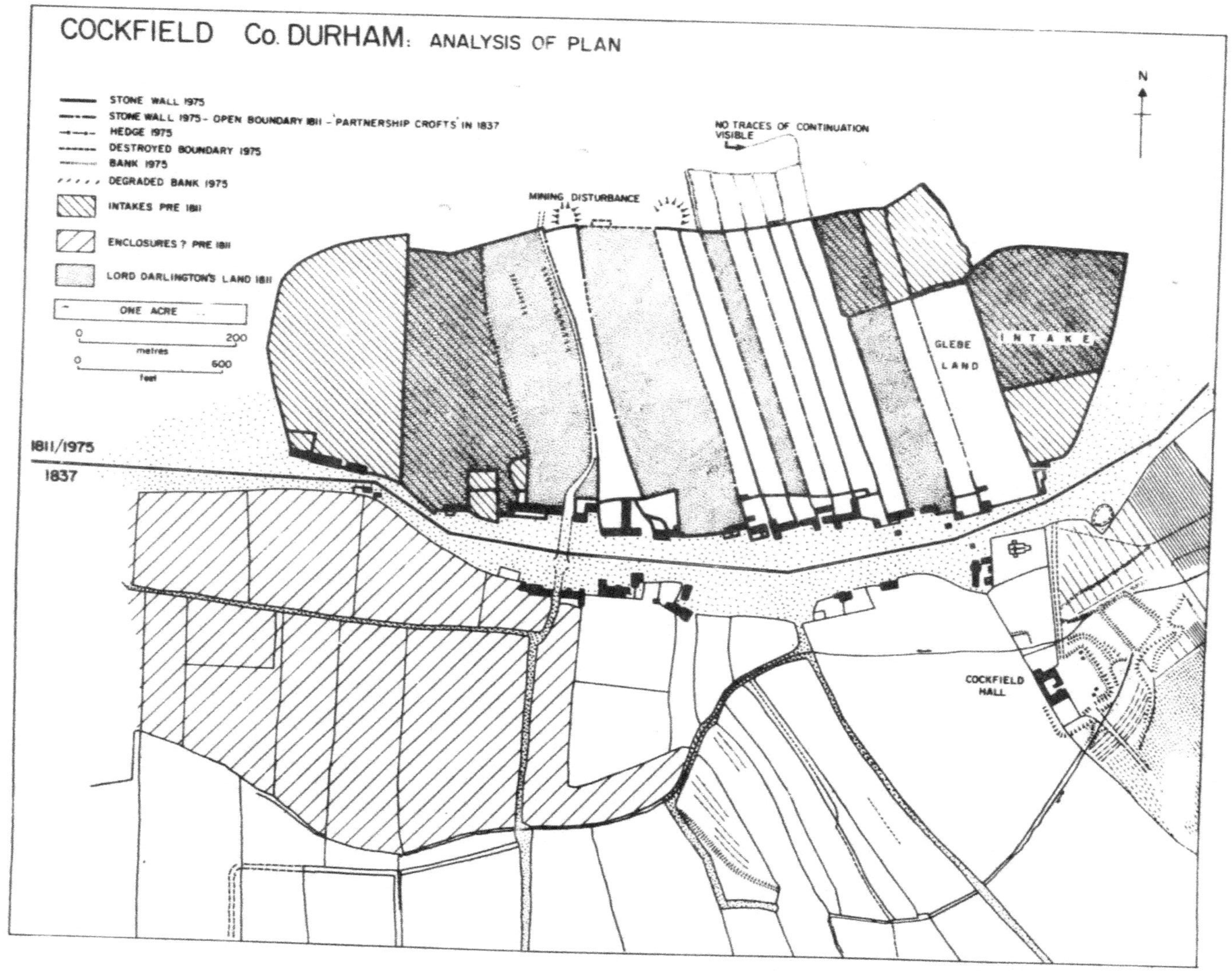

Fig. 2:2 Cockfield village: interpretational map.

1975).

A careful examination of the village suggests that beneath the landscape of rather grey and uniform nineteenth century buildings and stone walls there lies one of earthwork enclosures. The two are almost completely accordant. The earliest map of the settlement, in Lord Barnard's archives, is dated 1811 (BAKER and HARLEY 1973, 54) and an examination of this, the Tithe map of 1837 and the air photograph suggests the presence of three morphological elements within the village (Fig.2:2)

(1) the manor-house site, with a moated farmhouse, which clearly lies at the eastern extremity of a low domed oval field, almost bisected with a modern road, but with clear traces of lynchetted strip holdings at its western end (Roberts in ROWLEY, 1981, 145-161).

(2) a northern row, once very regular, with a building line which has been rendered irregular as a result of encroachment southwards onto the green edge. The map of 1811 suggests, by using pecked boundaries, that some of these long tofts or long crofts lay open. On the ground, and to a lesser extent on the air photographs, it is possible to see that they once extended beyond the present outer boundary wall onto the fell, and that, particularly at the eastern end, each long strip is lynchetted. In fact this northern row is made up of two units: the eastern unit contains the glebe and comprises strips some 250m in length, while the remainder are strips once just under 500m in length.

(3) there are in addition various subsidiary accretions, indeed one can postulate (for there is no clear evidence) that much of the western end of the present village represents accretion to an older, more easterly core, in-filling what was once a larger more open "green" area, an extension of the fell. The village is now dominated by nineteenth and early twentieth century buildings, and only here and there do rougher wallings and red pantiles hint at older structures, but even these may be no earlier in date than eighteenth century.

These points are no more than slight amplifications of the maps, which can be studied alongside the air photograph. It is clear that the present village is superimposed upon an older landscape, visible in a series of earthwork boundaries upon which, and along which, stone walls have been intruded. These latter cannot be dated from their visible forms: some are clearly characteristic of the late eighteenth and early nineteenth century; others, containing much field stone, are likely to be older, but the process of continuous rebuilding and adaptation has destroyed any coherent pattern, even if one ever existed.

COCKFIELD: AN EXPLANATORY DESCRIPTION

No documentary evidence exists which provides a secure

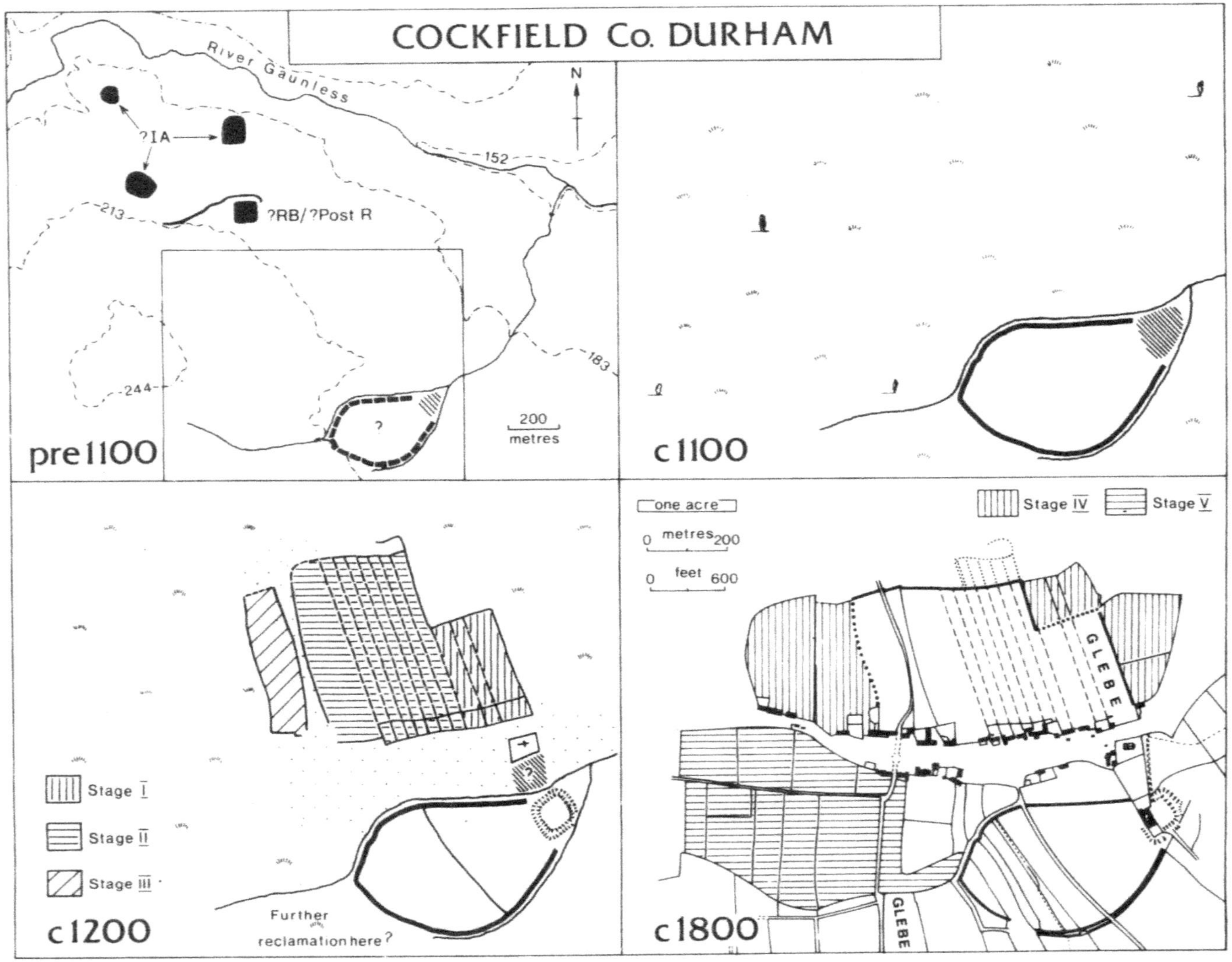

Fig. 2:3 Cockfield village: evolution.

date for the origins of the Cockfield plan, but sufficient exists to create a hypothesis presented in Figure 2:3: the location of the glebe toft offers a beginning, for its location suggests that the northern row developed after the establishment of the church, and a recent documentary discovery by Mr Martin Snape indicates that this may have been in the period 1228-1234 (Appendix to this paper) so that the date of c.1200 in Figure 2:3 proves not to have been wildly inaccurate - for the figure was created before the evidence came to light. A sequence of development is postulated, with growth from an initial field area of 25ha in extent, originally either demesne land or tenant strips: in the second or third decade of the thirteenth century a knightly family from Suffolk, the de Cockfield's, acquired the land, at that stage described as 'waste', but of course the oval enclosure may represent an even earlier phase of settlement which was re-occupied. Soon after, a church was established, with its glebe, and to that latter nucleus further tenant holdings accreted.

This sequence of events, and the detailed arguments, has been analysed elsewhere (Rcberts in ROWLEY,1981), but two points can be made here:

(1) this type of village is found throughout west Durham, a region experiencing active colonisation between 1050 and 1250. The long tofts or crofts provide a distinctive morphological feature separating villages of this type from those more common in Durham, with shorter tofts, less than 150m in length. In fact villages of the Cockfield type are normal in Cumberland, where they again appear in a context where active colonisation may be postulated in the period 1100-1250, although some Cumberland variants have strips 6-700m in length (ROBERTS 1978, Fig 2.7 above)

(2) Cockfield was a rural estate peripheral to a 'shire', a group of villages united by shared lordship and corresponding to the 'multiple estate' described by Glanville Jones. This is one context in which the overall distribution of village plan-types can be examined and in measure explained (ROBERTS 1977b,1978)

COCKFIELD: THE PROBLEMS

During the two years preceding 1981 there were tensions in Cockfield arising from a proposal to identify the fell as an Ancient Monument. From the viewpoint of the landscape historian this is an exciting prospect, but confusions, disagreements and simple misunderstandings by all parties have made the local community, for whom the fell remains an important economic factor, hostile to the proposals. The matter is currently being debated, but it should not be impossible for an arrangement to be worked out which is advantagous to all interested parties: a preserved ancient monument, suitably maintained and where necessary made safe and improved, could be managed by the local community, both for their own benefit and the good of the larger community of the country.

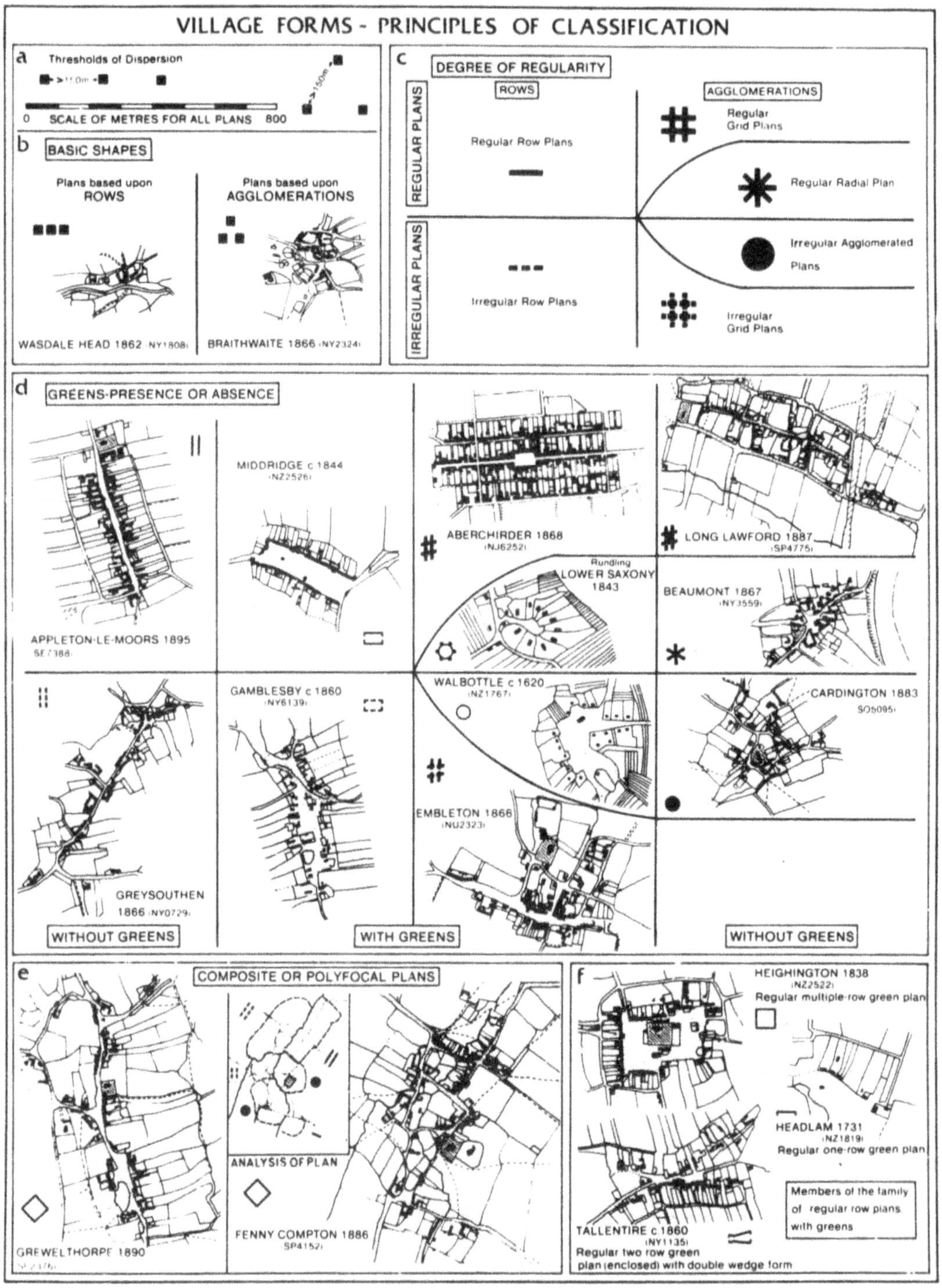

Fig. 2:4 Village forms: principles of classification.

The distinctive morphology of Cockfield raises many questions. The case of Wharram Percy (SE 8364) illustrates the uncertainties of attempting retrogressive analysis without excavation, but Hurst's identification of a head-row at the northern end, both introduces questions of terminology and serves as a reminder that even Wharram possesses a broader and distinctively northern English context. There are indeed some parallels between the plans at Wharram and Cockfield.

TERMINOLOGY AND CLASSIFICATION

The General Problem

The terminology and classification system adopted by the author for the study of rural settlement forms has its origin in discussions with June Sheppard in the early 1970's, and through her has some links with the work then in progress in Giessen. The original aim was to create a simple system which was not based upon German terms, drew upon observable morphological criteria and avoided references to supposed origins at any one time. A tabular structure originally devised by June Sheppard was gradually evolved by the author, into the grid shown in Figure 2:4d and this is in essence based upon an assumption (Fig 2:4b that rural settlement clusters (groups of houses, Haustättenverband) take two essential forms, a linear arrangement, or row, and a tighter more cohesive group, an agglomeration. The basic building blocks of the settlements are however in detail made up of three types of spaces, private spaces, communal spaces and public spaces. These have been examined in some detail elsewhere but these basic plan elements are shown in Fig 2:5 (ROBERTS 1982).

The morphologies of the two basic shapes are shown in Figure 2:6: a row may be defined as a morphologically distinct unit, in which the homesteads (Hausstätte) together with their associated outbuildings and enclosures, tofts and crofts (Hausparzelle, Hofparzelle) are arranged in a more or less straight line, which as Figure 2:4c demonstrates may be regular or irregular. These latter qualities cannot be easily defined, but 'regularity' involves an assessment of reality against a geometric ideal, a measure which in the context of general analysis must be largely subjective. At an early stage in the work, Professor Conzen drew attention to the important distinction between a row or farmsteads, architectually speaking not a row at all, and the serried arrangement or true row of wall-to-wall connected buildings, in eaves orientation, running parallel to a street or green. It would indeed be possible to distinguish between these by using the terms 'open' and 'closed' but the Durham evidence shows that, whatsoever the problems, there are sound historical reasons for applying the term row to a line of homesteads. In English the precise meaning can be made clear through the context of usage and any qualifying words. The word terrace can be applied to a serried row of buildings.

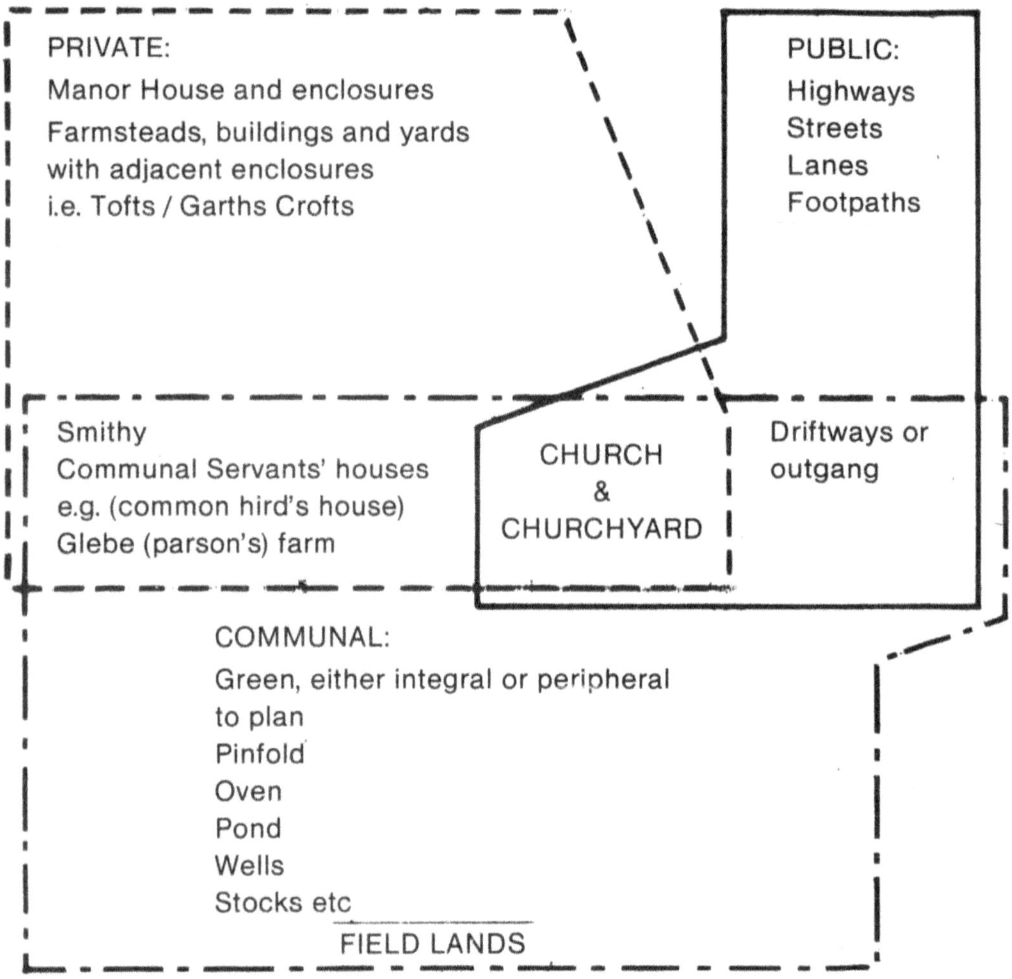

Fig. 2:5 Basic plan elements of settlements.

The word agglomeration basically means 'to gather together in a rounded mass' or 'to gather into a ball or cluster', the elements of circularity and tight clustering being the vital points. While irregular agglomerations are very common, corresponding to the German Haufensiedlung, regular agglomerations, either with a grid plan or a radial plan, also appear. The distinction between rows and agglomerations reflects Uhlig's distinction between Lineare Siedlungen and Siedlung mit flächigem Grundriss, but it is at this point the grid in Figure 2:4d diverges strongly from Uhlig's system, for instead of identifying 'green' villages, clusters arranged around a central common, as a separate type, the green forms are seen as subsets of the basic row/agglomeration contrast, and the six plan types in Figure 2:4c all either possess or do not possess a green, giving the twelve basic plan types of Figure 2:4d, of which only eleven are shown.

Thus far the grid distinguishes between basic plan types using three observable characteristics (1) basic shape, (2) degree of regularity and (3) the presence or absence of an integral green, but the result appears rigid, and further qualifications are necessary to allow it to be applied flexibly to an infinite number of particular cases. In essence the grid defines fundamental plan-type families, i.e. rural clusters, hamlets or villages and to some extent small towns, having characteristic arrangements of private, communal and public spaces. This essential flexibility of usage is given by three qualifications: first, each of the plan-types in the grid represents a characteristic member of a family, and Figure 2:4f is a reminder of the existence of 'cousins' to Middridge, related plans, made up of varying numbers of rows. A simple row, made up of a line of homesteads, is often surrounded by lanes and footpaths, and forms a toft compartment, a morphological unit intermediate between a plan element and a plan type. Such a distinctive group of houses would appear to correspond to the German Haustättenverbände; second, all of the individual plan types within the grid can be combined with similar or different types to create composite or polyfocal plans, as in Figure 2:4e (TAYLOR 1977). The German idea of a Haustättencomplex appears to embrace the idea both of a plan type and a composite plan; third, each of the types defined in the grid can be seen as representing one end of a catena of types: thus between Middridge, Co. Durham and Gamblesby, Cumberland, respectively a regular two-row green village and an irregular two-row green village, there exists in reality a range of minor variants. Such gradations are possible across most if not all axes of the grid and a series of theoretical topological transformations can be envisaged. In fact while Middridge is clearly very regular, Gamblesby, with a rather irregular disposition of homesteads has, nevertheless, traces of regularity in the plot boundaries to the rear of the village: thus it may be considered to lie closer to the central axis of the grid than Middridge. This accounts for the apparent

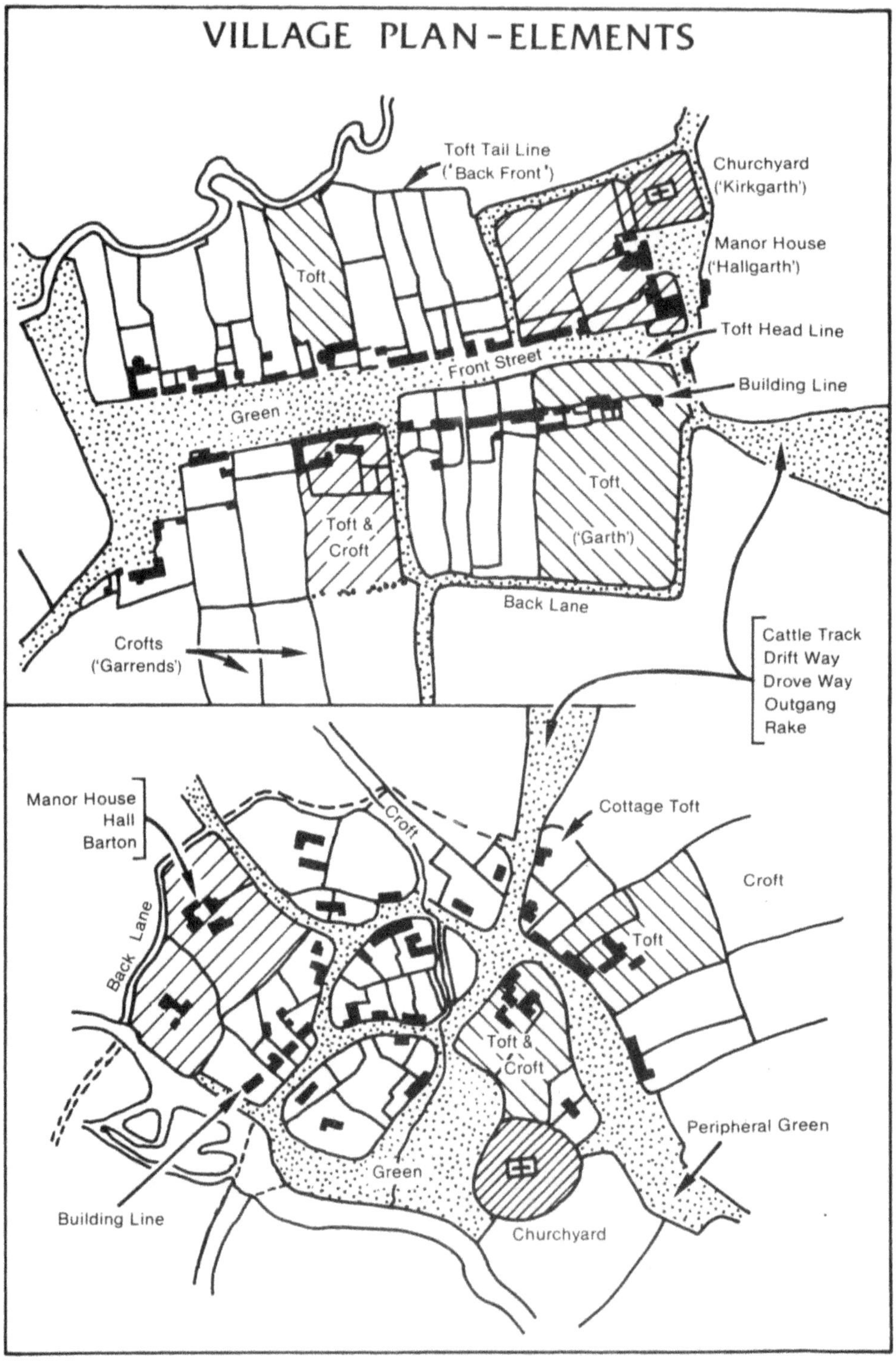

Fig. 2:6 Village plan-elements: two basic shapes.

discrepancy between Figures 2.4 and 2.7. Associated with each plan-type in Figure 2:4 there is an ideographic symbol for mapping purposes and these can be modified or aggregated as occasion demands. Standardisation is, however, essential for comparison.

The essential weaknesses of this system of classification are twofold: on one hand settlement size is not assessed, while on the other the density of homesteads for each unit of settlement area is not presented visually. In practice it has been found that the ideographic symbols can be size-graded using a simple measure of the circle most closely circumscribing the village homesteads, following the 150m or 'hailing distance' rule for those on the periphery. This produces effective maps (ROBERTS 1979). Although the density of dwellings per unit area can be measured, using a randomly imposed 50 x 50m grid, this measurement cannot easily be incorporated within the symbol system (ROBERTS 1982b). Size-grading of the symbols, together with the inclusion of all known deserted settlements, creates a valuable visual impression of the morphology of the cluster settlements within the system: the symbols for composite or polyfocal plans pose some problems but these can be made sufficiently large to allow the insertion of a qualifying symbol within the main diamond, although there is rarely space enough to incorporate symbols for all of the component parts.

To return to Harvey's point concerning means and ends: the classification presented here is sufficiently flexible to allow it to encompass all of the British plan types so far detected (see also BERESFORD & ST. JOSEPH 1979; HARVEY 1978; LOCKHART 1980; SHEPPARD 1966, 1974, 1976). Cockfield provides a useful illustration of the use of the grid and demonstrates its inherent flexibility. The village was originally mapped as an irregular two-row green village, with extended plots (provisionally termed 'long tofts' but as they are field strips 'long crofts' is equally appropriate) on one side (Roberts in DEWDNEY 1970, 51, second village north of Staindrop): these latter clearly represented field strips closely attached to each homestead and differed from the normal Durham homestead toft in that they were in excess of 200m in length. One hundred to one hundred and fifty metres of depth from the street frontage is more usual. As was noted earlier, these Cockfield strips may once have lain open, and this begs the question of the precise nature of the actual toft or house enclosure at the head of the strip. Such features were mapped simply by adding a small curved line to the appropriate side of the symbol for a regular two-row street green plan, and, to revert to Gamblesby, this village possesses long tofts on each side, necessitating to such additions to the basic symbol, although for simplicity those have not been incorporated in Figure 2:4.

Closer study, greater familiarity and more experience of the problems quickly showed that while Cockfield in the early mid-nineteenth century was indeed a two-row green village (the addition of the word street to the word green is optional) one side shows regularity while the other appears to be irregular, an observation which slides Cockfield along the catena of plan types theoretically extending along the Middridge-Gamblesby axis of the grid. Even the incorrect placing did not invalidate the grid as a useful research tool because comparisons were still made with other plans which show, in effect, the same partial regularity (Roberts in BAKER & HARLEY 1973, 54). The simplified versions of the morphological symbols used as the basis for Figure 2:7 show that the Cockfield plan-type has in Durham a distinctively western distribution and documented contexts associated with late colonisation, while in Cumberland such settlements, and variants with 600-700m strips, are the norm. This latter county was taken from Scottish hands as late as 1092 when William Rufus conquered the Carlisle district and sent 'many peasants thither with their wives and livestock to settle there and till the soil' (GARMONSWAY 1953, sub anno 1092), indeed large portions of eastern Cumberland were not taken by the Normans until the mid-twelfth century (KAPPELLE 1979, 200). These suggestions must at the moment remain hypotheses, for there is an absence of particular case studies to reinforce the line of argument suggested by the general distribution.

A cautious conclusion would be that there are grounds for believing that villages of the Cockfield type were developing in colonising contexts in the north of England during the second half of the twelfth and earlier part of the thirteenth century. If the southern field area at Cockfield is indeed the original unit then its limited size, 25ha, would imply that the settlement originated as a seigniorial farm with appendant tenant small-holdings. In a colonising context the author envisages the areas embraced by long tofts as 'foothold furlongs', and the southern field is seen as such a foothold unit with homesteads not located on each strip but probably concentrated at the eastern end. Thus Cockfield village demonstrates two morphological manifestations of the same feature. When the land was granted to the de Cockfield family, probably in the 1220s, it had a name, Statheleg, and was at a slightly later date known as Beaurepayr, 'the beautiful retreat'. It is possible to envisage rough but cleared land, used for hunting and grazing, with a succession of prehistoric and Romano-British enclosures visible, and it may be that the southern field represents a re-occupation of one of these.

COCKFIELD AND WHARRAM PERCY

It is at this point that comparison with Wharram becomes both necessary and difficult: Three points must be made:

(i) the dimensions vary substantially: the croft extensions to the homestead tofts at Wharram only run at a maximum for something under 200m from the toft head line fronting the street. By Durham standards these are short, a vital difference in spite of superficial similarity.

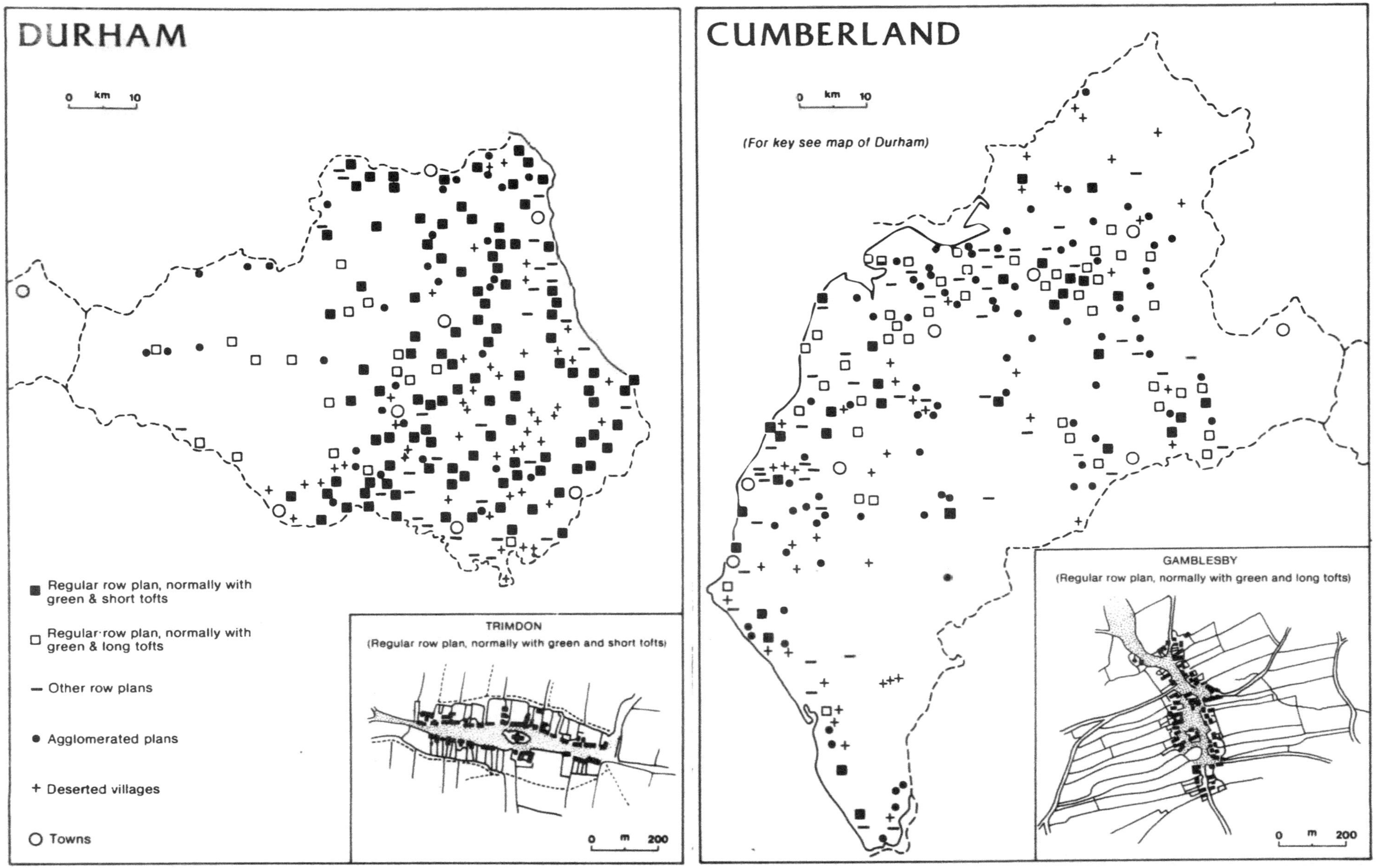

Fig. 2:7 Counties Durham and Cumberland: village types.

(ii) the actual tofts at Wharram appear in general to be somewhat smaller than those of Durham: if all of the Wharram tofts are indeed twelfth century then they are certainly about half of the size of those reconstructed at Kirk Merrington Co. Durham (ROBERTS 1972).
(iii) it is probable that both the Cockfield "long tofts" and the Wharram crofts represent a sharing of land in close proximity to the house for the purpose of more intensive land use. At Cockfield the owner of one long toft would be in possession of a plot not exceeding 0.8 ha. in area, hardly a full farm, while the largest Wharram crofts are of the order of 0.2ha. The latter are more suitable for garden cultivation, while the former hardly constitute a viable arable holding for a family. The enclosure of selions or field strips abutting the toft tail line at a late stage in the history of a village will, of course, produce a very similar morphological feature, but this seems not to have occurred in either Cockfield or Wharram. Such long homestead plots are occasionally documented in medieval records (PAGE, 1982, 58) and lie on an interesting threshold between the true village enclosure and an adjacent field strip: detailed studies using seventeenth century surveys from Northumberland reveal furlongs, with field strips running on from the tofts, but not forming a consolidated block to the rear of an appropriate toft and being sub-divided in a normal furlong sequence found throughout the fields (HODGSON 1899, maps facing 204, 376, 416 and original documents in Alnwick). To complicate matters these are always termed "crofts", the normal term applied to enclosures adjacent to the homestead plot or toft.

In Figure 2.6 the toft has been taken to be the immediate enclosure around the farmstead and its adjacent yards, the croft normally representing enclosed field land. Frandsen, in a personal communication to the author, makes the point that the Danish word Tofteagre means the strips next to the farm buildings, not enclosed and therefore technically measured as part of the fields. The enclosures on which the buildings were situated were called Gardstoften, but in regulated villages these did correspond exactly with the Tofteagre.

RETROSPECT AND PROSPECT

The aim of this paper has been to show how the questions raised by a study of the morphology of a single village can be enhanced by placing it within a broader context. To do this some consideration concerning terminology and classification is essential. It must be stressed, however, that the terms suggested and used here are by no means finalised or universally accepted, indeed constructive criticism is essential if they are to develop and become part of the fabric of the subject. The essential argument is that a point has been reached where studies of particular settlements and particular regions must be integrated in a wider pattern of work which identifies both contrasts and similarities: comparative studies necessitate a conscious attempt to define and agree a terminological framework. In the present stage of knowledge even basic terms such as row, agglomeration and toft and croft need attention.

Acknowledgements

My thanks are due to the Field Reeves of the Cockfield for allowing me access over many years, to Mr. Martin Snape for permission to reproduce his note on Cockfield as an appendix to this paper, to Mr. Don Wilcox for drawing my attention to the Cockfield Field Book, and to Lord Barnard for giving me access to private estate records. Without these kindnesses, and the willing help of the staff of the Department of Palaeography and Diplomatic, Durham, and the County Record Office, work could not proceed.

APPENDIX

Note on the possible origins of Cockfield, Co. Durham

(M. Snape
Dept. of Palaeography & Diplomatic,
Durham University)

It may be inferred from certain pieces of evidence that Robert de Cockfield was proprietor of Cockfield in Co. Durham in the early 13th century, and that he may well have been the founder or developer of the settlement there. The evidence is as follows:

Robert de Cockfield was a royal justice itinerant for Yorkshire and Northumberland in 1225.(1) He was sheriff of Yorkshire 1226-9, and constable of Scarborough and Pickering Castles(2), sheriff of Lincolnshire 1229-30(3) and briefly of Norfolk and Suffolk in 1232(4). He was married into the Yorkshire family of St. Mary; they were benefactors to the Cistercian nunnery of Sinningthwaite, a tradition which he and his daughter continued(5). Through his wife he was brother-in-law to Henry of Wales, who held Softley in Co. Durham and sold it to Finchale Priory(6). He was dead by 1242-3, when the marriage of his widow was declared to appertain to the bishop of Durham(7).

Robert de Cockfield witnessed charters of Bishop Richard Marsh in 1223, 1224, and 1225(8). From Bishop Marsh (d.1226) he had a grant of half an unidentified place called 'Statheleg' with some waste, which was confirmed by the crown in 1228 during the vacancy after Marsh's death (9). Between 1228 and

1234 Bishop Richard Poore granted or confirmed to him som^ unspecified lands and the advowson of a church he had founded 'in the place which is called 'Beaurepayr', likewise unidentified(10).

Robert, having acquired the wardship and marriage of John, son and heir of Robert le Vavasour(11), married him to his own daughter and heiress Alice(12). Subsequently the Vavasours were lords of the manor of Cockfield (and patrons of the church there), and it is a reasonable inference that their title was derived through Alice from Robert de Cockfield.

The Vavasours held no other possessions in Co. Durham, and it may well be that Cockfield represents 'Beaurepayr' with its church(13); the parish and the manor of Cockfield coincide, as befits such a proprietorial church. And although it would be natural to assume that Robert, as lord of Cockfield took his name from that place, it is quite as likely that he belonged to the family which took its name from Cockfield in Suffolk, in which case the Durham Cockfield takes its name from him instead of the other way round. It would be entirely understandable if a settlement proudly named Beaurepaire by Robert were to become known by the name of its founder instead.

1. Rot.Litt.Claus.,II (RLC), 77.
2. Calendar of Patent Rolls, 1225-32 (CPR), 38; Early Yorkshire Charters (EYC), I, 512, III, 224; RLC, 107, 117.
3. Pipe Roll 14 Hen.III; CPR, 275, & cf.337.
4. CPR, 480, & cf.489.
5. EYC, I, 511, III, 224; Burton, Monasticon Eboracense, 326.
6. EYC, III, 224; Priory of Finchale (Surtees Soc. 6), 105-9.
7. Miscellanea II (Surtees Doc. 127),45.
8. Durham Dean & Chapter Muniments (DCD), 3.3.Sacr.4, 3.3.Sacr.5,2.2.Pont.6.
9. Calendar of Charter Rolls, I, 66.
10. DCD, Misc.Ch.206 (confirmation by the chapter).
11. Norhumb. & Durham Deeds, 264; Excerpta e Rotulis Finium, I, 215.
12. PRO E368/15 m.7; Bodleian Library, MS Dodsworth 8, f.97v and MS Top.Yorks. e 7, f.149r-v.
13. And perhaps 'Statheleg' as well.

REFERENCES

BAKER, A.R.H. and HARLEY, J.B. (1973) Man Made the Land (Newton Abbot)

BERESFORD, M.W. and St. JOSEPH, J.K.S. (1979) Medieval England : an Aerial Survey (Cambridge)

COCKFIELD FIELD BOOK (CFB) This volume, in private hands, has been xeroxed and lodged in the Department of Palaeography and Diplomatic, University of Durham. Entries span the period 1857-1944.

GARMONSWAY, G.N. (1953) The Anglo-Saxon Chronicle (London).

DEWDNEY, J.C.(1970) Durham County and City with Teeside. British Association for the Advancement of Science.

HARVEY, D. (1969) Explanation in Geography (London).

HARVEY, M. (1978) The Morphological and Tenurial Structure of a Yorkshire Township Department of Geography, Queen Mary College, London, Occasional Paper 13.

HODGESON, J.C. (1899) A History of Northumberland V (Newcastle upon Tyne.

KAPELLE, W.E. (1979) The Norman Conquest of the North(London).

LOCKHART, D.G. (1980) 'The Planned Villages', The Making of the Scottish Countryside ed. M.L. Parry and T.R. Slater (London), 249-70.

PAGE, W. (ed.), (1892) The Cartulary of Brinkburn Priory Surtees Society XC

ROBERTS, B.K. (1972) 'Village Plans in County Durham: A Preliminary Statement', Medieval Archaeology XVI, 33-56

ROBERTS, B.K. (1975) 'Cockfield Fell', Antiquity XLIX,No. 193, 48-50

ROBERTS, B.K. (1977) The Green Villages of County Durham, Durham County Library, Local History Publication no 12, 56 pp.

ROBERTS, B.K. (1978) 'Rural Settlement in County Durham:Forms, Pattern and System', in Green, D,Haselgrove, C., and Spriggs, N. (eds.) Social Organisation and Settlement, British Archaeological Reports, International Series (Supplementary), 47 (ii), 291-322.

ROBERTS, B.K. (1979) 'Village Plans in Britain', in Recherches de Géographie Rurale, Numéro hors série du Bulletin de la Société géographique de Liège (Liège) 33-49.

ROBERTS, B.K. (1982a) Village Plans (Shire Publication No 27)

ROBERTS, B.K. (1982b) 'Village Plans in Warwickshire: A Preliminary Discussion' in Field and Forest : Essays in Honour of Harry Thorpe ed. T. Slater and P. Jarvis, (Norwich)

ROWLEY, T. (1981) The Origins of Open Field Agriculture

SHEPPARD, J.A. (1966) 'Pre-Enclosure Field and Settlement Patterns in a Yorkshire Township', Geografiska Annaler, 48B, 59-77

SHEPPARD, J.A. (1976) 'Medieval Village Planning in Northern England: Some Evidence from Yorkshire' Journal of Historical Geography 2, pt 1

TAYLOR,C.C. (1977) 'Polyfocal Settlement and the English Village', Medieval Arcaeology XXI, 189-93

UHLIG, H. (1972) Die Siedlungen des Ländlichen Raumes (Giessen): in German, English and French.

CHAPTER 3

POSSIBLE PLANNED MEDIEVAL VILLAGES IN THE NETHERLANDS

J.D.H. Harten & J. Schuyf
(University of Utrecht)

In recent years the work of Roberts and Sheppard has shown us the existence in England of a number of planned or replanned medieval villages, especially in the North (County Durham and Yorkshire). The existence of such villages in the Netherlands has been known for a number of years, but it appeared only on closer examination that there were many more than had been thought previously. In this article we would like to discuss some examples of medieval villages which show traces of either planning or replanning. It is by no means an exhaustive survey, but more or less an état de question. The most important question which we want to answer is whether these villages were primarily the products of careful planning or whether the visible regularity is mainly the result of a simple adaptation to physical circumstances.

STREET VILLAGES

Well-known to foreigners are the Netherland street villages of the fenlands. These were primary planned villages established in the wastelands from the 9th or 10th centuries onwards. (1) They were laid out along a reclamation-base - either a stream or a causeway - in such a manner that drainage would be easy. Strip parcellation was furthered by the Frisian common-law principle of opstrek, allowing the farmer to extend his strip into the waste until an obstacle was met. Such strips included both higher and lower areas, so that soils suited for arable as well as for pasture were involved (Plate 3:1).

Following the rise of local territorial lords in the 11th century the reclamation of the fens offered additional profit, with free men accepting grants of land. Farms were laid out according to a standard measure. This culminated in the innner fens of Utrecht and South-Holland, with their more uniform height and soil quality, in the so-called cope-reclamations (from kopen = to buy, viz. from the territorial lord) which were rigidly geometric. Here every plot was ca.110m wide and 6 furlongs (ca. 1200m) long (Plate 3:2 Hoenkoop). In these cases physical and political circumstances went hand in hand in creating a rigorously planned landscape. The system was widely exported, especially to Germany, where we find Dutch colonists around Bremen in 1106 (VAN DER LINDEN 1956,17) and in the Rhineland (KRINGS, this volume). Strip-like reclamations in wooded areas, Waldhufen (especially in more or less hilly terrain), are now known not to be only of 'Dutch extraction', as they were first mentioned in German sources of the 9th century (NITZ 1962,84). On the

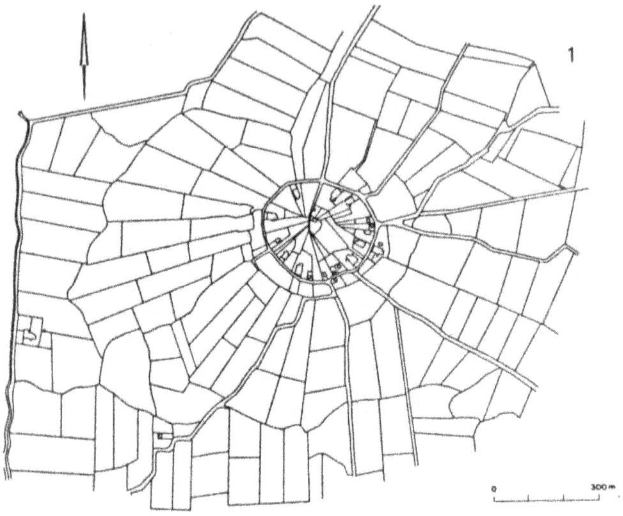

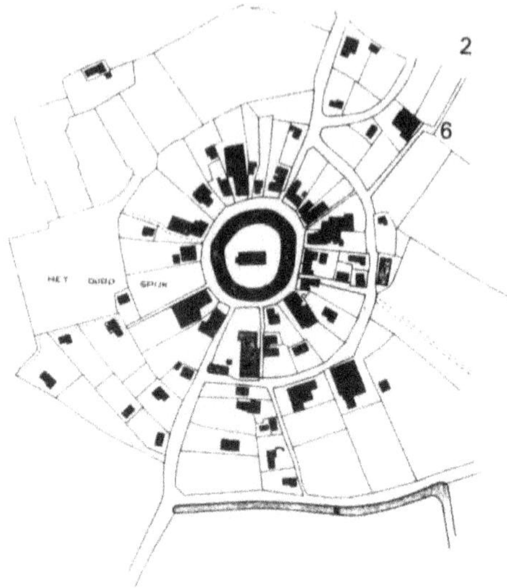

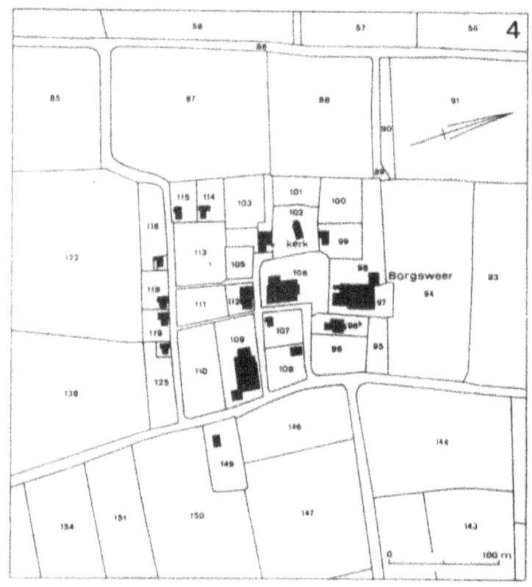

Different scales.

Fig. 3:1.1 Radial terp village Biessum near Delfzijl (Groningen) after the cadastral minute of 1830. (From Waterbolk and Boersma).

Fig. 3:1.2 Spijk: village on radial terp (Groningen). Church and churchyard are situated in separate ringroad and ditch. (From Halbertsma).

Fig. 3:1.3 "Trade settlement" Loppersum (Groningen), present situation. 1= castle site; 2 = rough limit of terp; 3 = water. (From Klok).

Fig. 3:1.4 Borgsweer near Termunten (Groningen): terp with rectangular parcellation, after the cadastral minute of 1830. (From Klok).

sandy areas of the Netherlands we find similar strips as assarts from the forest.

CONCENTRIC VILLAGES

In the marine clay area, both in the North and in the South-West of the country, we find concentric villages. The marine areas only became habitable at different times during the pre-Roman and medieval period and even then - until widespread use of dikes from the 11th century onwards - were under constant threat of submersion. In the provinces of Groningen and Friesland settlement was based on artificial mounds or terpen. Initial settlement was directly on the marsh, but very soon the continuous rise of sea-level forced the inhabitants to raise individual low house-mounds. At first such settlements were very small (1-3 farms), but as they increased in size it became difficult to maintain the unity of the three entities of the farm :meadow, pasture and arable. Even before the separate house-mounds grew into one village-mound the whole arrangement in the village could be re-planned. This led in some cases to a radial village wilth radial parcellation (Fig.3:1.1). In excavations such as those at Feddersen Wierde (GFR) and Ezinge (Groningen) this re-planning was dated shortly after the beginning of our era. (MAYHEW 1973,34).

Essential to the radial arrangement is the peripheral roadway or 'oxes way' along which the farms were placed. In some cases there is a second inner, ring-road (Fig. 3:1.2). These roadways were described in the 11th/12th century Frisian lawbooks (Schoutenrechten) and fixed measurements for the roads were proscribed. A certain amount of planning therefore cannot be excluded. At present the church is often found in the middle of the village on what was formerly an open space, which perhaps played some role in heathen cult. Thus, around the churchyard we find the wandeling or 'walk', another - narrower - encircling roadway around which, at funerals, the bier is twice or trice walked in the direction of the sun. This custom must be very ancient and can probably be regarded as the survival of an heathen rite-de-passage (GLAZEMA 1948,275 ff.) The terpen raised on newly emerged land in the 8th/9th centuries show much more rectangular planning (Fig.3:1.4). Moreover, some villages grew into trading settlements in the early Middle Ages, which led to the creation of an oval mound, with a single main street, in which merchants and artisans lived. Here the church is situated at one end and the castle (borg) at the other end (Fig.3:1.3) (KLOK 1975,147). In these cases replanning is probable.

In the south-west of the country, in Zeeland, we find a number of very small concentric villages whose origin is unclear (Fig.3:4.1). They are usually situated on small levée-ridges or in the basins of marsh deposits sedimented in the early medieval period. According to one author ('t HOOFT

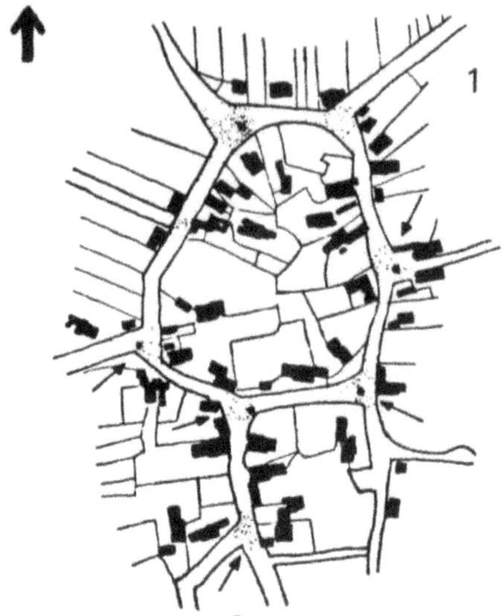

Fig. 3:2.1 Ringdorp Elp (Drenthe), ca. 1890. The village was created from separate "farm spaces" or little brinks (speckled). (From Harms).

Fig. 3:2.2 Cadastral minute (1830) of Ruinen (Drenthe) showing large central green. (From Brinkenboek).

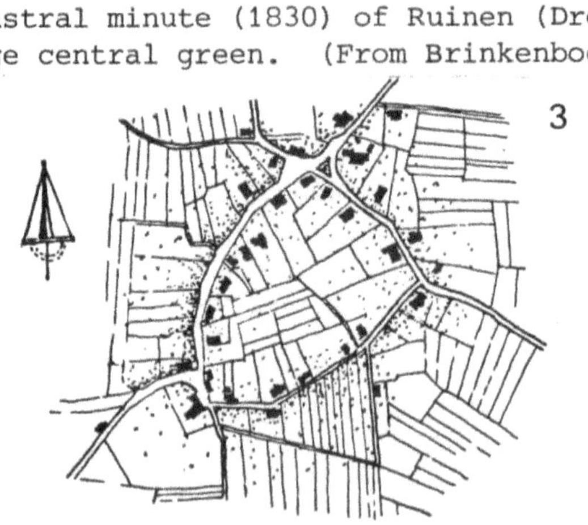

Fig. 3:2.3 Ringbrinkdorp Ballo (Drenthe). (From Kleijn).

Different scales

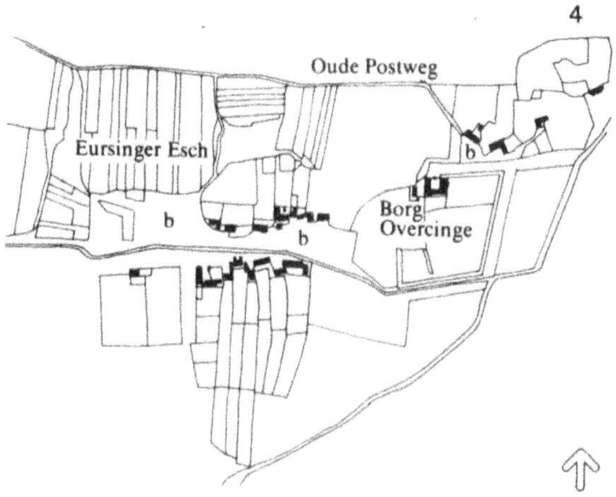

Fig. 3:2.4 Cadastral minute (1830) of Eursinge/Overcinge near Havelte (Drenthe). Secondary village "over the es" around central green(b). (From Brinkenboek).

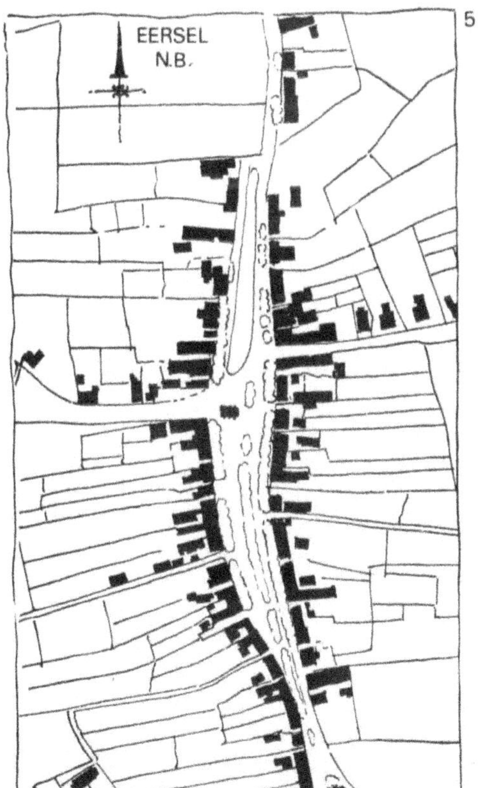

Fig. 3:2.5 "Trade settlement" Eersel (Brabant) around central green. (From Cleerdin).

Different scales

1944,22 ff.) the circular form is apparently accidental, caused by the construction of two side-roads next to the road-crossing which was the focus of the village and where the church, when present, was built. It can, however, be doubted whether this concentric form is really accidental. In some cases, such as 9th century Dreischor (2) the circle is so exactly round that there must have been some degree of planning.

Some larger settlements are based on a circular structure with a diameter of ca.200 meters found by excavations to belong to a 9th century fortification, analogous to Viking Trelleborg. They probably represent local defence against the Vikings and are found all along the North Sea coast from North-Holland to northern France (e.g.Middelburg, Souburg (TRIMPE BURGER 1973)).

Post-medieval reclamations in Zeeland very often show rectangular 'Renaissance' patterning. Newly reclaimed polders in both Zeeland and Holland were divided according to an ideal perceived pattern of squares or rectangles (the last with a length-width ratio of about 3:2, the so-called 'Golden Cut'). A perfect example is Borssele (Zeeland) laid out in 1616 on a scheme of rectangles measuring 89 x 140 m. with the village slightly askew to the parcellation due to its orientation on an older castle site.

BRINKDORPEN

The third main group of village types we want to discuss is that of the brinkdorpen (3) There are a great many different types of these in the Netherlands, but they all have in common that they are situated on soils well above Ordnance Datum (usually in the sandy eastern and southern parts of the country), that the arable used to be concentrated in one or more open fields, and that there were one or more (usually peripheral) greens present. The arrangement of the various elements in the village complex, i.e.farmsteads, arable, meadow and (extensive) pasture can be quite diverse leading to very different morphologies. Inquiries into the genesis of these brinkdorpen however have been largely concentrated on the history of the 'classic' nucleated brinkdorp such as we find it in the province of Drenthe. This brinkdorp was formerly thought of as 'disorderly' or, in the words of its first scholar, Keuning, 'a collection of stray farms put rather arbitrarily in the former common forest' (KEUNING 1936,83). This apparent disorder is, however, usually caused by the degeneration of the original form; recent excavations show that this is often the case. Waterbolk concluded from the results of the excavation of an early medieval village (ca.450-850) near Odoorn (Drenthe, Fig.3:3.1) (WATERBOLK 1973;1980) that a number of villages were originally planned along two axes, west-east and north-south, creating a kind of chequerboard pattern. Farmyards and adjoining parcels were separated from each other by means of these axial roads. These farms were removed during the first half of the 9th century, probably to the present site of the village, but the parcellation remained intact, as can be seen on the map, thereby suggesting a continuity at least from the 9th century onwards.

A number of other excavations have also revealed regular villages which moved slightly to a new location within the same territory during the middle ages. In Gasselte (in the North of Drenthe) excavations disclosed an elongated one-row village dating to the 9th-12th centuries. During the 12th-13th centuries this village moved to its present site slightly to the east (HARSEMA 1980) Waterbolk ascribed the marked regularity of these plans to an 'ordering force' in society, perhaps a local chief. Another influence could have come from Roman centuriation.

The excavations at Kootwijk (Gelderland) showed that the division into rectangular blocks could be found in other parts of the sandy area of the Netherlands too (Fig.3:3.2). As in Odoorn, the farms were all oriented in the same direction. In the course of its life, the village moved slowly from North-West to South-East, into the area of a former fen functioning as a kind of brink. The arable moved over the abandoned house-tofts. In the immediate neighbourhood, a little distance away, the remains of a late Roman and two Merovingian settlements were found; the main settlement belonging to the Carolingian period (HEIDINGA,N.D.) (3). The early medieval very regular village revealed by the excavations has all but disappeared in the present landscape. According to Waterbolk, the two-axial system of west-east and north-south crossroads can still be recognised in some extant Drenthe villages such as Emmen and Zeyen. Roman influence is again mentioned as an important agent in the formation of these plan-types (WATERBOLK 1973). In our opinion, however, a more simple solution is also possible.

Edelman and Vlam (1963) already pointed out that in many of the small villages in the sandy part of the country the centre of the village was formed by large block-like parcels, measuring ca.1-1½ ha (2½-4 acres) apiece. They regarded these blocks as initial reclamation units from the early medieval period, block parcellation being the most logical system of land division in a situation where cultivable land was still plentiful. (4) When arranging several of these blocks, forming the original arable of the farms, together, it is hardly surprising that the main structuring elements of a settlement run along two or more axes. One does not need an ordering force, with or without Romans. This theory is in accordance with that of Bouwer (BOUWER 1975). In his opinion neither the roads nor brinken were the main structuring elements in the village, but the original (rectangular) farm tofts. Their situation and that of the ecotopes determined the course of the roads, whereas the brinken were lying outside the village core (brink meaning edge).

1

Fig. 3:3.1 Excavation of early medieval village at Odoorn (Drenthe) showing large rectangular blocks between roadways and present parcellation and position of excavation in regard to present village (inset). (From Waterbolk).

2

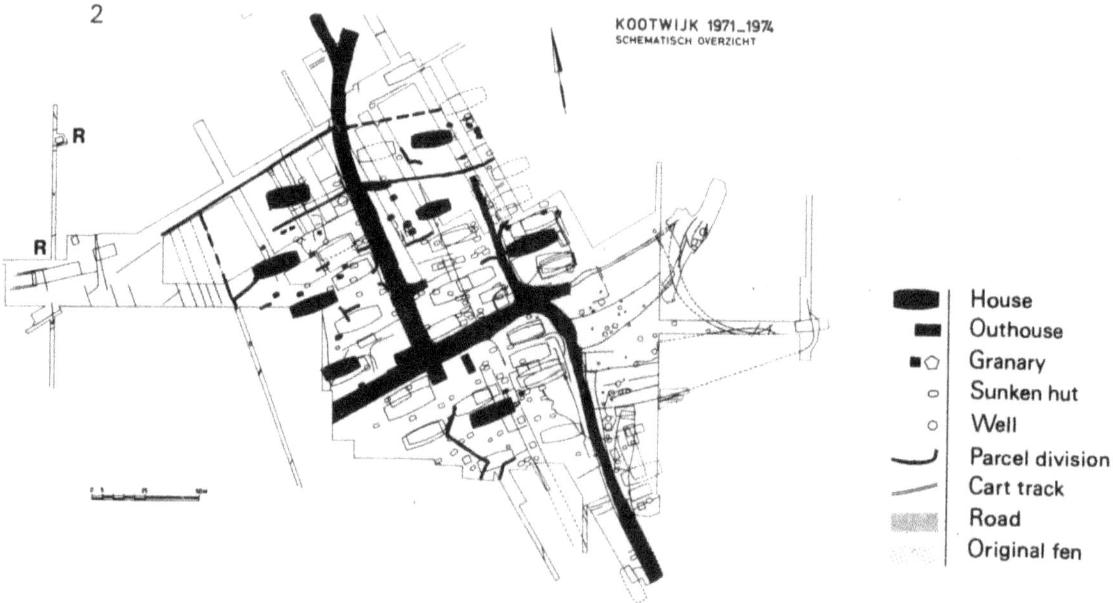

Fig. 3:3.2 Early medieval village at Kootwijk (Gelderland). General excavation results 1971-4; in black the first phase showing rectangular parcellation of village. Arable to the NW of village. (From Heidinga).

With regard to the brinken Bouwer is only partly right. Indeed the larger ones were originally situated on the fringe of the village, but this was not the case with the smaller ones, the so-called boerderijruimten. Harms was the first to discover that the 'seeming disorder' of most of the brinkdorpen disappeared if one saw the village as an agglomerate of farm groups, each situated around a small open space, the boerderijruimte, used to gather the animals at night. From these spaces the village grew. In some cases these boerderijruimten were more or less circularly situated. This led to village plans called ringdorp and ringbrinkdorp (with a larger brink in the ring) (HARMS 1957,146)(Figs.3:2.1 and 3:2.3). On toponymical evidence it appeared that the ring(brink)dorpen, which probably reveal a certain degree of planning, are younger than the normal brinkdorpen. Keuning already remarked on the foundation of daughter settlements of the old brinkdorpen on the other side of the open field from late medieval times onwards (KEUNING 1936,101). They had names like Eursinge, meaning 'over-the-es'. In the case of Eursinge near Havelte (Fig.3:2.4) this led to clear signs of regularity in the plan, rather reminiscent of long tofted plans in North England (ROBERTS 1972).

Several villages have a large central green (Fig.3:2.2). As this is somewhat unusual, the question can be asked whether this was the result of re-planning. Marked regularity, such as at Ruinen and Dwingeloo, suggests that it was; thus Dwingeloo was in historic times one of the main markets of Drenthe. Other evidence for planning has recently come to light in Gelderland. Barends found in the township of Geesteren three small nuclei with irregular block-parcellation and common land surrounding a remarkedly regular two-row village and also regular open field (BARENDS 1981). There are several English parallels for this type of village (e.g. ROBERTS 1978) all dating to the 12th/13th centuries. The reason for founding a new village in Geesteren is not known.

Re-planning of trading settlements seems likely in yet another part of the country where we normally find a regional type of brinkdorp:the sandy area of North Brabant. A number of these brinkdorpen were replanned into market villages, whenever they acquired a market function e.g.by situation on a land-water crossing. A market place was created either by widening the road (e.g.Eersel,Fig.3:2.5) or by creating a new market square next to the old nucleus of the village.

ELONGATED VILLAGE TYPES

Finally two other village types must be discussed. On sandy dune ridges along the western coast of the country we find so-called "geest"-villages (geest being the lowlying parts of the ridge on which these villages are situated). De Cock has shown that at least in the case of Rijswijk this type of village was structured like a brinkdorp with open field and brink (DE COCK 1975). Partly due to the elongated structure of the dune-ridge, the villages have a marked regular structure. They are usually one-row villages with a front and back row, both surrounding the arable which is thus in the middle of the village and divided into a large number of short strips diagonal to the ridge. The pasture is situated in the fens alongside the dunes. It is very probably that in almost all cases these villages were reclaimed under manorial control from the 8th century onwards; these manors were usually owned by the counts of Holland, whose earliest possessions are found on the dune ridges of Holland (DE COCK 1975,431;SCHUYF forthcoming).

Comparable villages are found in the western part of the riverine area in the middle of the country. On the elongated levee-ridges of especially the West-Betuwe we find so-called gestrekte rivieresdorpen. At first sight they show a degree of planning with their two or three rows, joined usually at the end or right in the middle by the church and/or brink (Fig.3:4.2;Pl.3:3). Most of the arable is concentrated into one or more open fields, the strips often slightly curved as a reversed-S. Toponymical evidence suggests that they were reclaimed during the Carolingian period (HARTEN 1976). Further to the east, in the East-Betuwe, we find few elongated esdorpen, but many irregular round ones (Fig.3:4.4) Toponymical evidence here suggests a large degree of continuity from Roman times onwards, maybe indicating that the round type of village is older than the elongated type. The area is situated somewhat higher above Ordnance Datum, therefore diminishing the influence of frequent flooding from the rivers. Furthermore the physical situation is clearly different in the way that the levée ridges are often wider that further west and the villages are usually situated on the crossing of two levées. An elongated structure was therefore not necessary (DEN UYL 1958). So it is probable that the physical situation was the main factor in the variation of the village plan types of the riverine clay area.

CONCLUSION

In this short survey we discussed some regular village types from the Netherlands:street villages, concentric forms, several types of brinkdorpen, elongated and market villages. In all of these cases a marked regularity was present, but this was due to no single cause: apart from social factors, usually stemming from reclamation as-a-whole, we found that physical geography played an important role. In the case of some of the terpdorpen and the market villages we find re-planning on the same site. In all other cases these villages were founded on new sites.

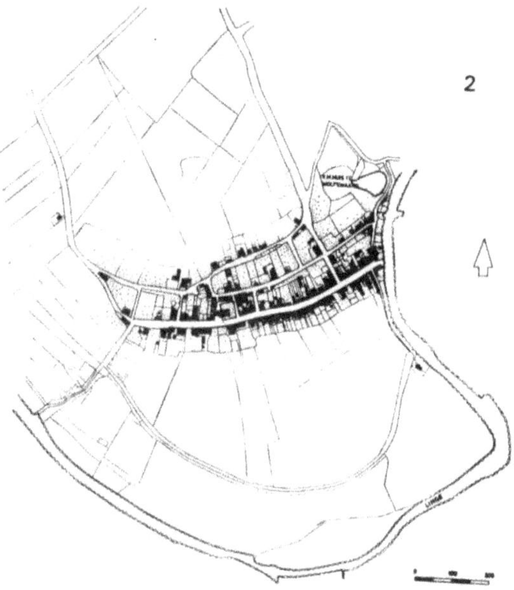

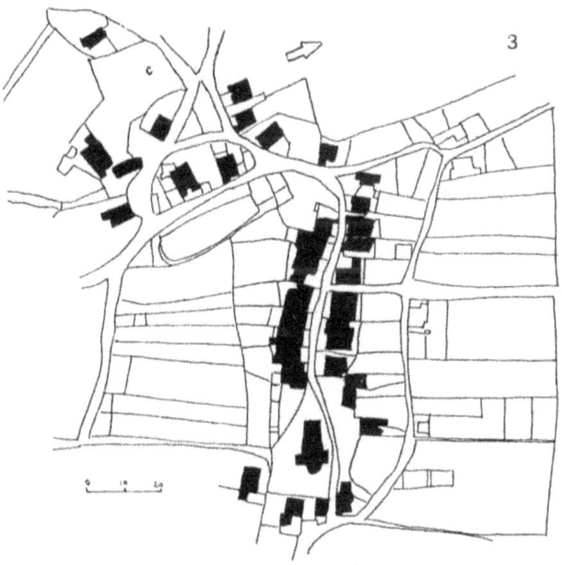

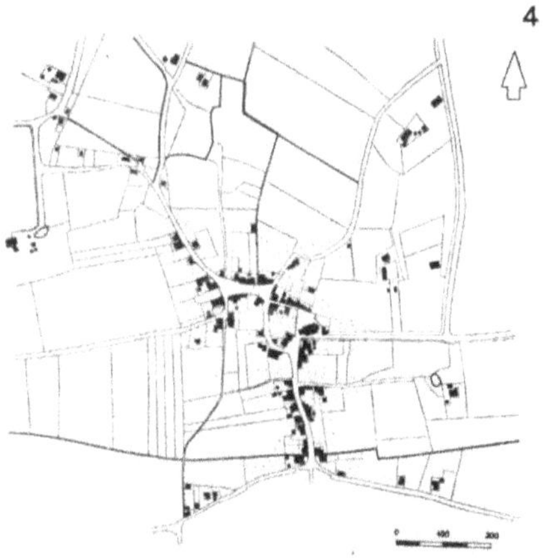

Different scales

Fig. 3:4.1 Concentric villages in Zeeland (Present situation). (From Lussanet de la Sablonière).

Fig. 3:4.2 Beesd (Gelderland): elongated riverine village; after cadastral minute 1826. (From De Beaufort and Van den Berg).

Fig. 3:4.3 Geesteren (Gelderland). "Replanned" village (right) and older nucleus (c = left). From cadastral minute 1830. (Barends).

Fig. 3:4.4 Lienden (Gelderland). Round riverine village, after cadastral minute 1826. (From De Beaufort and Van den Berg).

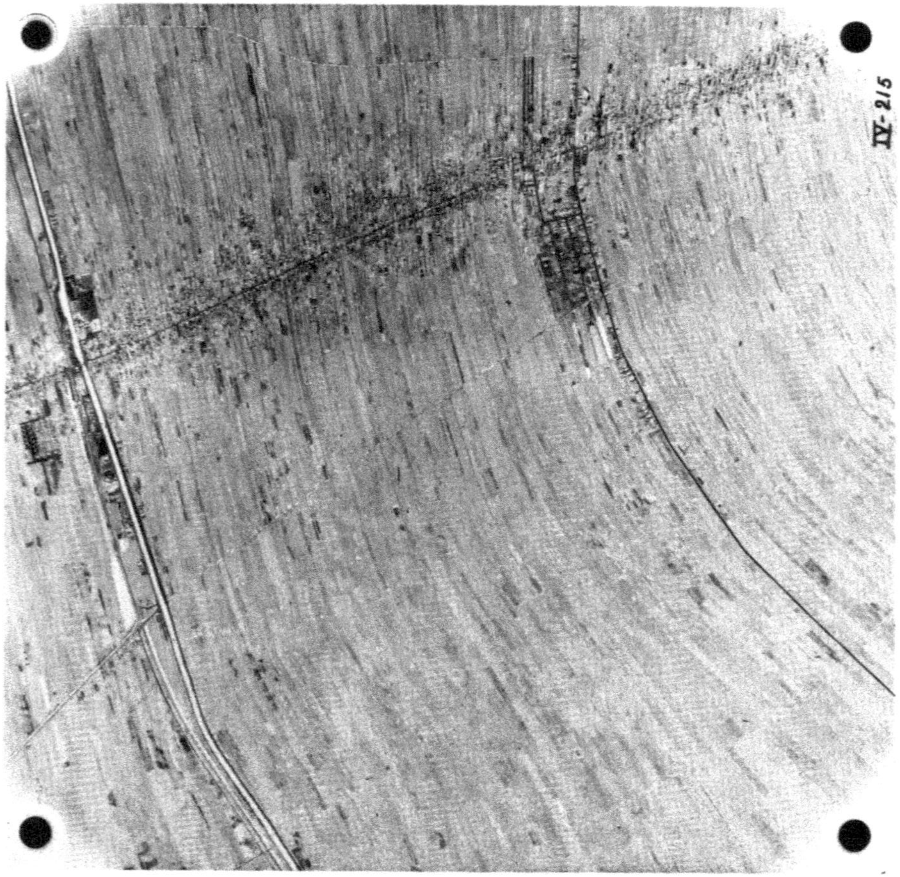

Plate 3:1 Vriezenveen (Overijssel). Fen-village showing original unlimited "opstrek" and very narrow parcels due to partial inheritance. Situation circa 1950; the area has now been re-allotted.

Plate 3:2 Hoenkoop (Utrecht). "Cope"-reclamation of probably 13th-century date. Situation 1955.

Plate 3:3 Wadenooijen (Gelderland). Elongated riverine village along the Linge. To the south the arable is divided into long and narrow strips. Further south the pasture was assarted from the fen in late medieval times. Situation 1950.

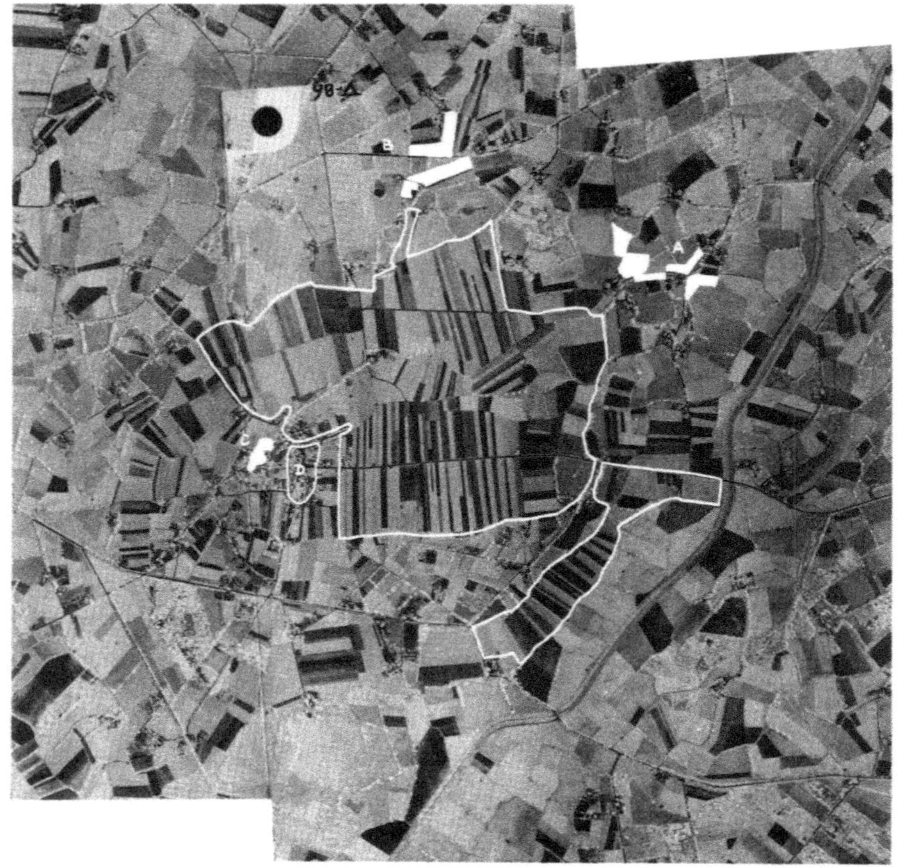

Plate 3:4 Geesteren (Gelderland), showing the older nuclei (a, b, c) and the present village (d) with the original arable (within black line). Situation 1950.

Photographs reproduced with kind permission of the Topographical Service of the Nederlands.

NOTES

1)Van der Linden found the earliest documentary evidence to striplike cope-reclamation dating to the 11th century. In some cases, however, there are indications that strip-like reclamation of the fens started earlier: the cope-parcellation is relatively late phenomenon. In the province of North Holland there were 8th century fen-reclamations with names ending in -more (=moor) (DE COCK 1965,254). In South Holland the earliest manors in the fen date to the 8th century. In this case it is not very clear at which point in time manorial economy changed from cattle-breeding to agriculture; the first alienation of single farms can be dated by documents to the beginning of the 11th century (SCHUYF, forthcoming).

2)This appears to be a mistake. The island drowned in 1287 and was reclaimed in the last quarter of the 14th century. If the village was replanned, it was a late-medieval replanning.

3)A brink has all the functions of a green, but differs in one major aspect from a green: originally it was very seldom situated in the centre of a village, but always on the edge. Usually a village had more than one brink (Cf.VAN DER VAART, this volume).

4)These blocks can still be recognised on the map bearing the field-name worth. (See for this phenomenon a.o.GLÄSSER 1967 and HAMBLOCH 1962.)

REFERENCES

BARENDS 1981	Barends, Sonja, Historisch Geografisch onderzoek naar de gevolgen van ruilverkaveling (Wageningen 1981).
BEAUFORT and VAN DEN BERG 1968	Beaufort, R.F.P.de and H.M.van den Berg, De Betuwe (Monumenten van Geschiedenis en Kunst III/1) Den Haag 1968.
BOUWER 1975	Bouwer,K.'Het Drentse Esdorp' Geografisch Tijdschrift N.R.IX,1975, 297-309.
BRINKENBOEK 1981	Werkgroep Brinken, Brinkenboek, een verkenning van de brinken in Drenthe (Assen 1981).
CLEERDIN 1944	Cleerdin,V,Het Brabantsche Dorp (Amsterdam 1944).
DE COCK 1965	Cock,J.K.de, Bijdrage tot de historische geografie van Kennemerland in de middeleeuwen op fysisch-geografische basis (1965).
DE COCK 1977	Cock,J.K.de, 'De middeleeuwse ontginningen Rijswijk en Rijswijkerbroek', Geografisch Tijdschrift N.R.XI 1977, 428-433.
EDELMAN-VLAM 1963	Edelman,C.H. en A.W.Edelman-Vlam,'Bijdrage tot de ontginningsgeschiedenis van de Nederlandse zandgronden', Boor en Spade 13,1963,1-11.
GLÄSSER 1967	Glässer,E, 'Zur Entstehungsgeschichte der bäuerlichen Kulturlandschaft' Geographische Rundschau 1967,383-9.
GLAZEMA 1948	Glazema,P, Gewijde plaatsen in Friesland (Meppel 1948).
HALBERTSMA 1963	Halbertsma,H, Terpen tussen Vlie en Eems (Groningen 1963).
HAMBLOCH 1962	Hambloch,H.'Langstreifenfluren in nordwestlichen Alt-Niederdeutschland' Geographische Rundschau 1962,345-356.

HARMS 1957 — Harms,J.A.,'Het Drenthse kern-esdorp in zijn veranderde functie' Bulletin Kon.Ned.Oudheidkundige Bond 1957, kol.129-158.

HARSEMA 1980 — Harsema,O., Drents Boerenleven van de Bronstijd tot de Middeleeuwen (Assen 1980).

HARTEN 1976 — Harten,J.D.H., 'Historisch-geografische achtergronden van het landschap in het gebied van de ruilverkaveling Avezaath-Ophemert' Avezaath-Ophemert,Bijdragen voor een ruilverkavelingsplan (Geldermalsen 1976).

HEIDINGA n.d. — Heidinga,H.A.,Verdwenen dorpen in het Kootwijkerzand. Barneveld, n.d. (ca.1977).

't HOOFT 1944 — Hooft,P.J.'t,Dorpen in Zeeland (Amsterdam 1944).

KEUNING 1936 — Keuning,H.J.,'Nederzettingsvormen in diluviaal Nederland ten noorden en ten oosten van den Ijssel 'Tijdschrift voor Economische Geographie, 1936.

KLEIJN 1949 — Kleijn,A.,Dorpen in Drenthe (Amsterdam 1949).

KLOK 1975 — Klok R.H.J.',Terpen zullen ons een zorg zijn' Groningse Volksalmanak 1974-5,129-166.

KRINGS 1982 — Krings,W.,'Late-Medieval Dutch Settlements in the Lower Rhinelands'. This volume.

VAN DER LINDEN 1956 — Linden,H.van der,De Cope (Assen 1956).

MAYHEW 1973 — Mayhew,Alan, Rural settlement and Farming in Germany (London 1973).

NITZ 1962 — Nitz,H.J., Die ländlichen Siedlungsformen des Oden-waldes (1962).

ROBERTS 1972 — Roberts,B.K.,'Village plans in County Durham: a preliminary statement' Medieval Archaeology (1972), 33-56.

ROBERTS 1978	Roberts,B.K.,'The regulated village in Northern England: some problems and questions' Geografica Polonica 38 (1978), 245-252.
SCHUYF, forthcoming	'Het reconstructiegebied Midden-Delfland historisch-geografisch bekeken', Geografisch Tijdschrift, forthcoming.
TRIMPE BURGER 1973	Trimpe Burger,J.A. Oost-Souburg, Province of Zeeland: A Preliminary Report on the Excavation of the Site of an Ancient Fortress (1969-71) Berichten ROB 23, (1973), 355-365.
DEN UYL 1958	Uyl, R.G.den ,'Dorpen in het rivierkleigebied', Bulletin Kon.Ned.Oudheidkundige Bond 1958, kol.97-114.
WATERBOLK 1973	Waterbolk,H.T.,'De oorspronkelijke structuur der Drentse Zanddorpen' Archeologie en Historie, opgedragen aan H.Brunsting (Bussum 1973), 429-439.
WATERBOLK 1980	Waterbolk,H.T.,'Hoe oud zijn de Drentse dorpen?' Westerheem (1980), 190-212.
WATERBOLK and BOERSMA	Waterbolk,H.T. and J.W.Boersma,'Bewoning in voor en vroeghistorische tijd' W.J.Formsma (ed), Historie van Groningen.Stad en Land (Groningen 1976), 13-73.

CHAPTER 4

THE BRINK OR VILLAGE GREEN IN BRINK VILLAGES IN THE PROVINCE OF DRENTHE (THE NETHERLANDS); ITS DEVELOPMENT, CONTEMPORARY VALUE AND PROSPECT

J.H.P. Van Der Vaart
(Fryske Akademy)

INTRODUCTION

In a society which is changing at an ever-accelerating pace we are also witnessing dramatic changes in cultural landscapes. In this process regional variations in landscape and settlements are becoming scarce, a result of the continuous pressure of adjusting these elements of our habitat to today's needs. However, a growing awareness of this process of landscape impoverishment can be noted. In the Netherlands concern about it is evident in both national and regional environmental planning policies, and it is now considered necessary to preserve the essential characteristics of regional landscapes and settlement structures. Nevertheless, although concern and good intentions are often expressed, this does not automatically ensure that action is undertaken on every relevant object. In my opinion historical geographers can play an important role in stimulating action. By using their knowledge and insight, and combining these with the rules of environmental and land use planning, they can make a valuable contribution towards preserving the heritage of the past and regional cultural differentiation.

In this paper I want to show you a method how to deal with the combination of environmental planning and historical geography in a rural settlement. As an example of this method some results will be shown from a study that recently has been done on the brink or village green in the province Drenthe in the Netherlands (HOUTING, DE POEL and VAN DER VAART, 1981).The main objective of this study was to make inhabitants and local government aware of the threats towards this regionally characteristic element of the village by providing them with information on the past, present and possible future of the brink. The brink village is very specific for the landscape of the Drenthe Plateau. Its value is 'felt' by many inhabitants, authorities and government bodies in this province, but up till now a systematic study of one of structurally the most important elements of this type of settlement was lacking.

Two major components of the expressed value of the brink can be distinguished: (1) the so-called 'village-scape' component and (2) the historical-cultural component. In combination they give the brink a special meaning. Places like this, 'places with a high sensory quality' as Lynch (1980,8.9) calls them, are scarce. Such desirable places are usually the remnants of a process of slow development, which occurred within sharp constraints of natural condition and cultural limitation and since then have been enriched by continuous habitation and reformation. The pressure of modern functions, for instance parking, poses a threat that is able to destroy in a short time something that has grown over the ages and is appreciated. Everbody cares, but not really. Someone has to take the first step of action: you as a geographer are able to do so.

THE QUESTIONS TO BE SOLVED

The contemporary functions of the brink (plural brinken) may be not the same as they were a hundred years ago, but this is no reason to abuse and neglect this part of public space in the village, as we see happen too often in recent years. Their history is too rich to let them wither away. Besides that the brink can have a function in today's village life in such a way that it serves the needs of the present and yet still shows the characteristics which originally belonged to it. From this point of view the next three questions were central in our study, reflecting past, present and future:

1. Which are the basic characteristics of the brink? This question should be solved by means of looking at the historical development.
2. What is the present state of the brinken, considering the characteristics and the village-scape value.
3. What needs to be done to integrate and protect the brink in the village of today and tomorrow.

SOME NOTES ON THE HISTORICAL DEVELOPMENT OF THE BRINK

To get full knowledge of the brinken in their present form it is necessary to study their history and development. Due to lack of documents much of the early history of the brink remains unknown, particularly the ultimate origins of this element of the settlement. On the basis of the agricultural evolution on the Drenthe Plateau it is possible to distinguish four stages in the development of the brink.

1) Although the origin of the brink or village green is not yet clear, most authors state that its origin must have been in the Early Middle Ages or the Dark Ages. It is not clear whether a brink existed in early settlements, and up till now excavations (e.g.Wijster, Odoorn and Gasselte) have not shown evidence of a brink. In the economic system of those ages however brinken may have existed as places where livestock could be driven together, as common grazing area on the immediate periphery of the settlement and as the place where communal activities and a market could take place. Because herding livestock, particularly cows, was one of the elements of the Dutch economy of the Early Middle Ages, it is likely that a place in the village was exclusively maintained for rounding up the animals when the herd returned from the grassland. The fact that a brink usually is situated on the side of the settlement

that is closest to the grassy meadows along the rivers where the cows were grazing during the daytime supports the hypothesis that a brink existed in these early medieval villages. In the description of the second stage further comment will be made on the location of the brink.

2) In the twelfth century the economy of the Drenthe Plateau entered a new era when a new hardy breed of sheep were introduced. This introduction probably took place under influence of two Cistercian monasteries (NAARDING,1947). These sheep were mainly kept for the manure that they produced. This supply of manure was very essential for fertilizing the poor sandy soil of the arable land. Even extension of the arable land, in Drenthe, called 'es', became possible. Because of this technique villages could grow, but only to a certain extent, because the area of the fell that belonged to a village limited the number of sheep that could be fed. This combination of mixed farming and herding large flocks of sheep lasted relatively unaltered till the second half of the nineteenth century.

The last two centuries of this stage provide information on the brink and its functions. This period also produced the basic features and structures which make up the characteristics of the brink as we see it today. At present we call these functions the 'original functions' and as such the following ones can be mentioned:

a) a place for rounding up cattle and sheep;
b) the production of high quality wood for construction and repair of homesteads;the brink was often partially or completely planted with oak trees;
c) the location of one or more village ponds; the water was used as drinking water for the animals and as a source of water to fight fires;
d) the ground of common meetings of the boerschap or entitled farmers;
e) the place where the local court assembled,
f) the place for markets and fairs;
g) the place for temporary storage of farm products during the harvest;
h) the place where the village school was built;

These characteristic functions are very similar to the functions of various types of greens on the North European Plain (THORPE, 1961, 115-116). The first function in particular calls for a comment on the location of the brink. The brink was the place where the flock of sheep gathered in the morning after having spend the night at the byres of their owners (in these byres the manure was collected). In the late afternoon the flock returned to the brink to be sorted out and return to their stables to perform their important task. Because of this continuous movement of the sheep the functional location of the brink should be on the fell side of the settlement. But in many villages the brink is situated on the grassland side, which is functionally far from optimal. We think that this indicates that

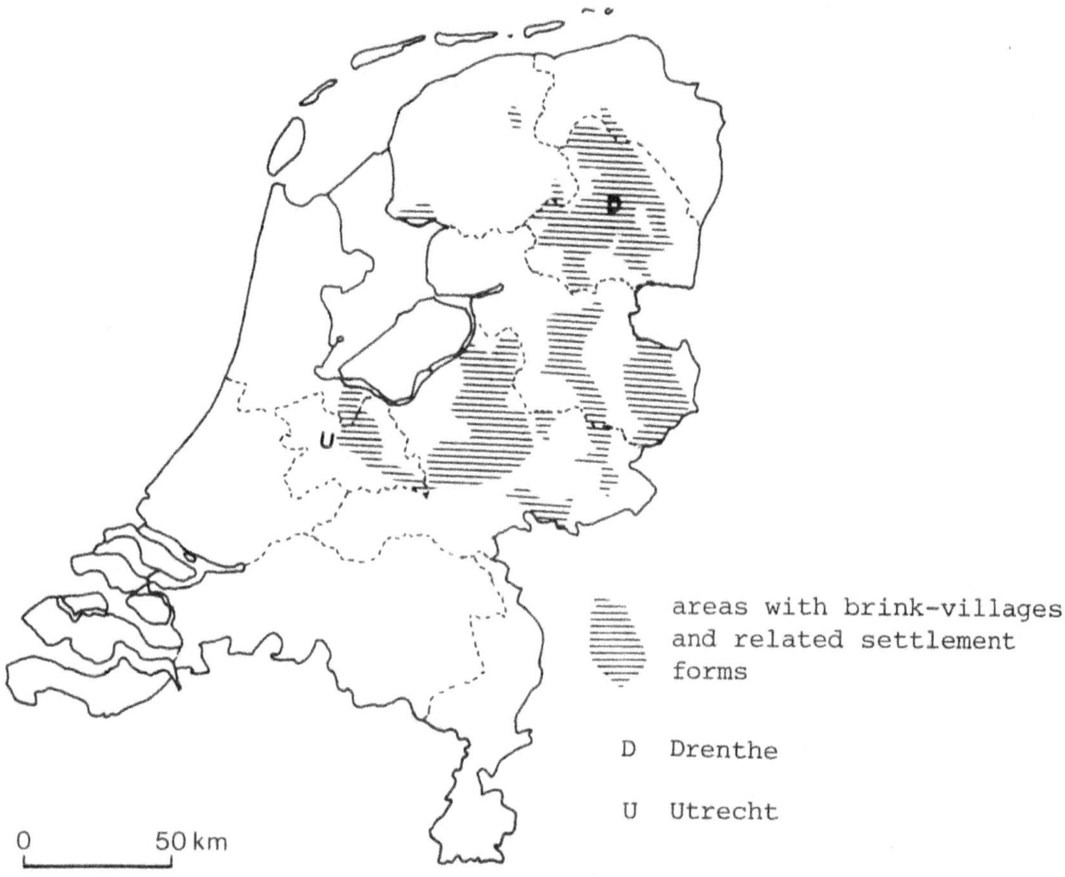

Fig. 4:1 The distribution of landscapes with brink-villages and related settlement forms in the northern part of the Netherlands.

brinken existed in Drenthe before sheep were introduced in the twelfth century and that the brink had a specific function for the cattle.

According to the archaeologist Waterbolk (1980, 197) there is evidence that most of the villages on the Drenthe Plateau existed in the tenth century. When we combine these two statements we think that there is evidence that brinken existed and had a function in the settlements of (at least) the Early Middle Ages.

3) In the period between about 1850 and World War II massive changes took place in the agriculture in Drenthe. Two major events started up these changes:

a) the 'Marke Wet' or Enclosure Act of 1868, under which the communal owned and used fell had to be divided and
b) the introduction of chemical fertilizer. Gradually the enormous herds of sheep disappeared, so the brink lost one of its main functions in the agricultural system. In many villages this resulted in filling in the ponds, dividing the brink to add it to the front yards of adjoining farms and houses, while in some cases houses were built on the brink. In most villages, however, the brink (or brinken, because some villages had more than one) remained the communal property of the boerschap.

4) A new stage of development was entered after World War II. The modernisation of agriculture in Drenthe began in the early fifties. The landscape changed in many ways to suit modern production techniques. Population growth, the building of many new houses, the introduction of small industries, supermarkets, tourism and an ever increasing traffic brought many changes in the old villages. Subsequently these new developments had their effect on the brink. The passage of traffic and the use as parking lot places an enormous burden on the most prominent elements of the brink. The oak trees, which are such a dominant and valued feature suffered a lot, frequently being damaged by cars. When the bark is badly damaged, it reduces the health of the trees substantially. Parking causes the soil to become too compact, so the trees can not get enough oxygen and water. A further threat to the trees is caused by mowing the grass with mechanical equipment, by which wounds on the foot of the trees are created. These bad spots reduce the vitality of the trees considerably.

Other threats to the brink are found in the careless widening of roads, and turning the brink into a neatly dressed public garden, introducing basically strange elements like rose gardens or hedges and levelling the micro-morphology.

THE SITUATION IN DRENTHE TODAY

The province Drenthe that covers most of the Drenthe Plateau, a sandy and loamy area of Pleistocene origin, has the best preserved collection of brink villages in the Netherlands. This type of settlement can also be found in small numbers on

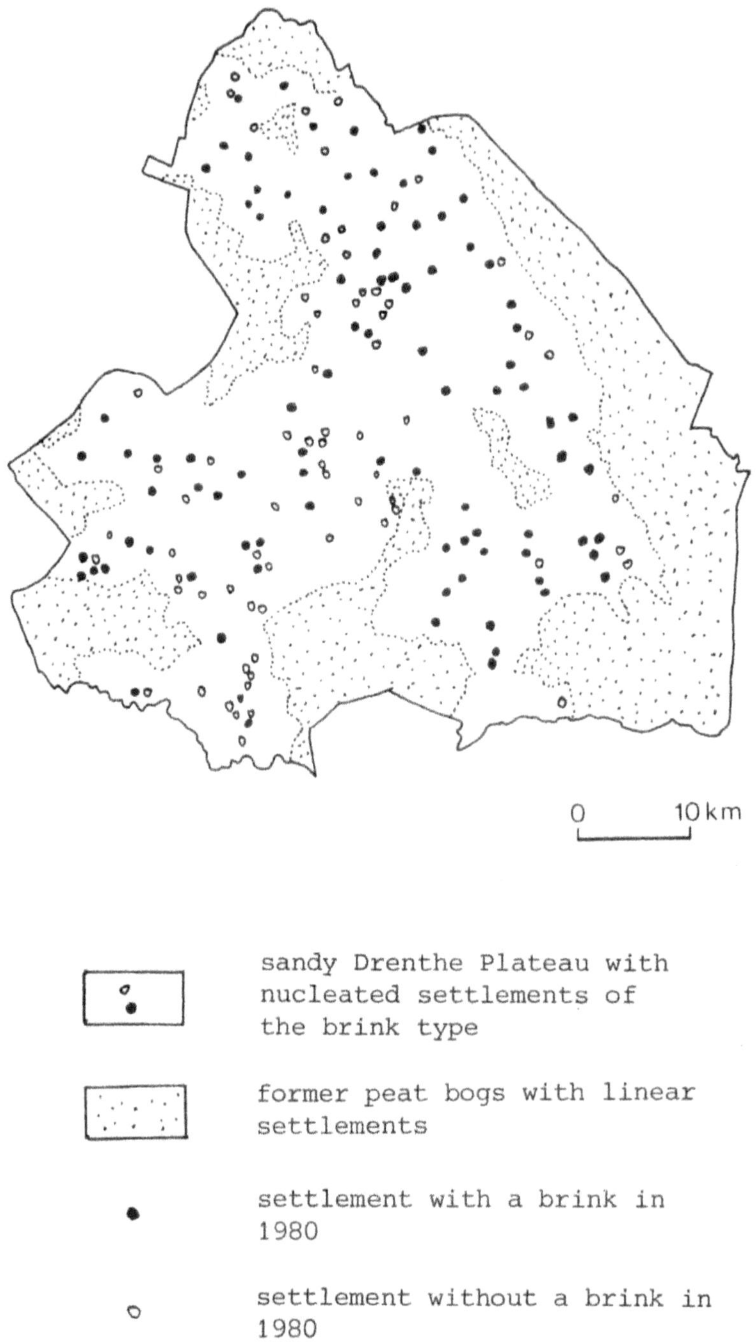

Fig. 4:2 Drenthe: villages with or without a brink, situation 1980.

the fringes of this plateau, in the provinces Friesland and Groningen (see also Fig.4:1). In a part of the province Utrecht some good examples of brink villages can also be found. (U on Fig.4:1). In the other shaded areas on this map settlements which are based on an originally similar agricultural system are present, but hardly any "brink-space" is to be found in these villages. In parts of neighbouring Western Germany however, for instance in the Hummling area, very fine examples of brinken still can be found,e.g. in Vreese.

Our inventory in Drenthe in the years 1979 and 1980 resulted in finding 162 brinken in 89 villages (Fig 4:2) 58% of the villages have one brink, 24% have two, 9% have three and another 9% have four or more, the maximum being ten.

By comparing the present situation with maps from the first half of the nineteenth century we discovered that during the last one hundred years 110 brinken had disappeared. Many of the brinken that we find today are smaller than they were around 1850. The example in Figure 4:3 illustrates this. Although 89 villages in Drenthe have a brink at present, it is worthwhile pointing at the fact that since 1850 36 villages in this province lost their brink completely. In another 21 villages there probably has been a brink, but further investigation of documents is necessary to prove this, while still another 15 very small hamlets never had a brink. These figures indicate that it is time to act if we do not wish to lose any more brinken.

Not only the number or size should bother us, the quality of the brinken deserves attention as well.

In our study the following method has been applied to evaluate the present quality of the brinken (HOUTING, DE POEL, VAN DER VAART, 1981,p.220). Valuation methods which are completely objective do not exist. Therefore the principle of intersubjectivity was applied by letting three people give scores on the following criteria:

a) state of the trees (age, health, indigenous or exotic),
b) the buildings around the brink (originality, architecture, height, shape, materials, colours, situation),
c) roads, paths and parking (dominance, materials, 'soft' or 'hard' edges),
d) the surface of the brink (micro-morphology, ponds, quality of the grass cover, maintenance),
e) the relation between the above mentioned elements: harmony or disharmony.

The possible score on each of these criteria could be:

bad 1 point,
mediocre 2 points,
good 3 points.

Finally the scores were added, every criterion carrying the same weight. It turned out that almost 40% could be

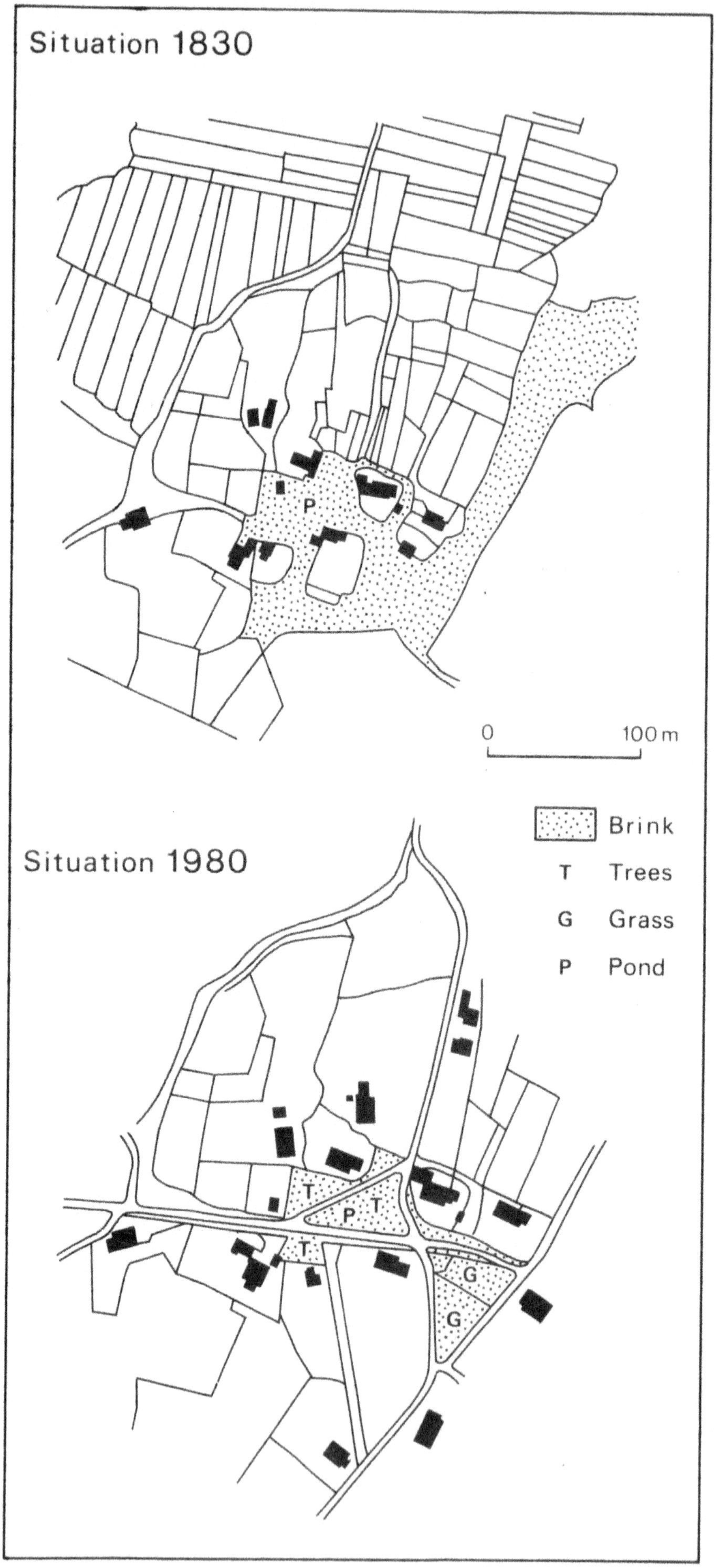

Fig. 4:3 Changes in the morphology of a brink; the example of the village Taarlo in Drenthe.

classified as reasonable to good, but about 60% of the brinken are at present in a bad to mediocre condition.

ENVIRONMENTAL PLANNING AND THE FUTURE OF THE BRINK

The brink, having lost its former agricultural function, still shows features that are a legacy of its past. Gradually new semi-urban functions enter the scene, adding new features. This is a very natural development in a society that is alive. There has always been change in the natural and man-made environment under the influence of mankind's actions. Most of the time these changes are welcomed as an enrichment of our environment. But there is a threat as well, specially when desirable places which are the product of evolution during the ages are damaged through sheer neglect. Open spaces are the most vulnerable to this threat, because they are considered to be 'idle land'. The structure of a settlement and open public spaces in this structure like the brink up till now did not get enough attention in land use planning. The aesthetical and historical value of village morphology is often overlooked. We as geographers have a task to pay attention to this subject.

The most important part of a study is the presentation of the results, particularly if one tries to help more people to gain a better understanding of their very own environment and tries to open the eyes of planners and authorities for the cultural and landscape value of a historical element. It is important to be understood. This means that we have to speak "the language" of the party that we want to reach. Towards environmental and land use planning authorities it is necessary to show in which way your 'historical' element does fit in their technical and legal procedures.

In my opinion a historical element must be able to absorb new developments in a living society so that it remains a living element of the environment. In the case of the brink or green the following general rules should be applied in environmental planning and management:

a. in a land use plan the brink (or green) should legally relate to the specific status of the brink (or green) and not the very general status of 'public space', which allows for a lot of non-characteristic influences and historical neglect.
b. The main structures of the brink should not be abused by new elements by letting them overrule this stucture. This allows only minor adjustments to modern function.
c. When new structures or elements are added to the brink, a state of harmony between old and new should be pursued.
d. Depending on the degree of rurality of the village, urban features like rose gardens, hedges and bitumen pavement should be avoided. Local materials should preferably be used.
e. A brink should be an informal public space like it always has been, so formalizing, e.g. straight paths, sharp

boundaries, has to be avoided.

f. As the buildings around the *brink* are very important to the *brink* itself, attention *should be* paid to the architecture. *With* new developments, like new buildings or in case of rebuilding, the architecture should be in harmony with the scale of the brink.

This paper is based upon a longer study - the Brinkenboek (HOUTING *et.al*, 1981) - but has, within the constraints of a *limited wordage*, sought to demonstrate the importance of a combination of historical geography, contempory geography and environmental planning in the preservation and management of *brinken*.

REFERENCES

HOUTING,E., K.R. DE POEL, J.H.P.VAN DER VAART, *Brinkenboek : een verkenning van de brinken in Drenthe* (Van Gorcum, Assen, 1981)

LYNCH,K., *Managing the sense of a region* (MIT Press, Cambridge Mass. 1980)

NAARDING,J., 'Vroegere Drentse schapenteelt en over het ontstaan der essen.', *Tijdschrift van het Koninklijk Nederlands Aardrijkskundig Genootschap* (1947) 669-708.

THORPE.H., The green village as a distinctive form of settlement on the North European Plain.', *Bulletin de la Société Belge D'Etudes Géographiques* (1961) 93-134.

WATERBOLK,H.T., 'Hoe oud zijn de Drentse dorpen'. *Westerheem*,vol.XXIX, no.2, (1980) 190-212.

CHAPTER 5

REGULAR RURAL SETTLEMENTS IN 19th-CENTURY TRANSCAUCASIA AS A CHARACTERISTIC FEATURE OF GERMAN COLONIZATION

G. Tiggesbäumker
(University of Bamberg)

The study of the handwritten documents transmitted by Baron Haxthausen (1) directed my attention to German settlements which had been founded in Russian Transcaucasia of the 19th century. At the invitation of the Russian government Haxthausen had travelled all over Transcaucasia in the middle of last century. He visited some of the German settlements and recorded the observations he made (2). Because of the lack of extensive studies which describe the settlements in detail Haxthausen's account is of particular importance. Using this material, as well as other contemporary travel descriptions and modern literature (3), I examine the following three complexes of questions:

1. origin of the colonists and colonization of Transcaucasia

2. formal arrangement of the settlements as a result of the methods applied in connection with the allocation of the land and the organization of municipalities

3. functional character of the settlements resulting from their location.

The settlements of German farmers from Wurttemberg founded in Transcaucasia between 1818 and the end of the 19th century are, in the broader sense, an after-effect of the systematic recruiting of German colonists, a measure initiated by Catherine the Great about the middle of the 18th century (4). Settlements around the Volga and the Black Sea had originally been founded to colonize the South-Russian steppes. In settling German colonists in Transcaucasia (Fig.5.1), the government hoped to give the agriculture fresh impetus in an ethnically heterogeneous region which had been annexed by Russia only shortly before. The colonies in Transcaucasia were thus conceived as model colonies; they were to give a good example for the purpose of extending and improving the agriculture and its ramifications. In stimulating this endeavour the Russian government complied with the desire of Protestant separatists to emigrate from Wurttemberg in South-West Germany, where they were regarded as an irksome group of population. With the approval of the King of Wurttemberg in 1816 and 1817 Russian agents were allowed to recruit about 500 families consisting of almost 4000 persons. These families hoped for unhindered worship in Russia, but the Russian government also promised to grant the emigrants numerous privileges such as exemption from taxes, military service as well as religious liberty, free land, and financial support during the first years.

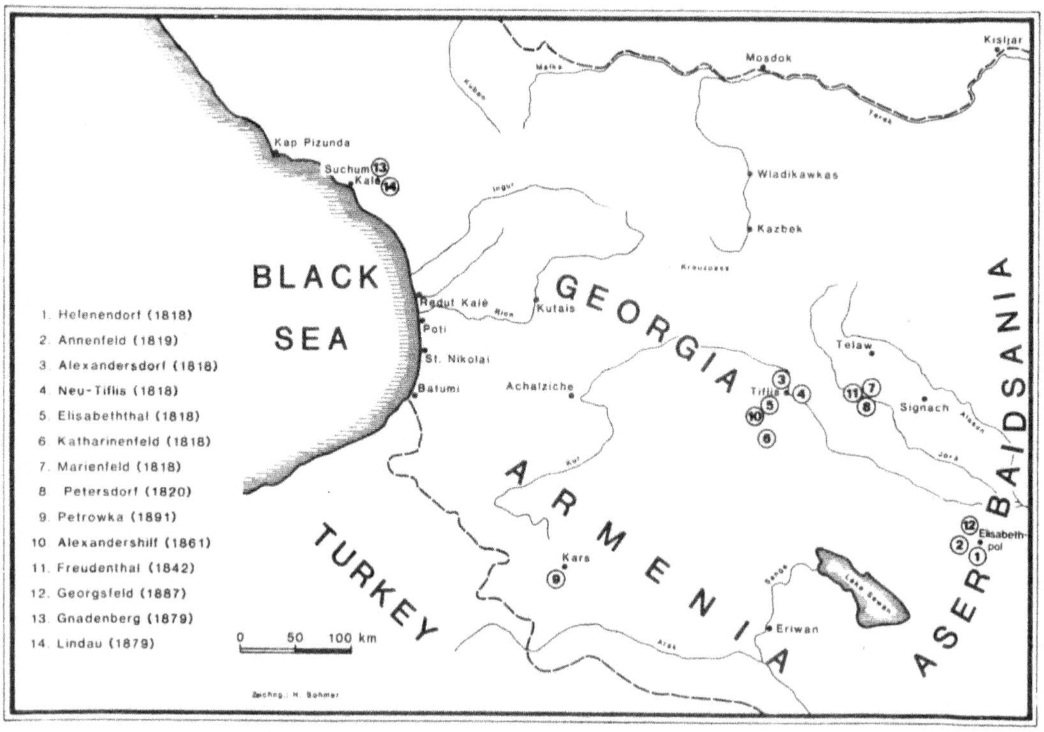

Fig. 5:1 German settlements in Transcaucasia founded between 1818 and 1900.

The emigrants gathered in Ulm, crossed the Danube by ship up to Odessa at the Black Sea. After they had been in quarantine for one year, they continued their way to Tiflis, the military headquarters of the Caucasian governorship. In Tiflis, the craftsmen were separated from the farmers: the group of craftsmen, very small in number, was settled in the local craftsmen's quarter, whereas the farmers were allotted land at the northern outskirts of the city (Fig.5:2)

Altogether about 12% of the German emigrants settled in Tiflis. The remaining colonists were allocated land in seven different places. For several reasons not all of the settlements were laid out in 1818. Near Annenfeld and Petersdorf foundation was delayed by one to two years. Thus, by 1820 only eight of the fourteen German colonies which were set up in Transcaucasia were in existence. These eight colonies were Helenendorf, Annenfeld, Alexandersdorf, Neu-Tiflis, Elisabeththal, Katharinenfeld, Marienfeld and Petersdorf (Fig.5:1).

This first period of colonization was followed by a second, in 1840: in 1842, 560 persons founded a settlement in the neighbourhood of Elisabeththal and Katharinenfeld. The lst emigrants from Germany, to be more precise, from Wurttemberg, reached Transcaucasia in 1879 and established two relatively small settlements in close vicinity: Gnadenberg and Lindau near Suchum-Kale (Fig.5:1).

Because of the rapid increase in population, the result of a healthy economic structure, Elisabeththal and Helenendorf - two settlements dating back to the first stage of colonization - were forced to establish the subsidiary colonies of Alexanders - hilf and Georgsfeld in 1861 and 1887. Measures of this kind had to be taken because they were not able to extend the boundaries of their community territories. The same process was repeated in 1891 when Petrowka, a subsidiary colony of Alexandershilf, was founded (Fig.5:1). Thus in Russian Transcaucasia 14 German settlements had been established between 1818 and 1891, and they constituted four settlement complexes of varying density, concentrated around a larger Russian administrative town (Fig.5:1).

The actual centre of the German settlements, consisting of 6 primary settlements and of two subsidiary colonies, was round Tiflis. A secondary centre was set up around the governmental capital of Elisabethpol. It consisted of two primary settlements and of one subsidiary colony. Gnadenberg and Lindau, which were established not far away from the Russian fortress and seaport of Suchum-Kale, consituted a third centre of minor importance. Finally, the isolated colony of Petrowka near Kars, a town which was in Russian hands till 1917 and is now in Turkish possession, must be mentioned.

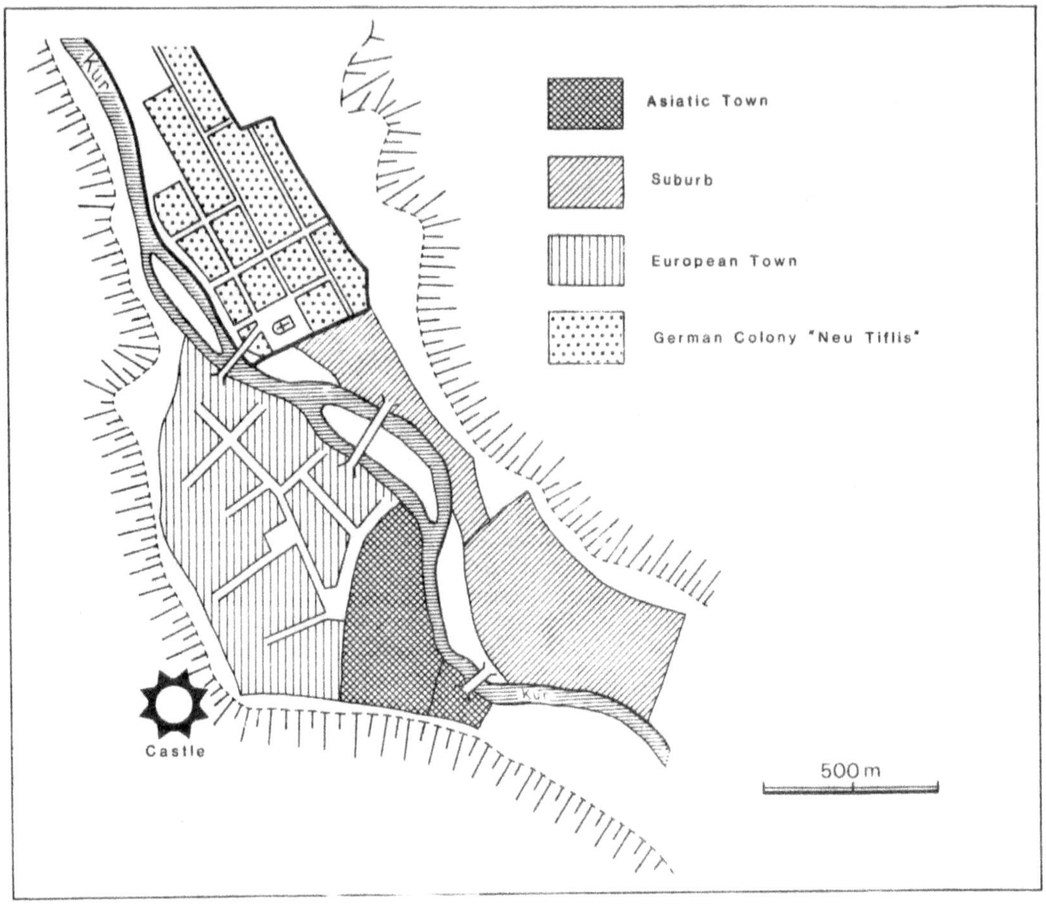

Fig. 5:2 The German colony "Neu-Tiflis" within the quarters of Tiflis about 1850.

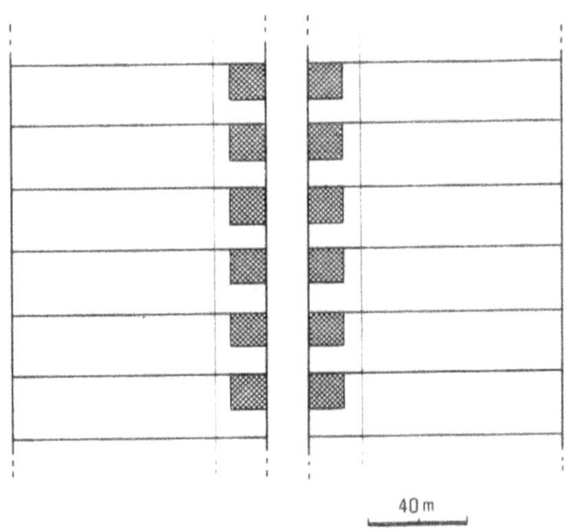

Fig. 5:3 Ideal ground-plan of a German colony (scheme).

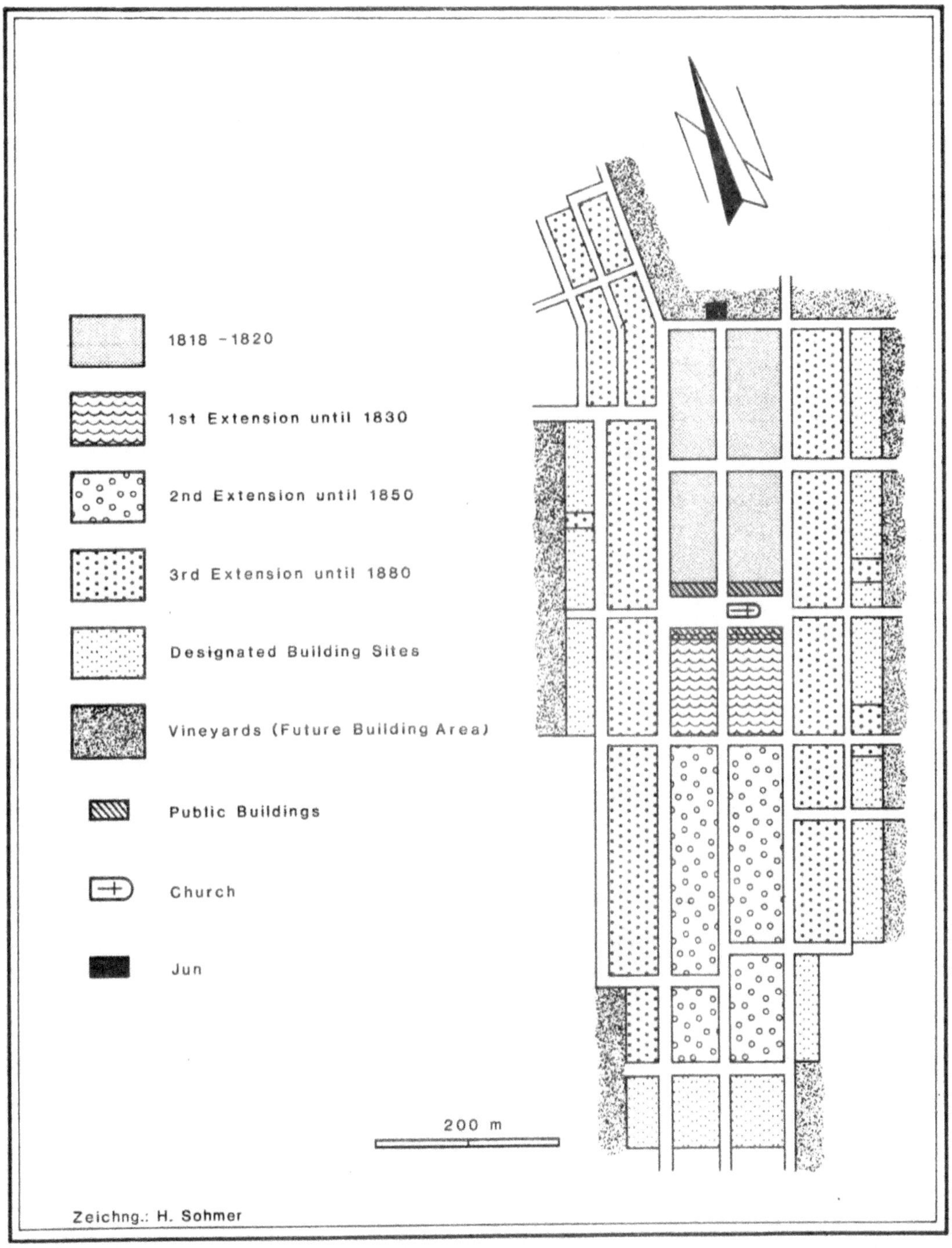

Fig. 5:4 Growth of a German colony from 1818 to 1880, illustrated by the example of Helenendorf (after Haxthausen and Schweinitz).

Without exception the colonies were allotted land in valleys with both favourable climate and good soil as well as with sufficient water, supplied by rivers or brooks. Moreover these valleys were situated close by main roads which permitted, on the one hand, fast connections with a larger town inhabited by Russians - to be more precise, with a garrison and an administrative centre - and, on the other hand, close connections with a number of non-German rural settlements. Each group of colonists which was to found a settlement obtained a limited territory of crown-land, which passed into posession of the community. A ground-rent, however, had to be paid for it, i.e. each community was liable to a collective duty assessed on the colonists in accordance with their economic situation. On account of the facilities guaranteed by the governement, the German settlements were exempt from payment of the customary poll-tax called Obrok. Within the territory of a community, officials of the immigration authority showed them where to settle and where to cultivate; the land was surveyed and distributed to the settlers. This distribution conformed with the regulations valid for all Russian colonists. The land was divided into so-called Wirthschaften (agricultural units), each of which contained a portion of land of approximately 11 acres. Because on average each agricultural community consisted of three independently operating families, constituting a large-scale family, each individual family possessed approximately only 3 acres of land. Some 0,6 acres of this lay within the settlement, the rest within a pattern of fragmented holdings outside the village. The areas of the settlement, to be more precise, the farmsteads, were distributed to the settlers only once, by lot, whereas the parcels within the pattern of fragmented holdings were allotted for cultivation for a limited lapse of time. Following the example of the constitution of the Russian Mir, officials distributed the parcels anew at irregular intervals.

Regulations made by the government for colonists dictated not only the distribution of land but also the ground-plan of the settlements. A regulated street village with regular areas of similar size had to be built. Parallel with the straight, 56 ft broad street and directly annexed to the farmstead, the parcel of each settler had to be 74 ft. wide and 335 ft. long, while the farmstead, which faced the street, took up a space of 49 to 92 ft. The rear part of each parcel was exclusively intended for gardening. For fire-preventive reasons a safety zone had to be reserved between the adjacent buildings (Fig.5:3).

When the colonists arrived, they first laid out a street with two rows of houses so as to lodge themselves as soon as possible. A market-place, with public buildings around it and arranged at one end of the street, tended to appear only much later. In Helenendorf, for example, a church was not built until 1855, nearly 40 years after the village had been founded. The arrangement of the market-place was often decisive in

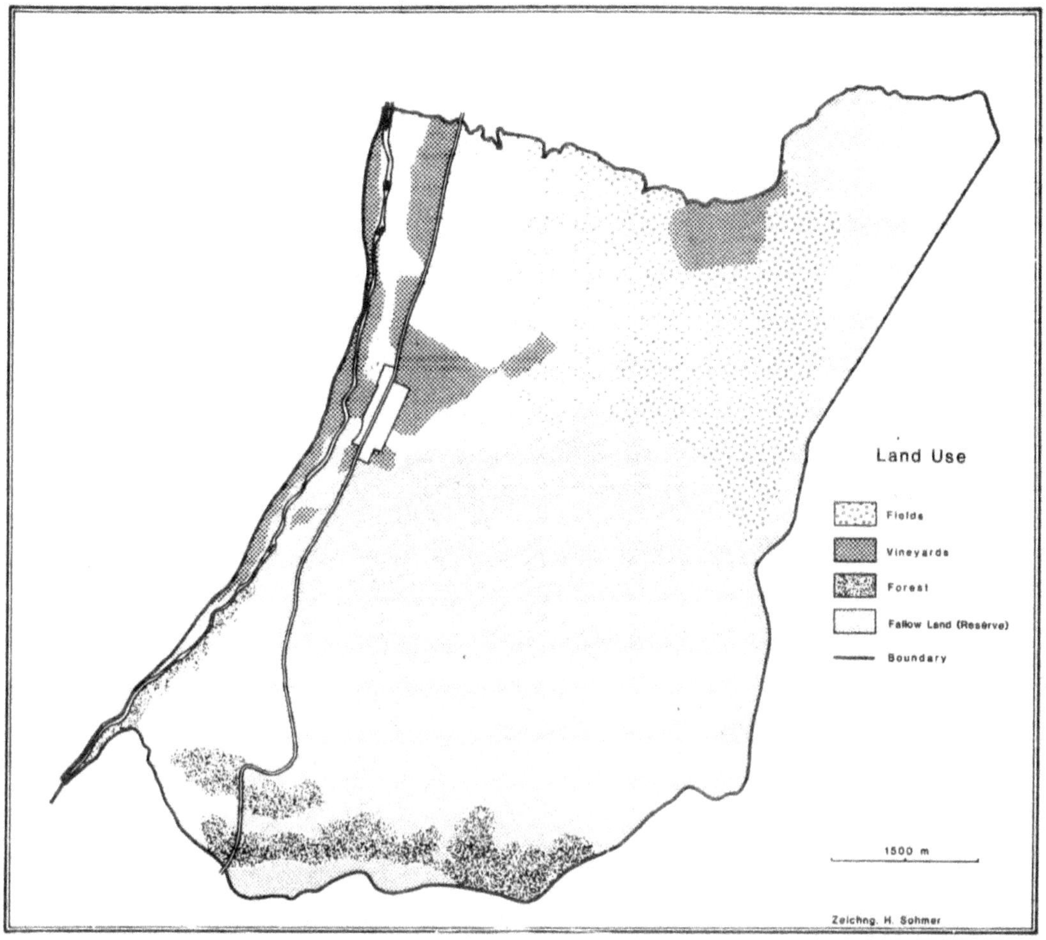

Fig. 5:5 Community territory of a German colony about 1880, illustrated by the example of Helenendorf (after Schweinitz).

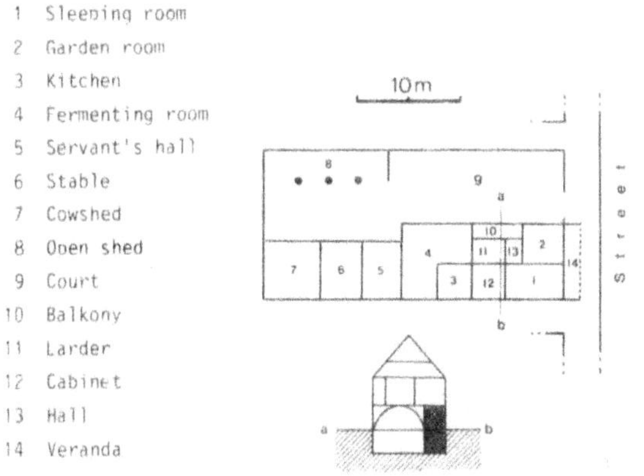

Fig. 5:6 Farmstead of a German colonist: ground-plan and front elevation (after Schweinitz).

controlling the direction in which the settlement was extended because the main street was normally continued on the other side, so that the market area situated in the middle of the village, and thus formed the centre. To limit the length of a settlement, extensions comprised the building of streets running parallel to the existing main axis. These were connected by ancillary streets and laid out according to the same regulations that were valid for the first row of houses. This way, the basis was given for a settlement to develop "concentrically" around a centrally located place (Fig.5:4). As the agriculturally used area bordered immediately on the housing area, arable land had to be given up when the village was extended. In exchange, areas located at far distance from the village, on non-cultivated land, were utilized in order to protect cultivated land against the extending settlement (Fig.5:5). The non-cultivated land within the total territory of a community, however, served primarily to supply new settlers and not heirs apparent of the resident population. As a rule, the non-cultivated areas were large enough for the expansion of a settlement. As mentioned before, only in three cases subsidiary colonies had to be founded. From 1840 on another method of land acquisition consisted in dividing estates. Due to the fact that in the meantime the government had consented to a partition of the real estates and to the transfer of land into private ownership, it was possible to distribute land among several heirs and to sell land. However to prevent the German settlements from being gradually dissolved, the government insisted that only German settlers should be allowed to purchase the land. Consequently, from about 1840 on, agricultural units which had been of equal size up to then became more and more heterogeneous: the spectrum ranged from giant agricultural units to estates consisting of an eighth part. It was forbidden to divide an estate into more than 8 parts.

While the ground-plan of the settlements was precisely stated by law, the settlers were free, though only to some extent, to build the farmsteads according to their own requirements. Special regulations made by the government only dictated the physiognomy of the village: the colonists had to construct stone buildings with one floor and with gable-sides facing the street. However, the ground-plan of the farmstead was entirely their decision. A certain standardization, however, cannot be denied: it is quite obvious that the settlers endeavoured to build a kind of Streckhof in the Wurttemberg tradition (Fig.5:6). From the street was access to a courtyard on which the farmhouse containing dwellings, cellar as well as quarters for animals (Wohn-Stallhaus) was erected on the one side and a shed on the other. In principle, ground-plan and front-view of the farmsteads did not change during the 19th century; in the course of time, however, the quality of the buildings was improved and more money was spent for the enlargement. During the eighties of the last century, travellers still observed buildings of poor quality which dated back to the time of foundation. The constructions did not form complete lines any longer but they mingled numerously with the new buildings. As the German settlements were founded on Russian crownland, they were subject to the administration and jurisdiction of the civil governor of Transcaucasia in Tiflis. The inner organisation of the colonies was based on municipal regulations which were common practice in Wurttemberg and had been granted to the colonists in 1818. Apart from the political community, a church community existed which was in most cases identical with the political one. The church congregation was organized according to a law valid for the Protestant church in Russia.

During the first years after the German colonies had been founded, their function as model colonies was to introduce the achievements of European farming into the Transcaucasian villages, which were generally poor. After a short period of time, however, it turned out that the native population was unwilling to adopt any innovations introduced by the German settlers. They preferred to adhere to their traditional, orientally influenced, way of life, indeed travellers report that it was extremely difficult for the colonists to make contact with the neighbouring population at all. Even years after the colonists had established their settlements, they were mistrusted by the native population.

The government soon realized that the original purpose they had intended in settling German emigrants failed. For this reason, the function of the German settlements changed quickly. The gradual conquest of the Caucasus by the Russians until 1866 resulted in a sudden Europeanization of the larger towns. As early as 1840 the European population of Tiflis e.g. outnumbered the local inhabitants considerably. The increased number of Europeans required a sufficient food supply of European quality, and as domestic farms were not able to satisfy these demands, the German colonies very soon took over this function. According to concurrent reports given by travellers of the 19th century, horticulture was completely unknown in Georgia. Thus it is not surprising that the German colony Neu-Tiflis, which was in close proximity to Tiflis and had specialized in horticulture, and showed an extremely healthy economic structure shortly after its foundation. Neu-Tiflis supplied the daily market of Tiflis with horticultural products.

The colony of Alexandersdorf, located some kilometres northwest of Tiflis, was economically isolated and it had to specialize in the growing of potatoes for lack of an appropriate irrigation system. According to the statements made by travellers, it was the German colonists who introduced the potato into Transcaucasia. On account of unfavourable conditions of cultivation, production was low in Alexandersdorf so that poultry farming, which was exceptional for German colonies, became the second main source of income. Potatoes and poultry were commercialized directly in Tiflis. The German colonies within a distant radius of Tiflis supplied the town to the same degree with agricultural products, but not primarily on the daily market. Intensive growing of wheat, barley, and oats on irrigated fields as well as productive vineyards guaranteed a long-term supply. Although viticulture - by the way, vines from Wurttemberg were cultivated - had only a small share in the agricultural acreage, its production was important and contributed considerably to the demand of Georgia and even to that of all Transcaucasia. In contrast to the native population, the German settlers applied improved methods and thus achieved better results. As early as 1840 wine was already refined, stored, and bottled in a central depot near Tiflis, which was run on a co-operative basis.

As a matter of fact, in the 19th century the German colonies near Tiflis provided to a great extent the food supply of the Transcaucasian capital. The three colonies close to Elisabethpol had similar economic relations but were of less importance. The colony of Petrowka sold its horticultural products on the market of Kars. It is not clear to what extent the colonies near Suchum-Kale sold their products, but parallel relations must be presumed.

Despite the fact that the results, based chiefly on the reports of Baron Haxthausen, are preliminary and incomplete, one can summarize that the German settlements in Transcaucasia were planned as new agrarian settlements. They could be clearly distinguished from the surrounding native settlements as well as from all the German colonies in Russia which dated back to the period of Catherine. Not only did they have a different pattern but also a diferent function, which, in my opinion, is very significant. The characteristic features of the German settlements established until the end of the 19th century are: a regulated ground-plan, a uniform physiognomy, a high economic level, the supply of 'Europeanized' towns with agrarian products, and the arrangement of land-use systems varying in spatial distance from the centre.

REFERENCES

1) Baron August von Haxthausen (1792-1866), a Westphalian noblemen; for further information see: HALLER, B. and G. TIGGESBÄUMKER, Die Kartensammlung des Freiherrn August von Haxthausen in der Universitatsbibliothek Muenster,- Beihefte zu Westfaelische Geographische Studien,2. Munster 1978. Manuscripts and handwritten correspondence by Baron Haxthausen, dating from 1842 to 1866 are to be found in the University-Library of Muenster (West Germany), Autograph-Division, August von Haxthausen Papers.

2) Apart from the handwritten documents the following printed publications made up by Baron Haxthausen served as an additional source:
HAXTHAUSEN, A. v., Transcaucasia: Sketches of the Nations and Races between the Black Sea and the Caspian (London 1854)
HAXTHAUSEN, A. v., Tribes of the Caucasus (London 1855)
HAXTHAUSEN, A. v., The Russian Empire, its People, Institutions and Resources (London 1856)
HAXTHAUSEN, A. v., Transkaukasia: Andeutungen ueber das Familien und Gemeindeleben und die socialen Verhaeltnisse einiger Voelker zwischen dem Schwarzen und dem Kaspischen Meere.. Vol. 1 (Leipzig 1856).

3) The most important writings on German colonies in 19th century Transcaucasia are:
HOFFMANN, P., Die deutschen Kolonien in Transkaukasien (Berlin 1905).
KOLENATI, F.A., Die Bereisung Circassiens - Reiseerinnerungen, 2 (Dresden 1859).
SCHWEINITZ, H.- H.v., Helenendorf: Eine deutsche Kolonie im Kaukasus. (Berlin 1910).
WAGNER. M., 'Die deutschen Colonien jenseits des Kaukasus', - Der deutsche Auswanderer 4 (1850).
WAGNER, M., Reise nach Kolchis und nach den deutschen Colonien jenseits des Kaukasus. (Leipzig 1850).

4) For further information on the history of German colonization in Russia and, in particular, in Transcaucasia see:
MATTHAEI, F., Die deutschen Ansiedlungen in Russland. Ihre Geschichte und ihre volkswirtschaftliche Bedeutung fur die Vergangenheit und Zukunft. (Gera 1865).
SCHRENK, F.M., Geschichte der deutschen Kolonien in Transkaukasien (Tiflis 1869).
STUMPP, K., Die Auswanderung aus Deutschland nach Russland in den Jahren 1793-1862 (Tuebingen 1960).

CHAPTER 6

REGULAR SETTLEMENTS IN FRANCONIA FOUNDED BY THE FRANKS IN THE EARLY MIDDLE AGES

F. Eigler
(Werner-von-Siemens Gymnasium)

At the Roskilde meeting June 1979 it was pointed out that the heights of the Franconian Jura (about 50 kilometres south of Nuernberg) are characterised by extremely regular green villages and field systems resulting from late-medieval colonisation (A.D. 1180-1320). Only a few miles away, however, there appear a group of eighteen regular street villages, clearly of a different morphology. The aim of this paper is to examine the relative chronology of these two types and the contexts within which each developed. The analysis can be broken into several sections:

SETTLEMENT FORMS

The main axis of the characteristic street-village is a roadway, never completely straight, but a little bent, and usually about 200-500 metres in length. At each side of this street is a line of farmsteads. Most villages have a further characteristic, a street or lane leading away at right angles from the main axis towards the adjacent forest. This can be clearly seen in Burgsalach, Raitenbuch (Fig. 6.2 and 6.3), Reuth and Kaldorf (Fig. 6.1) and today this is usually called the Holzweg or Holzgasse, i.e. 'the wood lane'. Secondary dwellings, small-holders cottages, are normally concentrated along this lane. This distinctive morphology is the first indication of the presence of planning, for while in most cases the villages extend parallel to the forest edge, with the 'wood lane' leading towards the woodland, in those cases where a village is not parallel, there is no 'wood lane'.

FIELD PATTERNS

The first accurate maps, of about 1820, show that the field patterns associated with the street villages, are the same as those of agglomerated villages (Haufendorfer) that can be dated to as far back as the seventh century. There are a great number of small furlongs (Gewannen) with short strips within a pattern of fragmented holdings (Kurzstreifen-Gemengeverband) and there are groups of block-shaped parcels. If one examines these field systems retrogressively, shorter but wider strips can be seen near the village: further work is still required upon these structures.

LANDOWNERSHIP

The structure of ownership is characterised by a great number of different manorial and territorial lords. This is a

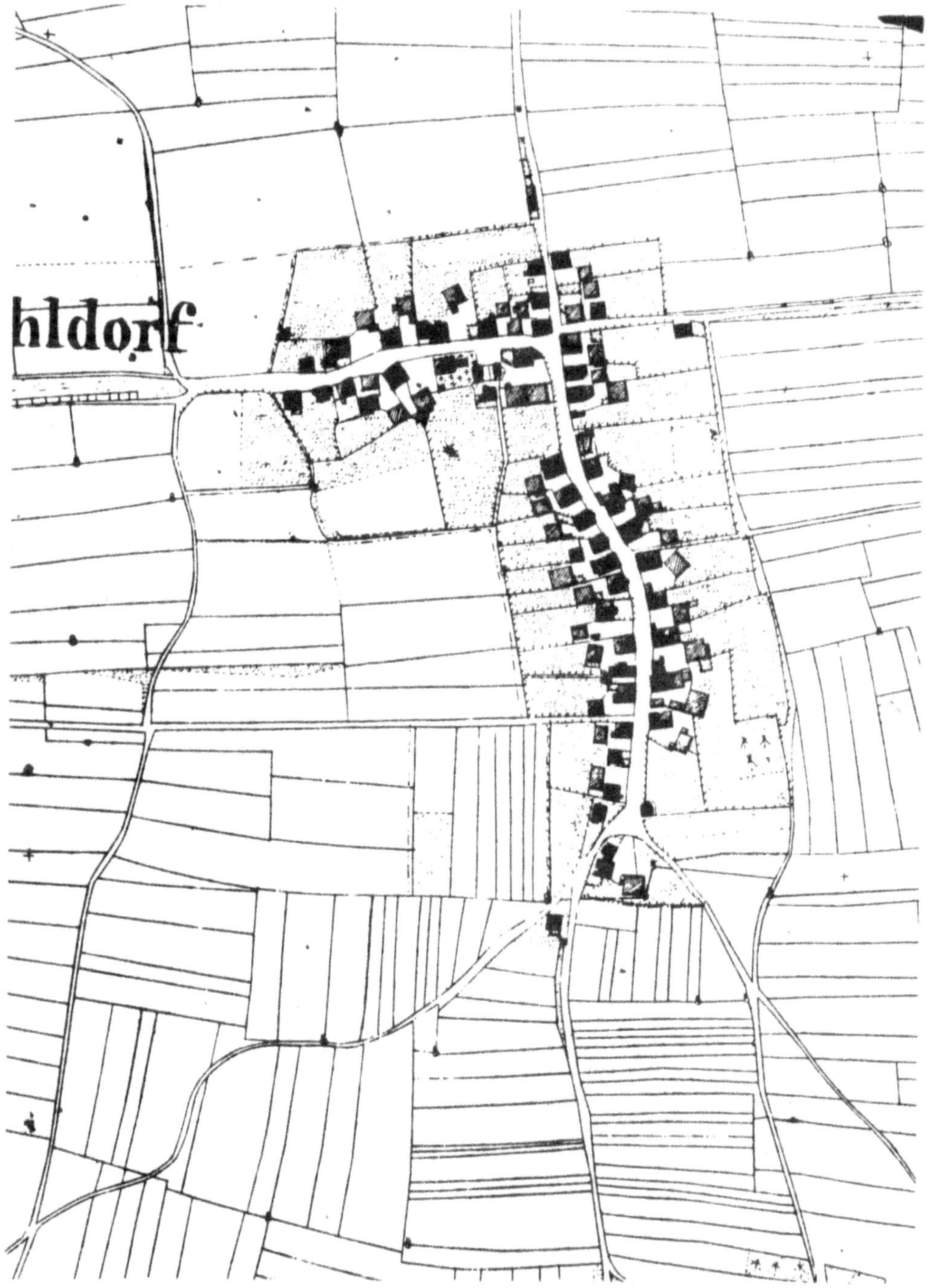

Fig. 6:1 Kaldorf: map of 1820 (scale 1:5000) showing north-south street with farmsteads, and 'wood lane' with smallholders cottages leading off the main axis.

markedly different situation to that found in the late-medieval green village, where single lordship is more usual, but is similar to that of the agglomerated villages (Haufendorf) of the neighbouring agricultural lands known to have been settled between the 6th - 9th centuries. If one examines the available sources carefully it seems that the street settlements originated as a small number of big farmsteads which lay at least 100 metres away from each other. Thus, in Raitenbuch there were in the 14th century only seven big farmsteads, and in Reuth in the 14th century only four. The gaps between these original farmsteads were gradually filled in with a great number of small-holders' cottages which gradually developed from the houses of farm-labourers.

PLACE-NAMES AND CHURCH DEDICATIONS

The place-names of these villages, the date of their first appearance in the documentary record and their church dedications are as follows:

(Ober -) Hochstatt	899	St. Martin
Indernbuch	1119	14- Nothelfer
Kaltenbuch	1435	St. Nikolaus
(Burg)Salach	1057-75	St. Koloman
Raitenbuch	1087 (867)	St. Erhard
Reuth am Wald	1057-75	
Kaldorf	1119	St. Andreas
Petersbuch	1119	St. Peter
Heiligenkreuz=Neuenkirchen	1140	Hl. Kreuz
Seuversholz	1080	St. Nikolaus
Sperberslohe	1153	
Workerszell	1305	
(Langen)Sallach	11th century	
Rupertsbuch	1002	St. Michael
Sappenfeld	1305	St. Sebastian
Schernfeld	1068	St. Georg
Eberswang	about 1250	St. Alban
Bieswang	889	St. Martin

The suffixes -buch, -holz and -lohe are synonyms for 'wood', while Reuth means 'clearing of woodland'. The ending -feld indicates a 'thinly wooded area for grazing cattle' or 'a pasture within a forest'. Collectively these place-names are signs that the villages have come into existence by way of clearing woodland. As most are first documented between A.D. 1000 and 1150 they are clearly older than the late-medieval green villages, but as in this region even the oldest village types are not documented before AD 1000, the street villages could have come into existence well before that year. Bieswang is of special importance, for it is mentioned as early as 889, over a century before 1000. The suffix -wang means 'grassland on a slope appropriate for grazing'. Significantly, the dedications of the churches of this group tend to be 'early', the oldest being to St. Martin (in Oberhochstatt and

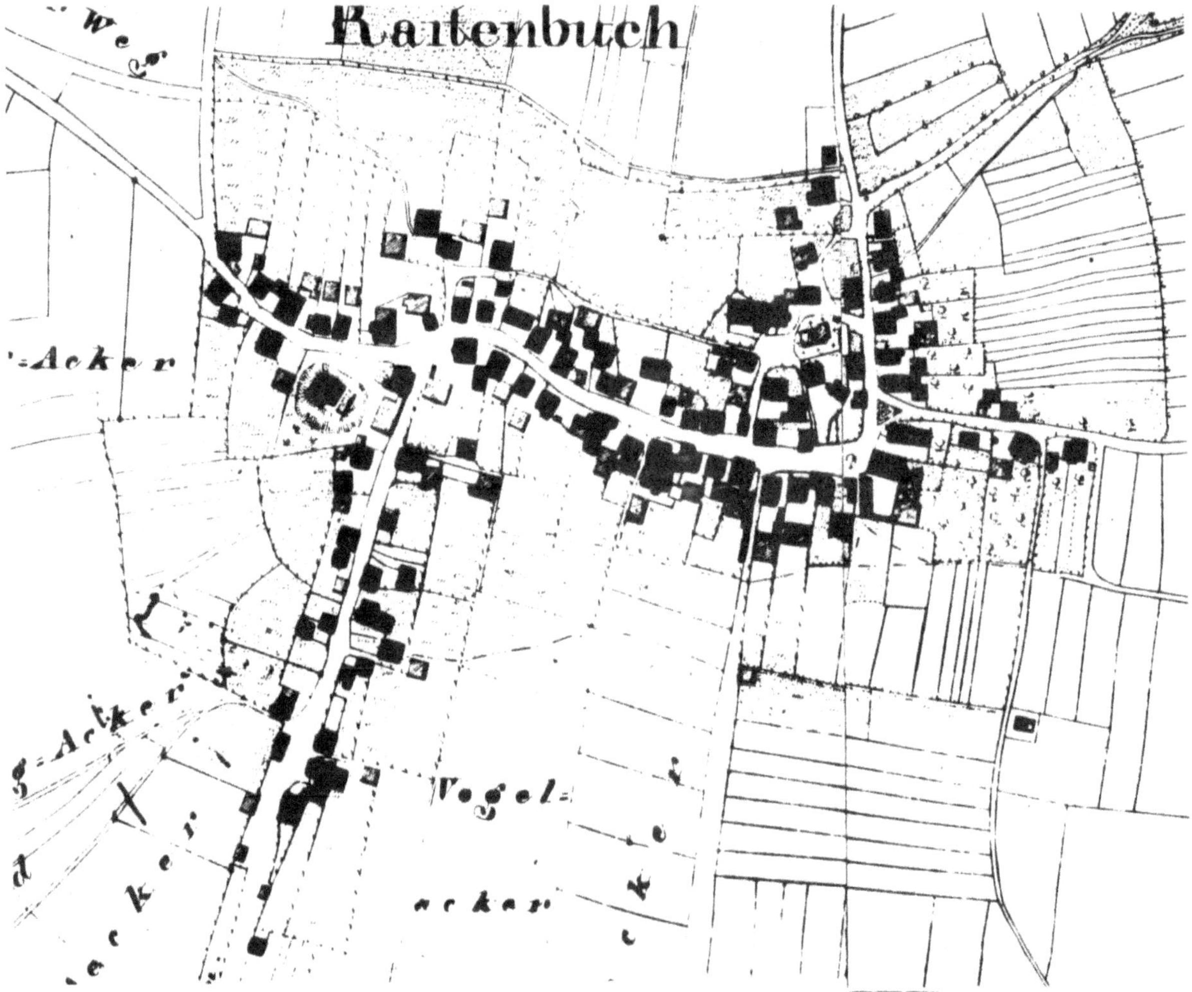

Fig. 6:2 Raitenbuch: map of 1820 (scale 1:5000) showing main street east-west and 'wood lane' on south side.

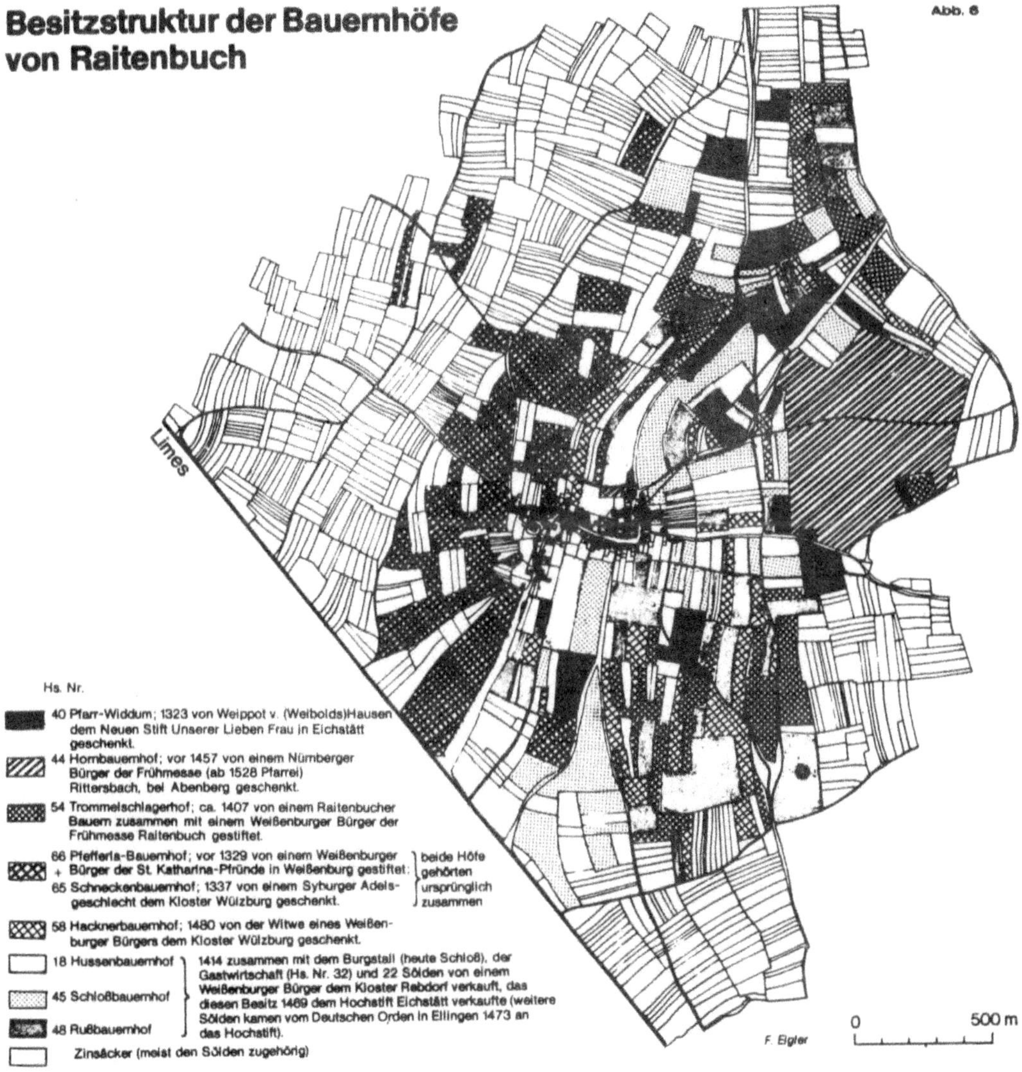

Fig. 6:3 Raitenbuch: showing large number of furlongs of varying size and shape.

Bieswang), followed by St. Michael, St. Nicholas, St. George and St. Alban, etc.

IMPERIAL PROPERTY

Together with their fields these villages form an uninterrupted zone of arable land, bow-shaped, and extending for about 20 kilometres between the towns of Weissenburg and Eichstaett, and many of them are linked by the presence of imperial property. In 1153 Sperberslohe was still in royal hands, but others, for example Kaldorf, Petersbuch and Heiligenkreuz, were called 'royal villages' even when administered by the 'Free Imperial City of Weissenburg'. The presence of certain jurisdictional features in Seuversholz, Kaldorf and Petersbuch, together with the levying of Reuthmuth, a type of tax levied in oats, from Raitenbuch, Indernbuch, Reuth, Sperberslohe, Burgsalach, Oberhochstatt, Seuversholz and the three 'royal villages' were both royal prerogatives taken over by Weissenburg. Reuthmuth is a helpful indicator, for muth means 'sixteen baskets of a certain size' while reuth means 'cleared woodland'. As the clearing of royal forest was not allowed there is only sense in this levy if it was taken from that region where the woodland had been cleared - i.e. where we now see the fields of the eighteen villages.

It is probable that the area of royal forest, together with the farmsteads existing within it, was granted by the crown to vassals, but under the ban not to make further clearings. To make the existing settlements viable it was necessary for the peasants to be allowed rights within the royal woodlands, an arrangement which surely must precede the subinfeudation to vassals. Thus the peasantry could take firewood and timber, and graze cattle and pigs, and to this day the 'wood lanes' lead to the so-called Viehtrieben (or cattle droves). Such grazing rights were of vital importance because of the absence of communal woodlands and relative scarcity of bottom-land suitable for hay-meadows within the region.

In summary, settlement morphology, field systems, landownership and the presence of royal property suggest that there is an underlying uniformity within this group of villages and the differences between the street villages and the green villages are summarised in the following table:-

Fig. 6:4 Dettenheim: map of 1820 (scale 1:5000). Street village with -heim place-name, possibly of sixth-century origin.

Tabular Summary of Certain Settlement Characteristics in Franconia

	Street Villages	Green Villages
Settlement Morphology	Street village forms	Green village forms
Field Patterns	Irregular field patterns	Planned furlongs with long strips
Size	Relatively small	Relatively large
Landownership	Fragmented lordship	Unitary lordship
Church Dedications	Relatively old	Relatively new
Royal Property	Present	Absent
Woodland	Grazing rights in Royal Forests	Communal woodland present

THE HISTORICAL CONTEXT

The region in question was a border zone: in Weissenburg a Frankish royal court had been established by about A.D. 500, but to the east was Bavarian territory which the Frankish kings tried repeatedly to annex. In the eighth century they were able to establish a bishop's see in Eichstaett, indicative of secure conquest. The royal capital at Weissenburg (whose continuing importance is attested by grave finds) was separated from Eichstaett by extensive woodlands, and it makes sense for the Frankish crown to have secured this frontier by establishing 'defensive' villages. When was this done? The first reference to Bieswang is in A.D. 889, and the suffix *-wang* is the only example found within this royal forest area: nevertheless, it is one known to have been in use in Carolingian times, and the dedication to St. Martin is certainly in accord with Carolingian usage. It is possible that Bieswang can be regarded as a kind of field base *(Etappenort)* between the royal property at Pappenheim and Eichstaett.

In the year 788 Charlemagne liquidated Tassilo III, Duke of Bavaria, bringing Bavaria under Frankish administration and creating a royal court at Neuburg. The map suggests that a similar zone of street villages is to be found between Eichstaett and Neuburg and it may well be that the establishment of the whole zone of distinctive linear settlements can be attributed to Charlemagne after 788; he certainly tried to link the Rhine and the Danube with a canal, the *Fossa*

Carolina, and as this failed, defensive villages might have been seen as a viable substitute. The village of Dettenheim (Fig.6:4) is situated in the early-settled agricultural land south of Weissenburg, where Merovingian graves of the sixth century have been found: thus, it is possible that the Carolingian settlements were modelled on even older types.

REFERENCES

EIGLER, F. Die Entwicklung von Plansiedlungen auf der suedlichen Frankenalb (Muenchen 1975).

EIGLER, F. 'Weissenburger Reichsforst und Pappenheimer Mark', Zeitschrift fuer bayerische Landesgeschichte, Band 39, Heft 2 (1976) 353-377.

EIGLER, F. Weissenburgs ueberregionale Bedeutung zur Zeit Karls des Grossen, villa nostra IV (Weissenburg 1975).

EIGLER, F. Late-medieval manorial colonization of the Franconian Jura, Permanent European Conference for the Study of the Rural Landscape. Collected Papers, Denmark Session 1979, (Copenhagen 1981), 63-67.

CHAPTER 7

LES ESPACES COLLECTIFS VILLAGEOIS: L'EXEMPLE DES 'USOIRS' LORRAINS

J. Peltre
(Université de Nancy II)

Les rues des villages lorrains (montagne vosgienne exceptée), notamment les rues principales, se distinguent par leur largeur inusitée qu'on ne retrouve ni en Bourgogne, ni en Alsace. Cela tient à la présence d''usoirs', c'est-à-dire d'espaces libres compris entre la chaussée et les façades d'immeubles. Naguère on y entassait le bois, le fumier, les instruments de ferme; aujourd'hui cette utilisation intensive est très exceptionnelle (Plate 7:1).

Ces espaces intermédiaires dénommés dans les ancient textes: 'usuaires', 'aisances', 'pairs' ou 'parges'[1] sont le plus souvent dépourvus d'appellation précise dans le langage populaire actuel. Ils n'en posent pas moins d'intéressants problèmes d'ordre scientifique et pratique, notamment en matière de genèse et d'aménagement. Les usages locaux, codifiés dans chaque département, les définissent sensiblement de même façon et pourtant leur diversité est grande, même si l'on s'en tient aux seuls usoirs frontaux, en façades d'immeubles.[2]

- diversité de statut foncier: en règle générale, les usoirs sont des espaces collectifs. Ils appartiennent à la commune et ne sont pas cadastrés. Accessibles à tous, ils sont cependant plus spécialement réservés à l'usage des riverains qui peuvent en disposer à leur convenance sans toutefois faire obstacle au passage. Telle est la règle en Lorraine centrale, dans la quasi-totalité des départements de la Meuse et de la Moselle,[3] notamment aux confins de la Belgique, du Luxembourg et de la Sarre. Toutefois, en Lorraine du Sud et du Sud-Est, sur les auréoles gréseuses de la plaine et des collines sous-vosgiennes, ils sont en majorité appropriés aux riverains et donc cadastrés.

- diversité de formes et de dimensions: inclinés vers le caniveau qui borde la route pour assurer un meilleur drainage, ils sont généralement rectangulaires mais éventuellement trapézoïdaux ou triangulaires selon la disposition des façades par rapport à la rue. Leur largeur varie de quelques décimètres jusqu'à 10, 15 mètres et davantage, au point d'évoquer parfois ces espaces indivis, non bâtis et gazonnés connus sous les noms de 'coudercs' dans le Massif Central, 'trieux' ou 'rejets' en Wallonie, 'green' en Angleterre, 'anger' en Allemagne.

- diversité d'utilisation. Dans la Lorraine du Nord-Est, aux limites de l'Alsace et de l'Allemagne où se maintiennent les petites exploitations, les usoirs ont gardé leur fonction traditionelle de cour de ferme. A ce titre, ce sont des lieux d'entrepôt, de stationnement, de passage et même de véritables 'tournières' pour les longues et lourdes charrettes pénétrant dans les granges. Ailleurs, ils sont généralement sans utilisation définie sauf au droit des fermes, peu nombreuses aujourd'hui.

Plate 7:1 Ommeray (département de la Moselle) (Photo A. Humbert).
Dans ce village-rue aux maisons jointives, on distingue nettement de larges usoirs gazonnés et plantés d'arbres d'ornement.

Cette diversité multiforme ne se réduit pas aux seules oppositions régionales. Elle peut jouer à l'intérieur d'un village déterminé. C'est ainsi par exemple qu'au voisinage de la Haute-Marne, le statut foncier peut varier à partir du centre de la localité où les usoirs sont collectifs vers la périphérie où l'appropriation privée, fréquente, engendre des constructions adventices en équerre sur la rue.[4] Même au niveau de la parcelle bâtie, des complications surgissent[5] faisant deux parts dans l'usoir non cadastré: l'une collective et banalisée, l'autre de 1 à 3 m de large, attenante à la maison et juridiquement liée à celle. C'est le 'tour de volet' des usages de la Moselle.[6] Cette division en deux parts est souvent concrétisée par l'existence d'ouvertures extérieures de caves et d'un pavement jouxtant les façades, destiné autrefois à faciliter la communication entre les différentes travées de l'immeuble, voire entre les immeubles eux-mêmes.[7]

A la diversité spatiale se superpose une diversité temporelle. A la suite d'aliénations et d'usurpations, la situation enregistrée au XIXe siècle par l'ancien cadastre n'est plus nécessairement celle d'aujoud'hui.[8] Une ambiguïté existe d'ailleurs au sujet de l'appartenance des usoirs collectifs au domaine public ou au domaine privé communal tant par l'ignorance des municipalités que par les interprétations divergentes des juristes.[9] Dans l'ensemble, le décompte statistique des arrêts tend à situer les usoirs dans le domaine public communal. Des aliénations ont été consenties cependant au cours des deux derniers siècles, au profit d'élargissements répétés des voies de circulation, jusqu'à faire disparaître complètement les usoirs de rues secondaires. Les bouleversements occasionnés par les guerres ont engendré des changements profonds: après la guerre de 1914-1918, les usoirs des villages ruinés ont été considérablement élargis par suite du déplacement des fermes sur les meix aux dépens des jardins. Ils sont donc aujourd'hui en grande partie et très logiquement propriété des riverains (Fig.7.1).

D'autres mutations de statut, de sens opposés, sont plus déconcertantes. Nous disposons pour le village de Réning (Moselle, environ d'Albestroff) de quatre documents parcellaires échelonnés sur deux siècles. Sur trois d'entre eux, datés de 1689, 1812 et 1978, les usoirs ne sont pas représentés. Au contraire, en 1738, c'est-à-dire en pleine période d'explosion démographique des campagnes lorraines et de multiplication des maisons jointives, les usoirs sont de statut privé comme si la "fermeture" du village-rue engendrait alors une réaction de défense tendant à garantir aux riverains l'usage d'un espace de service plus que jamais indispensable. Mais alors pourquoi ce retour à l'appropriation collective après quelques décennies? On peut y voir à nouveau l'effet de la croissance accélérée de la population: 1738 était l'époque de construction de maisons individuelles. Il fallut ensuite, pour répondre à la demande, partager nombre d'entre elles, loger plusieurs familles sous le même toit sans pouvoir attribuer à chacune d'elles un lambeau d'usoir, d'où le retrour forcé à l'indivision.[10]

L'usoir était en effet une nécessité pour les fermes lorraines qui ne disposaient pas d'issue charretière à l'arrière, en

LANEUVEVILLE-AUX-BOIS état ancien (XIXᵉ siècle)

LANEUVEVILLE-AUX-BOIS (arrondissement de Lunéville, Meurthe et Moselle)

village reconstruit

Fig. 7:1 Dans le village reconstruit après la 1ère guerre mondiale les usoirs sont cadastrés sur la même parcelle de propriété que la maison riveraine.

direction des champs. Il leur fallait un espace de service frontal, sorte de cour où transitaient nécessairement les animaux, le fumier, les récoltes sous la surveillance attentive des voisins. Toutes les manifestations de la vie agricole trouvaient leur expression dans la rue, sur l'usoir auquel s'appliquait la règle d'appropriation collective comme pour les terres cultivées, périodiquement remises au service de tous, après la récolte.

Cette idée de l'usoir élément du système agraire des pays de champs ouverts est confirmée en Lorraine par le fait que sa non-cadastration originelle, son statut originel de bien communal sont courants dans les pays d'assolement obligatoire classique où jouaient à plein les contraintes collectives. Au contraire, l'usoir cadastré, propriété des riverains, est de règle dans les secteurs qui ont échappé à ces contraintes, notamment en matière d'habitat. Ainsi dans les Vosges gréseuses et dans la plaine sous-vosgienne toute proche, l'usoir collectif disparaît en même temps que la compacité du village et le caractère jointif des maisons.[11] Nous sommes là sur des sols siliceux, relativement pauvres, où les règles strictes de l'assolement obligatoire ne pouvaient s'appliquer.[12] Sur le muschelkalk plus favorable aux céréales, au contraire, l'usoir indivis réapparaît.[13]

Il faut donc distinguer, en Lorraine:

- des usoirs 'vrais', fonctionnellement liés à l'ancien système agraire et non seulement à l'obstacle continu des façades jointives du village-rue.[14] On peut donc les rencontrer dans les villages en ordre serré ou en ordre plus lâche. Ils sont soumis à l'appropriation collective, libres de constructions et non enclos. Des changements de statut ont néanmoins pu intervenir à la suite d'usurpations ou de reconstructions;

- des 'pseudo-usoirs', simples portions de <u>meix</u> privés sans fonction ni obligation strictement définies. Souvent associés à l'habitat desserré, éparpillé même, ils se prolongent souvent sur les côtés, voire à l'arrière de la ferme. Ils sont propriété des riverains qui peuvent y bâtir ou les enclore à leur gré.

Les autres caractères des usoirs, leur largeur entre autres, dépendent de facteurs variés. Le site par exemple qui, par son étroitesse ou son incommodité à flanc de coteau peut interdire toute extension en largeur.[15] Quant à l'importance des exploitations, à la condition sociale des propriétaires, leur influence semble plus réduite qu'on ne l'a dit. Que la petite ferme du manouvrier ne justifie pas un usoir étendu est un argument logique mais très souvent non vérifié,[16] sauf aux extrémités de villages où d'autres explications ont d'ailleurs été suggérées.[17] L'influence des types de maisone, de leur plan, n'apparaît guère plus convaincante. En revanche, on n'a pas suffisamment dit l'influence de l'âge de construction. En règle générale, les usoirs les plus larges sont ceux des fermes traditionnelles les moins anciennes. Ce rapport est amplement vérifié dans les villages planifiés établis au nombre de plusieurs dizaines dans la Lorraine du Nord-Est aux XVIe et XVIIe siècles.[18] De même observe-t-on dans les villages-rues dont les espaces libres ont été progressivement colmatés que les dernières parcelles occupées ont leurs bâtiments nettement en retrait par rapport à d'autres. Les décrochements de façades qui en résultent sont donc un précieux indice du façonnement d'un village-rue.

Mais les problèmes les plus immédiats posés par les usoirs sont d'ordre pratique. La plupart d'entre eux ont perdu tout ou partie de leur fonction agricole par suite de la réduction du nombre des fermes (plus des deux tiers en cinquante ans) et des règlements d'hygiène interdisant le dépôt des fumiers à proximité immédiate des habitations et de la voie publique. Ceux-ci ne sont tolérés que dans les secteurs de petites exploitations aux mains d'agriculteurs à temps partiel. D'autre part, les obligations découlant des anciennes pratiques agraires ne sont plus qu'un souvenir.

Dès lors, que faire de ces espaces vides? Comment les entretenir? Questions sans réponses dans les très petites communes aux budgets de misère. La tentation est grande pour les riverains d'aménager à leurs frais, mais aussi à leur profit exclusif la portion du domaine public qui les concerne, parfois avec l'accord tacite des municipalités. Le risque est grand de voir se multiplier anarchiquement des initiatives plus ou moins extravagantes notamment en matière de clôture.

Pour conserver la maîtrise d'un espace collectif, des municipalités en ont entrepris l'aménagement sur le modèle urbain: construction de trottoirs, engazonnement, plantation d'arbres. Les exemples les plus réussis résultent d'une répartition harmonieuse des responsabilités entre les autorités municipales et les administrés. Après une nécessaire période de sensibilisation du public, une réflexion collective définit l'utilisation de l'espace collectif villageois: les issues des fermes demeurent libres d'accès pour le bétail et les machines tandis qu'ailleurs sont aménagés sur les anciens usoirs des plages fleuries ou gazonnées dont l'entretien est confié aux riverains. Toute clôture érigée par des particuliers est interdite mais il est paradoxalement plus difficile de freiner les initiatives intempestives des grands services publics: Electricité de France, Postes et Télécommunications, etc...

Entre tous les éléments de l'habitat rural de type lorrain, l'usoir a été jusqu'alors le moins étudié, son importance étant apparemment secondaire. Pourtant, à l'heure de la promotion du cadre de vie et d'un tourisme rural de qualité, ce retard entretient une indécision fâcheuse des pouvoirs publics vis-à-vis de certains villages qu'un aménagement intelligent, entrepris à bon escient et peu coûteux aurait pu sauver de la médiocrité.

NOTES

1 Dans les terriers de la période de reconstruction consécutive à la guerre de Trente ans, les usuaires sont mentionnés mais, à de rares exceptions près (Coincourt (1696), Bezange la Petite (1687), Gondrexon (1714), Reillon (1695)), ils ne font pas l'objet d'un arpentage séparé si bien que leur largeur, leurs dimensions, ne peuvent être appréhendées, non plus que l'emprise exacte du bâti sur la parcelle foncière. Il est certain néanmoins qu'à cette époque de nombreux usuaires sont non seulement devant mais aussi sur le côté des maisons.

2 Cf. par exemple: Bull. officiel du département de la Moselle, n° spécial du 20 avril 1961, chap. VII, art. 57: 'on comprend sous la dénomination d'usoir l'affectation spéciale d'une bande de terrain, mais aussi cette bande de terrain elle-même le long des routes à la traversée des localités jusqu'aux immeubles construits... L'immeuble peut être attenant à cette bande de terrain par la façade principale ou par les côtés et même par l'arrière.

3 Bulletin Officiel de la Moselle du 20 avril 1961. art. 58: 'En règle générale le terrain qui se trouve devant les maisons appartient à la commune, à moins qu'un titre spécial ne prouve le contraire'.

4 P. de Martin. La dégradation de l'habitat rural de type lorrain vers la Haute-Marne, Nancy, 1965, Diplôme d'Etudes Supérieures de l'Université de Nancy II, Laboratoire de Géographie Humaine, dactylographié.

5 'Il était et est encore admis qu'une partie plus ou moins large du sol joignant la construction appartient au riverain' Arrêt de la Cour d'Appel de Nancy, 20 octobre 1966.

6 Bulletin Officiel de la Moselle, art. 58 (op.cit.): 'Les propriétaires ont le droit d'en user mais ne peuvent revendiquer que le tour du volet'. P. de Martin (op.cit.) rapporte à ce sujet qu'en Lorraine méridionale les usoirs cadastrés sont souvent plus étroits que les usoirs réels.

7 ex. de Servigny-les-Sainte-Barbe (Moselle).

8 Le jugement de la Cour d'Appel de Nancy (note 5) le laisse entendre: 'que si ces sols figurent actuellement sur les plans cadastraux sans affectation ou rattachement à l'immeuble riverain et paraissent de ce fait rattachés à la voie publique, il n'en résulte pas, compte-tenu de l'utilisation qui en est faite, preuve d'une possession de la part de la commune, en l'absence de tout acte explicite établissant dépossession régulière du riverain'.
A l'inverse, sur son terrain d'étude, P. de Martin (op.cit.) estime que, depuis le XIXe siècle 'l'administration s'est aperçue qu'elle avait intérêt à cadastrer au maximum pour augmenter le revenu de l'impôt foncier et pour diminuer la charge des communes...'.

9 Un arrêt du tribunal de Colmar en date du 27 avril 1892 précise que 'ces places... doivent être considérées comme

faisant partie du domaine public de la commune et sont, par conséquent, imprescriptibles'. Un autre arrêt du même tribunal (27 février 1901) reprend le thème de l'appartenance au domaine public communal mais en soulignant la désaffection susceptible d'être induite, soit d'un état de fait tel que 'l'absence prolongée de protestations de l'autorité contre les entraves apportées par les riverains à la circulation du public', soit de toutes circonstances indicatives de la commune 'de ne plus entretenir les usoirs'. En revanche, le Tribunal Régional de Metz (12 février 1897) nie expressément qu'il y eut domaine public dès lors qu'il n'existe plus d'affectation à la circulation générale. (cf. Ministère de l'Equipement: Centre d'Etudes Techniques de l'Equipement de l'Est. Statuts et règles d'usages des espaces collectifs: les usoirs lorrains. Metz; 1978; offset, 52 p., p.9.)

10 J. Peltre. L'évolution des meix et du bâti dans un village lorrain depuis le XVIIe siècle; Géographie historique du village et de la maison rurale; Paris, 1979, édition du CNRS, pp.113-20.

11 L. Lopez. Les limites vosgiennes du village et de la maison de type lorrain, du côté du massif cristallin. Diplôme d'Etudes Supérieures, Université de Nancy, 1964.
J.P. Giron. L'habitat rural de la plaine sous-vosgienne, Mémoire de maîtrise, Nancy, 1969.

12 A. Dufour. Géographie d'un secteur rural en Lorraine méridionale au début du XVIIIe siècle. Diplôme d'Etudes Supérieures, Nancy, 1972.

13 ex. à Vomécourt, Padoux, Sercoeur près de Rambervillers (Vosges). (cf. L. Lopez, op.cit. p.37).

14 Cette dernière explication, avancée par A. Demangeon, a été souvent reprise après lui. Cf. La France économique et humaine (Géographie Universelle), t. VI, 2e partie, 1er vol., p.189, Paris, 1946, A. Colin.

15 Ex. St Baslemont, Provenchères, Monthureux/Saône, Siersdorf, Cottendorf, Hemmersdorf.

16 Sur la frontière belge par exemple où de petites maisons ont de profonds usoirs alors que des fermes à quatre travées n'en ont pas.

17 P. Fénelon. La structure défensive du village lorrain; Actes du 84e congrès national des Sociétés Savantes, Dijon, 1959, section de Géographie, pp.29-42. M. Fénelon expliquait par des nécessités de défense le rétrécissement des rues aux issues des localités.

18 J. Peltre. Une génération de nouveaux villages en Lorraine. Revue géographique de l'Est, 1966, 1-2, p.17.
A Tanconville (Meurthe-et-Moselle, arrondissement de Lunéville) créé en 1732, l''arrest d'appointé' précise: 'sexto: que chaque emplacement de maison de laboureur aura 6 toises de largeur à la face et 8 toises au moins de profondeur, et celle de chacun manouvrier moitié seulement de

la dimension marquée ci-dessus. (La toise lorraine mesurait 2,86 m.)

'septimo: que chacun des emplacements laissera 6 verges de façade de chaque côté de la rue au devant de sa maison pour servir d'aisance, place au fumier et passage du chariot' (cité par J. Nicod: Un texte sur la structure du village lorrain. Information géographique, 1950, n° 4, p.161).

CHAPTER 8

GLEBE AND VICARAGE ON THE ISLAND OF FALSTER, DENMARK, IN THE 17th CENTURY

K-E. Frandsen
(University of Copenhagen)

During the last 11 years a group of scholars within the Danish section of the Scandinavian Deserted Villages Project has been studying the development of settlement on the island of Falster, where 20 of the 112 villages mentioned in Kong Valdemars Jordebog, from about 1240, disappeared before 1569 and 10 more before 1660, while 28 new villages were established in the same period. Apart from this very important information the medieval sources are very few, and for this reason the project has laid much emphasis on morphogenetic analysis of evidence from the 16th and 17th centuries in order to discern earlier stages. The rich material included in the land register of 1682 (Kristian 5th's matrikel) permits us to study the agrarian structures of that time in great detail, but as no maps were made in 1682 (unlike in Sweden), most of the retrogressive analysis has to be made by reconstructing maps and using the enclosure-maps from about 1800 as the main cartographic source.

Regarding village-and field-structures the investigations have shown one particularly striking feature: a marked regional differentiation between the northern and southern parts of Falster, in spite of the fact that the island is physically very homogeneous. To the north and west of the island almost all villages are strictly regulated street-villages (Angerdorfer) with farms situated in two parallel rows. The land-division was regulated according to the solskifte-rules (GÖRANSSON 1976, 22ff.) and the furlongs were divided among the farms according to fixed quotas (bymal) corresponding with the width of the tofts and the manorial rents (landgilde). In contrast, in the eastern and southern parts of Falster most of the villages have irregular forms. In the east the land-division was according to the bolskifte principles (GÖRANSSON 1976, 27), but in the south both the village-forms and the field-divisions were completely irregular.

We have for years been trying to explain these facts but without convincing results until a preliminary investigation by Dr. Svend Gissel, leader of the Danish section of the project, on the structure of the glebe land inside the open-fields (a problem until then totally neglected in Danish research) gave some interesting results, which encouraged me to examine all

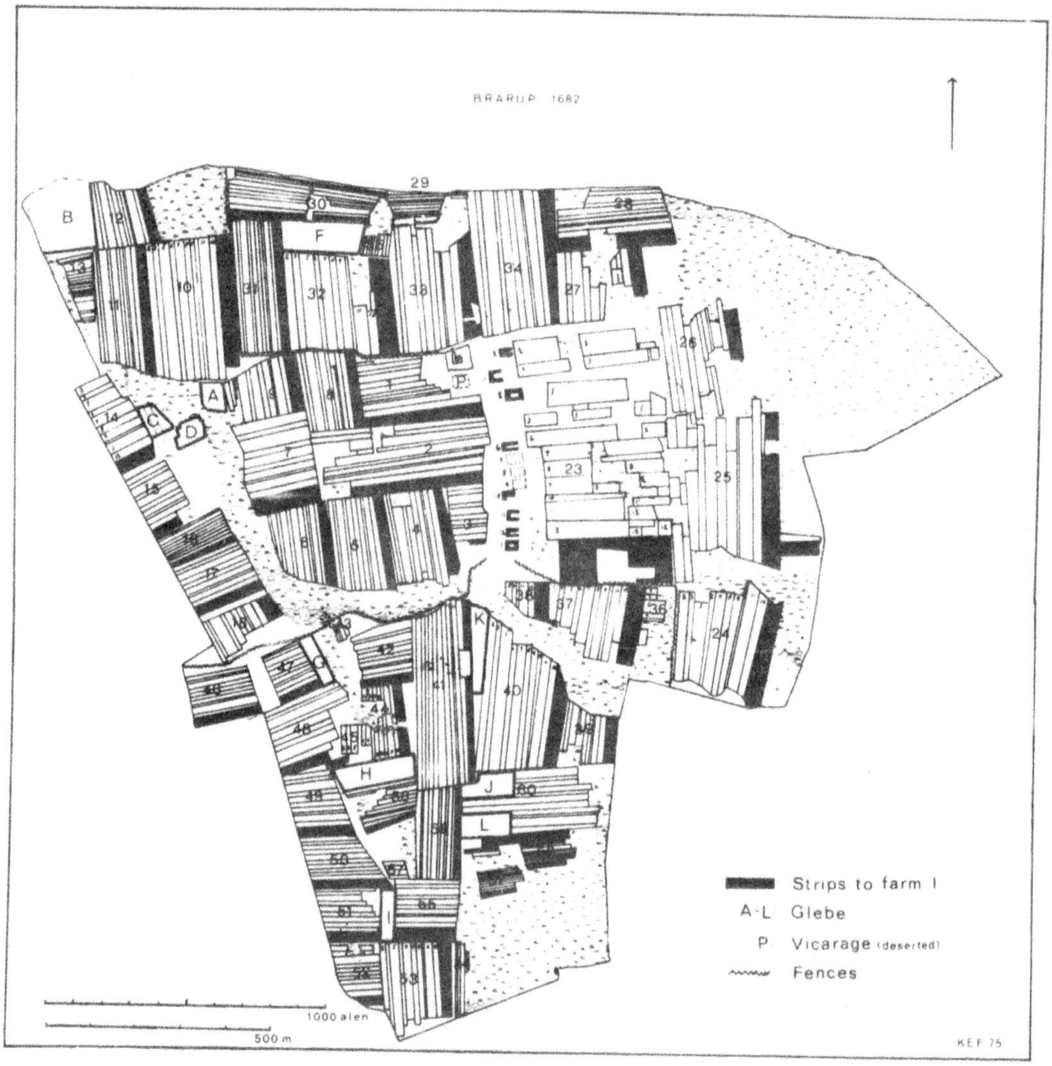

Fig. 8:1 Brarup 1682

the 27 villages in Falster which in 1682 had a vicarage. In marked contrast to Sweden the Danish medieval laws have very few provisions for vicarage and glebe. Only from the end of the 14th century does it seem to be normal that in every parish the vicar should have a vicarage (in which he should reside) and a glebe exemptam de omnibus oneribus regiis ('Handfaestning' 3/5 1376 para. 2) and in regional laws (Lolland 1330 and Falster 2/1 1417) it was laid down that if a vicar had no glebe it should be given to him on church land (in fundo ecclesiae sitam).

After the Reformation in 1536 all clerical lands belonging to the bishops and monasteries were confiscated, but the parochial clergy kept their glebe and bona mensalia pastoris, farms from which the vicar got manorial rents. During the next century it can be observed that the crown supported the clergy by enlarging their possessions so as to make it possible for the vicar to live of his own, including the tithe.

In the following I shall illustrate the three different types of glebe structure on Falster in 1682.

1) The village of Brarup (Fig.8.1: reconstructed on the basis of the measurements of 1682 combined with the enclosure-map of 1792) consisted in 1682 of 11 farms of which 3 (nos.7, 8, 9) disappeared before 1792. The farms were situated in a single row with irregular tofts to the east. The village had its arable lands collected into 61 furlongs, organised in three big open fields (as most of the villages in Falster) farmed with a three-year rotation (rye-barley-fallow). The parish of Brarup was, since 1536, annexed to the parish of Kippinge, and the vicarage in Brarup was abolished (its supposed former position is indicated by P on Fig.8.1). The church owned as fabrica the farms nos.2 and 3. Besides his own farm the tenant no.3 used some scattered pieces of land (farm no.4) called 'the glebe of the vicar' (Danish, Sognepraestens annexgard), for which he paid rents to the vicar.

The land-division in Brarup was according to the solskifte principles, the sequence of strips in every furlong being: 2-3-5-6-7-8-9-10-11-12-1 corresponding with the sequence of farms in the village row. There was, however, no correlation between the widths of the tofts and the widths of strips, but we know that in 1634 the furlongs were remeasured according to a standard of 4 bol (C.A. CHRISTENSEN 1970). The most interesting feature, however, is that the glebe land was completely independent of this field-division. Farm no.4 had no tofts, and its land was dispersed in the 3 fields in 12 rectangular blocks (Danish, Stuf-jorder) presumably unfenced but farmed in severalty. Below is listed the names of the blocks in 1682 and the corresponding names on the enclosure-

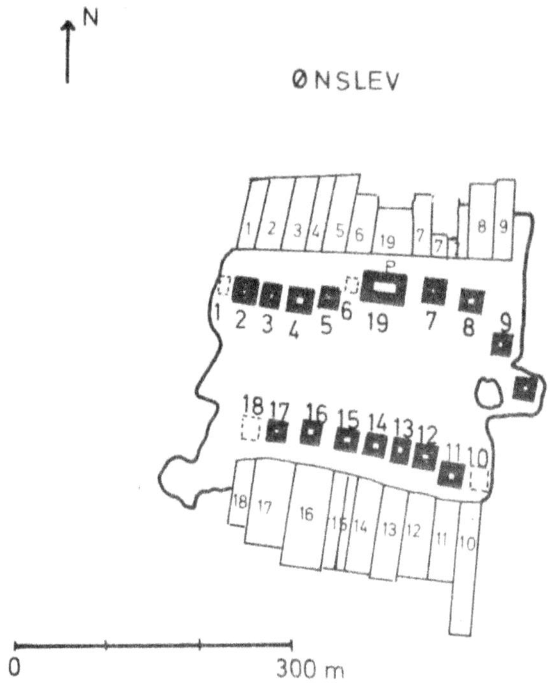

Fig. 8:2 Ønslev.

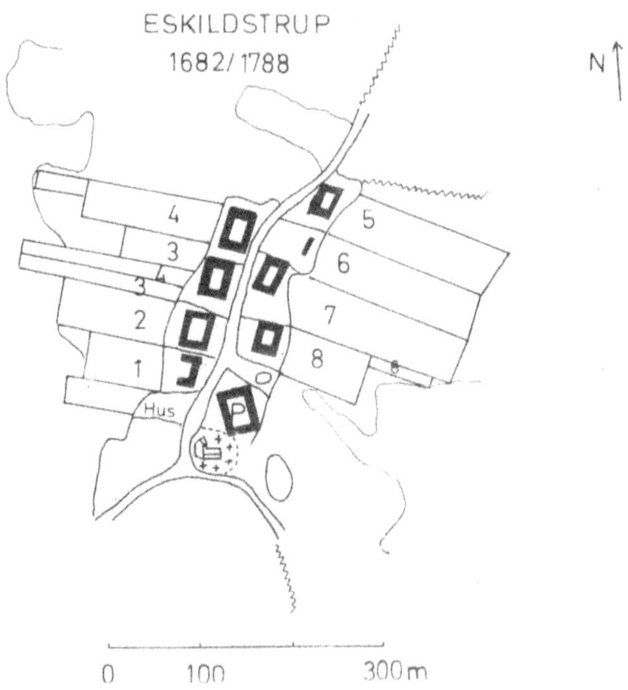

Fig. 8:3 Eskildstrup 1682/1788

map of 1792

Markbog 1682	Map 1792
Westerwang:	
A: Kelle Stude Aas	Kelde-Leeds Agre
B: Korsse Stude	Kors Stuv
C: Kierke Stude	Praeste Stumper
D: Becks Stude	-
E. Strip no.10 in Eppe Stene Stockerne	Eble Steens Stykker
Kirkemark:	
F: Trinne Stude	Praestestuv
Syndermark:	
G: Ellebechs Stud	Stuv Agr.
H: Den store Stud bag Ellebech	Stuv Agre
I: Holm Stud	
J: Den trinde Stud	Stuv Agre
K: Kierlinge Bro Stud	Kierling-broe
L: Støffuls Stud	Huule Agre

A small but interesting point is that block C was divided between the glebe of the vicar and a strip belonging to the church of Brarup (but used by a tenant from the village Alstrup).

The fundamental question is whether these blocks represent a primary or secondary structure, in relation to the strip-and furlong-system, but before trying to answer the question it is necessary to examine more examples.

2) The village of Ønslev, 4 km SE of Brarup, is a very representative example of the regulated street villages of North Falster (Fig.8.2). The land-division too was strictly regulated solskifte with the sequence in every furlong being the farm nos.: 10-11-12-...18-1-2-3...6-19-7-8-9. The vicarage (on Fig.8.2 marked p = no.19) was situated in the northern row of farms. It had tofts and strips in the furlongs twice as wide as the other farms. In a county register from 1690 (Herredsbog) a document from 1477 is cited: concerning Praedia et possesiones curiae sacerdotorum in Ønslev it says 'It has part in all fields of 11 ørtug (1 ørtug = 1/24 of a bol) according to the standard of the village'. This means that before 1477 this glebe was integrated in the village fields. It must however be stressed that the church is situated 500m to the east of the village, and it is possible that the village had been moved to its present position in the middle ages, and in this connection a regulation of the village and field-structure could have been undertaken.

3) A third type of vicarage and glebe-structure is

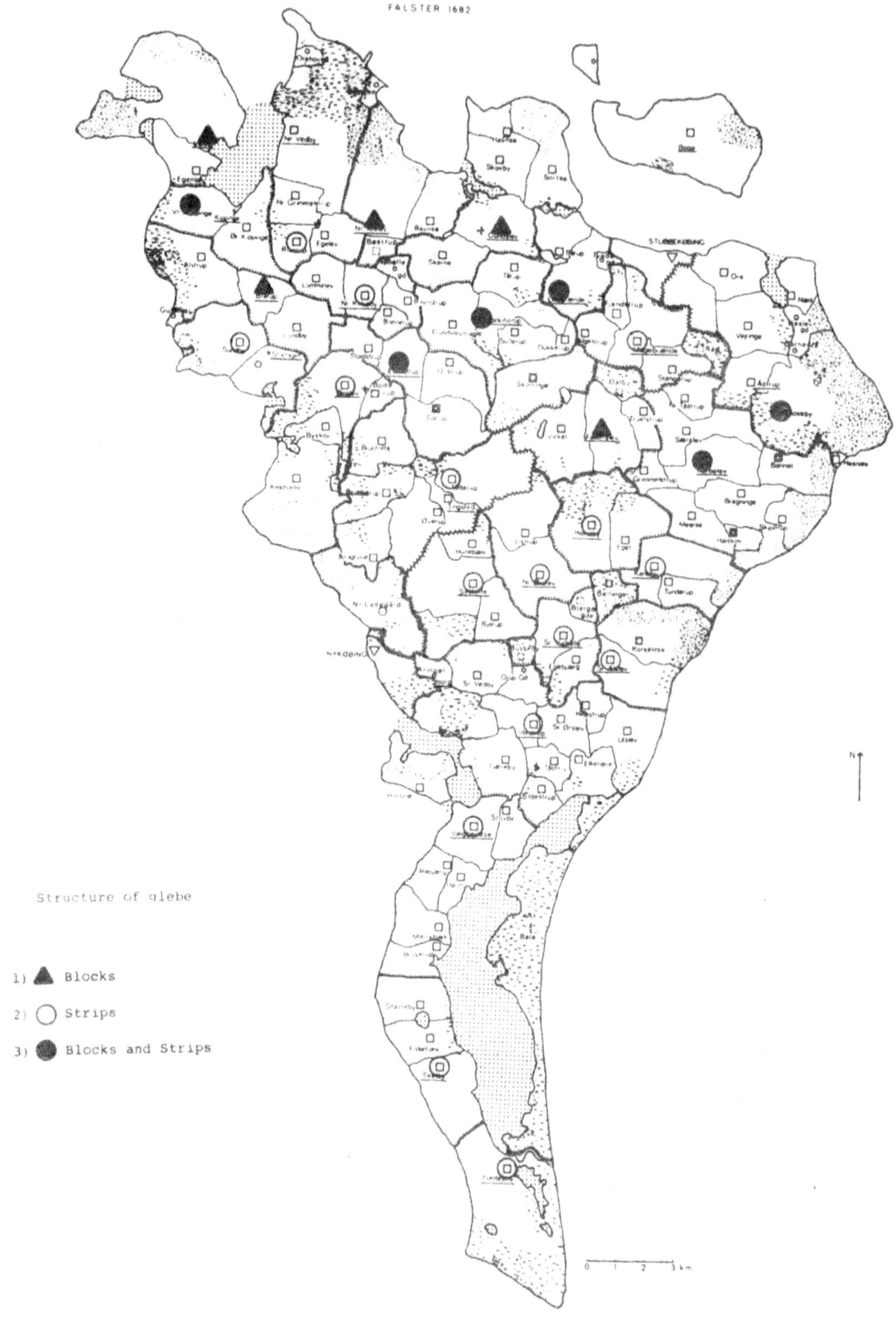

Fig. 8:4 Falster 1682: type of glebe holding.

demonstrated in the case of Eskildstrup (Fig.8.3) just 2 km to the east of Ønslev. The village consisted in 1682 of 7 farms and a cottager plus the vicarage (marked P) by the church. The farms were situated in two rows with crofts to both sides but the layout of settlement was not very regular.

In contrast to the other farms the vicarage had no tofts in connection with the farm buildings, but it did possess the toft no.6. The glebe consisted of two elements: a regular position in the solskifte-divided furlongs, as farm no.6, and a great number of blocks scattered in the three fields. In this particular case we are able to trace the origin of the first element. In a document from 10/8 1575 (Fynske Registre 1, 586) we are told that because the vicar has complained that his glebe is very small, the King Frederik II has given him a desolate holding belonging to the crown to the north on the eastern street in Eskildstrup. There can be no doubt that this holding is identical with no.6 in 1682, and that the original glebe consisted of scattered blocks as in Brarup, to which, in 1585, were added the strips belonging to farm no.6.

For Falster the structure of the glebe in relation to the peasant land can be summarised in 3 groups:

1) scattered blocks (as in Brarup);
2) fully integrated with the peasant land (as in Ønslev);
3) a combination of blocks and strips (as in Eskildstrup).

Glebe Structure on the Island of Falster, Denmark

Glebe-structure	(1) blocks	(2) integrated	(3) 1+2	Total
Position of vicarage:				
Separate	4	3	4	11
Integrated	1	13	2	16
Village-form:				
Regular	4	7	3	14
Irregular	1	9	3	13
Field-division:				
Solskifte	5	8	4	17
Bolskifte	0	3	1	4
Irregular	0	5	1	6

The statistical analysis of the different subjects suggests that glebe in the form of blocks only occurred in regulated villages (Brarup is an exception) with solskifte-division, and in the same villages the vicarage was situated outside or on the outskirts of the village. When the glebe was integrated in the fields, the vicarage was integrated in the village layout (as in Ønslev).

Fig.8.4 discloses a clear regional dispersion of the different types of glebe. The blocks and block-and-strip structure are found exclusively to the north of Falster, whereas the integrated glebe is absolutely dominant in the middle and southern parts of the island.

The explanation of these structures is not very simple. First the existence of glebe in the form of blocks can be interpreted in two ways. One possibility is that the vicars according to the laws and privileges of the 14th century were given glebe mainly in the form of existing peasant farms (belonging to the church), but where this was not possible untilled blocks between the existing furlongs were made available. But another (and as it seems to me more convincing) explanation is that the blocks are a relict from a former widespread field system of Blockgemengefluren well known from Germany (KRENZLIN 1961) being 'a system which seems to lie half way between separate, fenced closes and extensive subdivision into... strips, half way between severalty and comprehensive common rights' (FOX 1981, 86). This type of field system corresponds very well with the expressions in the laws of the 13th century.

It seems very possible that the organisation of the fields in the form of a regulated Gewannflur, with furlongs and elongated strips, was made as a part of the reconstruction of the rural landscape after the great depression in the 14th century. In the southern and eastern parts of Falster the church participated in this process, and the fabrica-land (some of it later transformed into glebe land) acquired the structure of a normal peasant farm. In the northern parts of the island the crisis was more severe and protracted. Here many villages disappeared, some were moved to new sites (as has been proved archaeologically) and the rebuilding was late and conformed to very strict rules in regard to village layout and field-divisions. But at the time the vicars had assumed a great deal of autonomy and exemption from taxation for their glebe, and therefore in 11 parishes the vicars did not participate in the regulations but kept their glebe in the old block-form as well as their vicarages beside the churches instead of moving them to new places together with the peasants. Later on, mostly in the 16th century, the vicar in 6 of these 11 parishes got in possession of a peasant farm to enlarge his glebe lands.

It is obvious that this theory needs testing on a broader national, Scandinavian and even European basis, and I shall finish this paper by confirming the statement of Harold Fox: 'In this connection there is an urgent need for detailed studies of the disposition of glebe land, for the glebe tended to be the most stable of holdings and the most likely to take us back to early arrangements' (FOX 1981, 90).

REFERENCES

Sources:

FYNSKE TEGNELSER: Rigsarkivet, Copenhagen, Danske Kancelli.

MARKBOGER 1682: Rigsarkivet, Copenhagen, Rentekammeret 312.

HERREDSBOG 1690: Landsarkivet for Sjaelland: Fortegnelse paa Praesternis... i Norre Herridt i Falster tillagde visse og u-visse Indkomster.

Works:

C.A. CHRISTENSEN, 1970: Det falsterske landstings erklaering 1484 om skyldjord og rebning. Landbohistoriske studier tilegnede Fridlev Skrubbeltrang.

H.S.A. FOX, 1981: 'Approaches to the Adoption of the Midland System', The Origins of Open-field Agriculture, ed. by Trevor Rowley (London 1981).

SVEND GISSEL, 1979: Falsterske praestegarde og andre ejendomme som agrarstrukturelle indikatorer. Non-printed paper to Det nordiske Odegardsprojekt.

SOLVE GÖRANSSON, 1976: 'Solskifte: the definition of a confused concept', Fields, Farms and Settlement, ed. R.H. Buchanan (Ulster).

ANNELIESE KRENZLIN, 1961: Die Entwicklung der Gewannflur, Berichte zur deutschen Landeskunde 27.

II: VILLAGES AND FIELDS

INTRODUCTION

'In this year King William went north to Carlisle ... he drove out Dolfin who formerly ruled that district ... thereafter he returned thither southwards, sending very many peasants thither, with their wives and livestock, to settle there and till the soil.' Anglo-Saxon Chronicle, sub anno 1092, trans. G.N.GARMONDSWAY (1954).

The morphology of the individual village is the result of the interaction of many forces, some environmental, some economic and others social, but all act through time, and the second section of this volume draws together papers concerning the relationships between settlements, villages or hamlets, and the economic areas supporting them. HALL examines the way in which early field arrangements can be reconstructed and demonstrates some of the evidence for remarkably elaborate arrangements for the allocation of landownership parcels within sub-divided field systems. In some settlements many re-allocations must have occurred, but in others the presence of important tenurial or manorial differences represented fixation lines for all subsequent developments, and, tenuous as lines of argument can become, allow some discussion of the earliest arrangements which may originate in the twelfth century or even earlier. SIMMS provides a broader context for such detail by a study of the settlement forms and functional variations on two royal manors in southern County Dublin, where feudal settlements and field systems were forcibly superimposed upon an Irish base. Such appraisals of Anglo-Norman colonisation in Celtic environments have much to tell us of complex processes of interaction which must have been common during the Norman conquest of Englan, and involve both cataclysmic change and subtle continuity. JONES uses Welsh sources to penetrate beneath visible, relatively superficial, aspects of settlement and to explore relations between these and the social and economic linkages in a lnd-conscious society. He reminds us that beneath Anglo-Norman, Anglo-Saxon and Scandinavian organisations there lie even more ancient roots and it is at this level that attachments to special places, sacred places, valuable places, have been forged, elements of continuity within otherwise widespread patterns of change.

In a wide-ranging and precise paper NITZ links together the three morphological components of an evolving settlement system, the emergent pattern, the morphology of individual villages, and the character of the field systems. He concludes by touching upon important questions concerning the origins of the model or 'village idea' manifest in village plans in Saxony: the regular two-row street village plan-type he identifies as characteristic (Fig. 12.6) could be exactly paralleled elsewhere in Europe, and the need for carefully defined criteria, morphological, social, organisational and economic, which distinguish and define particular settlement forms and field systems, is apparent. Comparison must rest upon secure foundations. KRINGS returns to the theme of Dutch settlement and the settlements established in the Lower Rhineland during the medieval period, and like VAN DER VAART considers the problem of retaining these ancient and characteristic field patterns - long and remarkably regular strips - within present and future landscapes. In her study of Polish villages SZULC begins by emphasising the problems of source material, continues with an appraisal of the morphogenetic levels identifiable within Silesian and Pomeranian villages, and looks, in conclusion, towards a broader survey of morphogentic types within Europe as a whole, a leitmotiv which emerges from many of the studies in this volume and one which is indissolubly linked to problems of definition and classification.

Papers by WINDELHED and ULSIG present different aspects of the study of Scandinavian villages and their fields: the former outlines a research project in progress within a marginal landscape of east central Sweden, building upon very limited documentary materials by using archaeological methods to treat problems of settlement evolution. In contrast ULSIG uses documents to study changes in landholding between the medieval period and the late seventeenth century in East Jutland, changes which took place in spite of underlying continuity in the unit of fiscal assessment, the bol, a measure which carries the argument back to HALL'S paper. With the final paper of this section, by FOUTREIN, we turn to Flanders, and the social and economic forces generating settlement aggregation and dispersion. The threshold between a nucleated and a dispersed pattern is, in many parts of Europe, by no means sharply defined, and as studies of settlement development reveal tendencies for substantial readjustment within regional settlement systems, leading to aggregation or dispersion, the study of areas characterised by a mixture of types may allow the identification of the factors leading to a balance, or stability, or imbalance, or change, in either pattern.

CHAPTER 9

FIELDWORK AND FIELD BOOKS: STUDIES IN EARLY LAYOUT

D Hall
(Cambridgeshire Archaeological Field Officer)

This paper will discuss the combined use of the evidence of fieldwork and historical records to identify early estates and tenurial arrangements. For effective analysis of parish structure three sources are essential: 1) an open field plan, 2) a field book, 3) adequate documentary records.

THE EVIDENCE

Open-field Plans

Some townships have pre-enclosure maps that detail the layout of the open fields. For those that do not the problem can be remedied by archaeological fieldwork. The techniques have been discussed recently, and need not be treated here in detail (Hall 1982 a). In summary, the action of the plough transferred soil from each strip to the ends, causing accumulation over the centuries. The edges of furlongs, where there had been a series of turning points, were demarked by considerable banks of soil. The effects of modern ploughing rapidly render a field of earthwork ridge-and -furrow planar, but leaves the furlong boundaries as long linear banks which are fairly easy to identify and survey. Examination of every modern field in a parish, and plotting the linear banks, allows a reconstruction of the furlong pattern of earlier subdivided fields. It is always worth checking parishes that do have open-field maps, because areas of ancient enclosure, that do not have furlong details mapped, usually reveal evidence of strip ploughing on the ground.

Field books

England does not have the benefit of a detailed strip-by-strip national survey of all subdivided holdings, such as exists for Denmark and Sweden. However, many individual parishes or townships do have such surveys, often called field books, town books, or terriers. They are descriptions of all the individual strips, and usually group information by field and furlong, and give the size of the holding, most commonly in an estimated form such as 'one rood' or 'half an acre', or 'one land'. In exceptional cases strip widths at each end, the length, and the area, are recorded, accurately measured in poles. The lands and furlongs are sometimes numbered for ease in identification. Owners and tenants are named, and often the nature of the holding identified, i.e. whether it is glebe, freehold, demesne, former monastic land etc. Many terriers were prepared immediately prior to enclosure, or partial enclosure, which would have necessitated exchange of lands:

sometimes these exchanges are recorded as marginalia in field books. Most field books state or imply the cardinal orientations of land by adding after the furlong names such comments as 'begin west' i.e. implying that the land runs north and south. Abuttals relating a furlong to its neighbours or to topographical features are often mentioned.

Documentary sources

The field book itself obviously constitutes a major source. It is necessary to add up the numbers of strips in each furlong and compile a table of names and data. As many additional furlong abbutals as can be found are required, obtained from charters, extents, surveys, deeds etc. The next type of information needed is a compilation in map form, of all the modern field names recorded on Tithe and estate maps, and sale catalogues. Many of these will be furlong names, which thus become accurately located on a modern 1:10,000 (or 10,560) map. Field names can also be identified for those parishes that have surveys of the enclosed fields, provided that acreages of each field are given. The locations can be worked out by reference to modern acreages, and the relationship of the old farms to the present day ones, and to the enclosure holdings. Primary research may also be necessary, i.e. asking present-day farmers what the names of the fields are.

From the charters, extents, and deeds etc, already used to search for topographical information, the descent of most of the holdings can be determined. Thus the field book may not state that a given person is the inheritor of former monastic land, or holds the tenancy of, for example, the demesne, but other sources will often make this quite clear. Published county histories can be used as a general framework for property descent. Any ecclesiastical or monastic land should be carefully studied: the accumulation of monastic estates will often be minutely detailed in a cartulary, and glebe disputes are frequently informative as to the nature of the land, i.e. who gave it and when.

Furlong identification

We are now in a position to attempt to identify furlongs using the three types of data compiled by the methods indicated above. The furlong map provides the physical landscape, the collected enclosed field names give locational data, and the furlong list gives the medieval field names, their sizes and abuttals. A simultaneous usage of all three should lead to an accurate and precise plotting of furlongs on the reconstructed plan.

In clear cut cases where there is a field book, and a good set of post-enclosure field names, and plenty of furlong abuttals, it is an easy matter to start from a known point, say on the parish boundary, and plot the furlongs directly on to a 1:10,560 plan. On this scale lands of average 7m width are 0.7mm wide, i.e. there are 14 lands to the centimetre.

In an area where the data are not good, a slightly different procedure has to be adopted. A strip of thin cardboard, no more than 2.5cm wide, is marked off in 'furlongs' measured to scale. These are then cut up, the name of the furlong written on with any abuttal information and the land orientation. It is then possible to arrange and re-arrange these 'furlongs' on the reconstructed plan until they form a reasonable interpretation of all the evidence. The sizes (widths) of the furlongs are known by either adding up the stated widths of the lands or by using an appropriate average value. For the English Midlands this works out to be 7 metres in most town-ships. Further details of how to identify furlongs have recently been given (HALL 1982a).

Having satisfactorily located the furlongs it is a relatively simple matter to plot whatever information the field book or other documents choose to impart.

ANALYSIS OF RESULTS

Four topics will be examined here to indicate the potential of a combined fieldwork - historical approach: 1) demesne estates; 2) early lay out of subdivided fields; 3) tenurial arrangements; 4) relationship between peasant holdings and the Domesday assessment.

The first two themes have been treated before, and so will only be summarized here.

Demesne estates

The land directly farmed by the lord, the demesne, is commonly specified and its exact whereabouts can be determined using fieldbooks. The Saxon demesne was called inland, and usually lies in a compact block next to the manor house. Wollaston, Northamptonshire, is a good example: the inland of one of the two manors represents a very ancient core of land. It contains a Roman villa (reflected in the furlong names as Walton acre, i.e. the ton of the Wala or British), an early Saxon site, and a manorial enclosure.

By the time that detailed documents survive the demesne is often disposed in a different way. At Ashby St Ledgers a field book of 1715 specifies pieces of land as paying tithe which can be identified with the demesne of c1210. It lies in blocks of complete or half furlongs, near but not actually adjacent to, the manor house. Later examples in other parishes show a uniform dispersal of the demesne throughout the parish grouped in parcells of just a few lands in each furlong. At Wollaston this had happened by 1430, and at East Haddon by

1598. There are numerous 17th-century cases showing the same uniform dispersal. (See HALL 1982b for plans and more details).

The above examples suggest the following model. First the demesne was a compact piece of land next to the manor house, and close to it for convenience, and probably contained some of the best quality soil. The lord did not intend to suffer the inconvenience of traversing the whole parish to attend to strips scattered in single units. Probably with passage of time, the demesne became over-exploited, and the villagers were getting better yields of corn in their scattered holdings than was the lord. There would then be a desire to disperse the demesne. Ashby St Ledgers shows a half-way stage with the demesne partially dispersed in half furlongs, but still reasonably accessible. The later examples, such as East Haddon, show the extreme case of complete dispersal, yet even here the demesne is distinguished by being in small parcels of land with groups of three lands lying together.

Such a model demonstrates the authority of the lord over a long period of time. He clearly had a lot to say in whether the open-field holding should be re-arranged. Subsequent events demonstrate the same authority. At Wollaston some of the inland had been dispersed before 1430, so that the lord's demesne lay throughout the parish, and one of his tenants, and tenants of another manor, shared strips in the inlands (Fig. 9:1). Yet by 1583 matters had been reversed, the demesne had been consolidated within the area of the old inlands and it was enclosed (Leicester Museum 1430, Henry Huntington Library).

Early lay-out of subdivided fields

Archaeological and physical evidence was put forward, in 1978, to show that there had been major changes in the furlong patterns of Midland England, notably that the patchwork of small furlongs with strips about 200 metres long, had been created by lateral division of furlongs with strips very much longer, up to 1,000 metres long (HALL 1981)

The physical evidence may be summarized as follows: aerial photographs show that some small furlongs have been created by changing parts of the strips of a larger furlong through 90°. In many other cases the strips of one furlong can be seen to go through (under) one or more boundaries of adjacent furlongs (HALL 1982a). Most striking are those cases which show that a massive furlong, forming a reverse-S aratral curve, is divided into several furlongs (e.g. at Raunds where six furlongs are so formed from a single one (HALL 1982a, Fig 29)).

Historical evidence has also been forthcoming showing the subdivision of such large units. At Hardingstone near

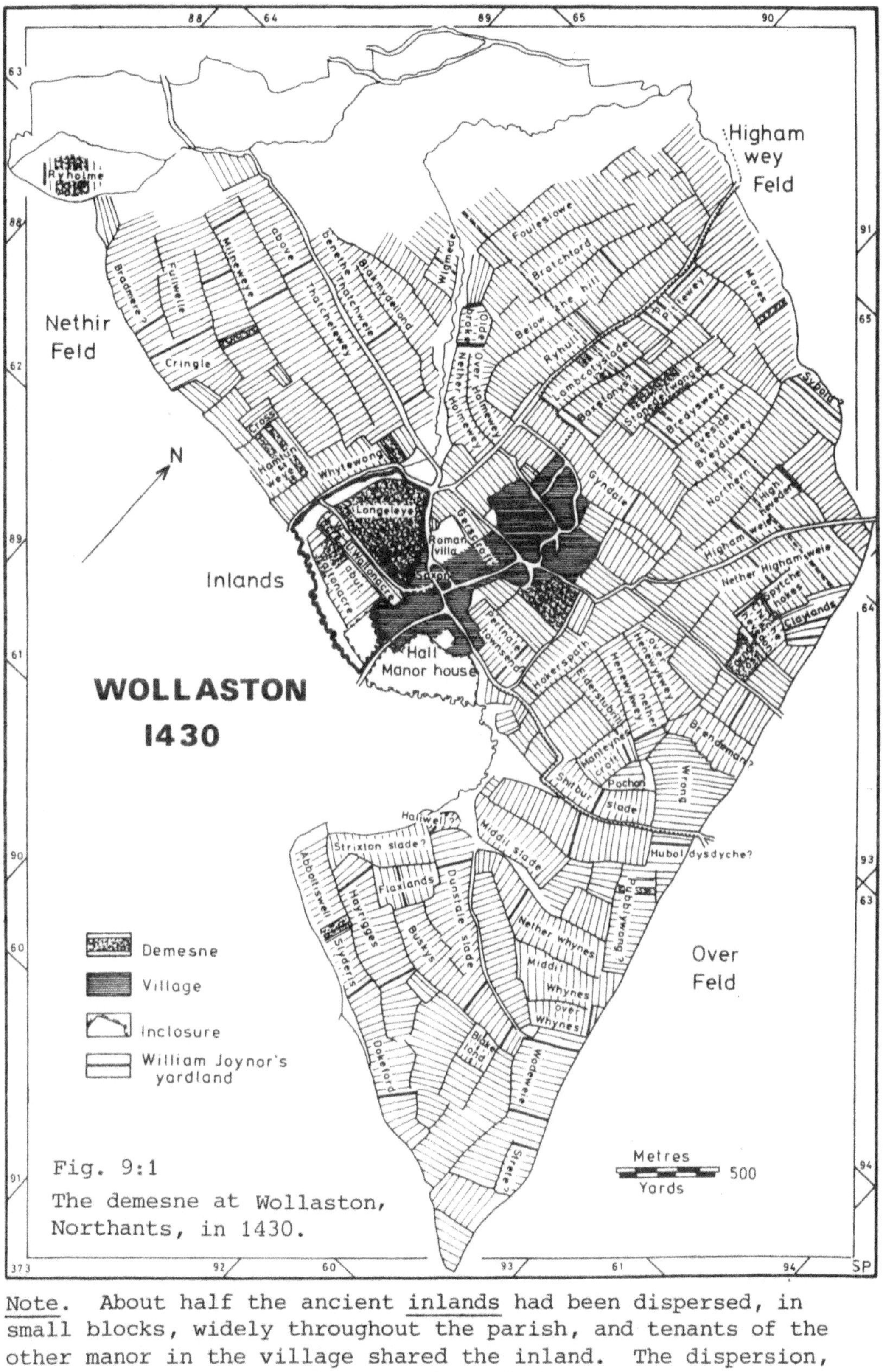

Fig. 9:1
The demesne at Wollaston, Northants, in 1430.

Note. About half the ancient inlands had been dispersed, in small blocks, widely throughout the parish, and tenants of the other manor in the village shared the inland. The dispersion, however, is not as complete as that of the peasant virgates, indicated by the plot of the single virgate held by William Joynor, dispersed in single lands.

The furlongs were surveyed on the ground and the names worked out by comparison of a map of 1774 (NRO) with the demesne terrier of 1430 (Leicester Museum). A few of the furlongs cannot be identified and a few of the parcels are therefore not plotted.

The furlong pattern illustrates a large-scale laying out of the fields, in that no less than 14 furlongs form a continuous alignment, 12,500m long.

Northampton, there were repetitive cycles of tenurial holdings intimately related to the furlong pattern and suggesting an older and simpler arrangement. (see below, HALL 1980). Generally the furlong names themselves point to an earlier larger scale unity in that they frequently occur with adjacent 'upper' and 'lower' elements, often with 'middle furlongs' between them.

Away from the Midland region there are many parishes which had simple 'long land' fields unchanged until enclosure, e.g. in Yorkshire (HARVEY 1980, HALL 1978c) and in the Cambridgeshire silt fens (HALL 1978). Lands of 1,000 metres length are very common. This is also the case in much of Germany and Austria (NITZ and EIGLER, these proceedings). It is, after all, the so-called classic English Midlands which are different. The dating of the Midland furlong subdivision is before the 12th century, according to terrier and charter evidence.

Archaeological-fieldwork results show that there was a drastic change of settlement pattern in the Middle Saxon period (c650-850), A landscape consisting of numerous small settlements was rearranged to form fewer large villages (the ancestors of the present-day ones). On pottery evidence this process occurred before the middle of the 9th century. The furrows of the open-field strips cut across these Saxon sites and must, therefore, be later. Saxon charters demonstrate that subdivided strips were in existence at least by the 10th century. There were thus two major landscape re-arrangements occurring at a similar date, it is likely that the two events happened together, as part of a large-scale replanning of the countryside. An 8th-9th century date is therefore probable for the establishment of sub-divided field systems based upon strips.

Such an early date is compatible with the German evidence, where planned villages and long strips can be dated to the period 775-850 (see NITZ in these proceedings).

Tenurial arrangements and field structure

Given that it was required to distribute uniformly all peasant holdings throughout the subdivided fields of a township, only two methods seem probable. Either there was random distribution by lottery, or there could have been regular cycles with each tenant always occupying the same position relative to his neighbours.

A lottery system could have been random in all the township at once, or, more likely, there would have been a fixed cycle consisting of the number of holdings to be dispersed, but with the position of each land in the cycle decided by lottery. It is very difficult to demonstrate mathematically the existence of lottery systems in fieldbooks, because the size of the cycle is unknown, and so is the whereabouts of its beginning or end in each furlong. Various mathematical tests would need to be applied to confirm the random nature of the arrangement.

In the Midlands there is usually a multiplicity of small furlongs making it difficult to work out sequences. Further problems arise when there has been even a limited amount of holding fragmentation or exchange. No Northamptonshire field system has yet been proved to have been arranged by lottery, but the majority are likely to be so, since they exhibit no regularity, although the possibility of complete devolution from an originally regular layout can never be excluded.

On the other hand regular cycles are easy to recognize, even if they have been partially disguised by fragmentation and exchange. Several Northamptonshire parishes reveal examples, such as Ecton and Great Billing. Göransson has already discussed Ecton (Göransson 1961) and further information will be given here. A given peasant holding, the yardland, is completely dispersed, occurring at intervals in the furlongs once in every 110 lands (see below). Numerous charters and deeds from the 13th to the 18th centuries reveal a rigid tenurial system, with the same two neighbours on each side. The system is very ancient, because an undated charter of the late 13th century grants:

> one messuage....and one virgate of arable land in the fields of Eketon between the land of Robert de Catteworth on one part and the land of John le Hirde on the other...(N.R.O. 1)

John le Hirde is also a neighbour of the messuage and so probably at one time both the order of houses in the village street and the lands in the fields were identical. Such examples are rare for Northamptonshire, although Göransson quotes cases for Yorkshire and Harvey has found many others in the same county. (HARVEY 1980).

Another parish with a regular tenurial system, until enclosure, is Great Billing, adjacent to Ecton. No map or fieldbook survives, but a glebe book of c1684 has a sketched scheme of two sets of 43 lands with occupiers' names written in (Fig. 9:2, N.R.O.2). The unit strips are called 'roods', but some of the holders with two adjacent roods had ploughed them together as one double unit, called a 'land'. The sketched plan does not state that it is a scheme for a cycle of lands, but it quite clear that this is the case, because the 43 lands are divided into groups of 20, 10, and 13, each group separated by a balk. The glebe on the scheme is marked as being the 19th, 20th, and 36th rood. A glebe terrier of 1684 states that two yardlands lay in half acres in the 20-rood hides, next to the common balks which divide the 20-rood hides from the 10-rood hides, and one yardland lay in the 13-rood hide, 5 roods from the balk on one side and 7 roods from the balk on the other (N.R.O. 3). This description agrees precisely

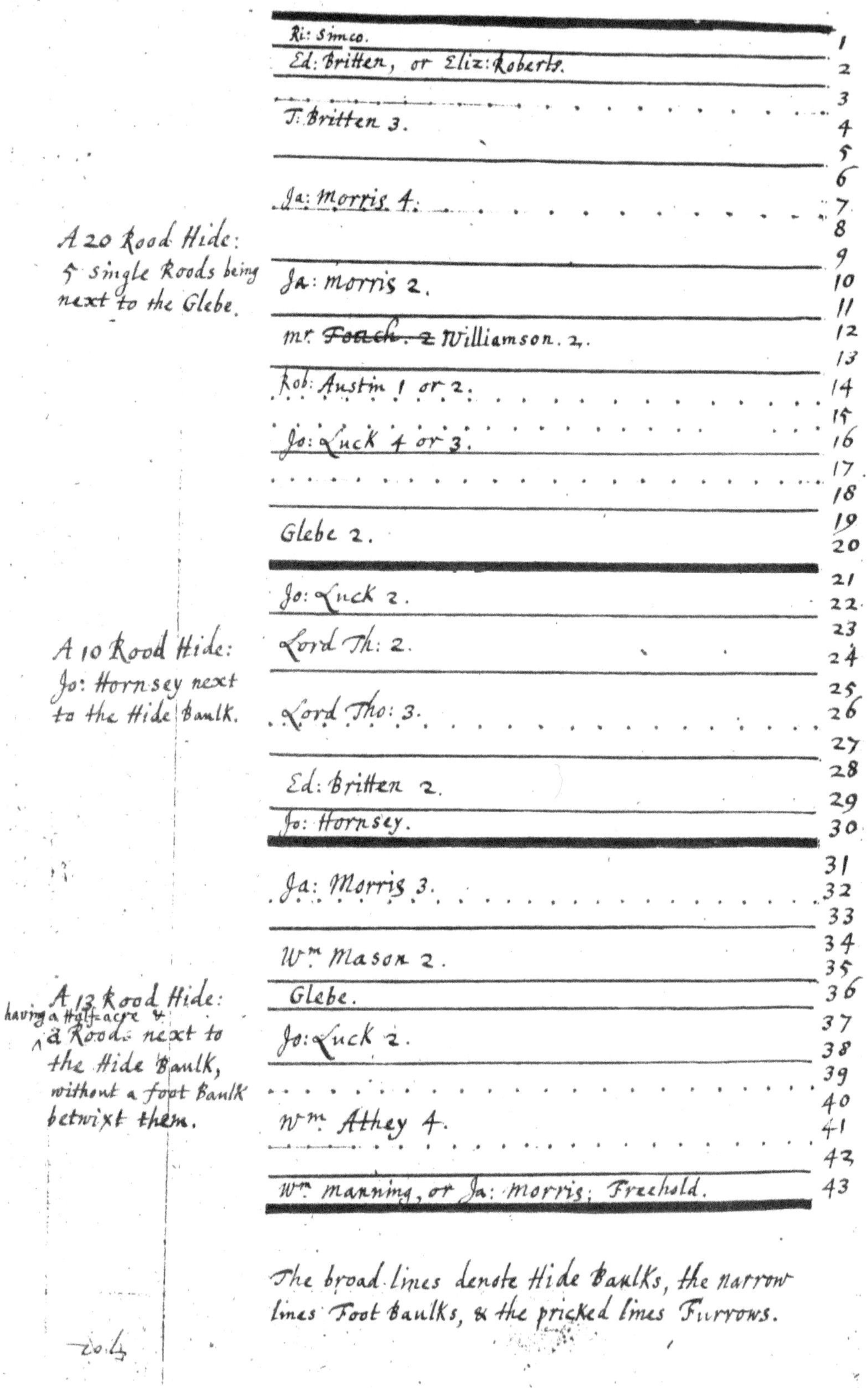

Fig. 9:2 A scheme for the regular disposition of tenurial holdings at Great Billing, Northamptonshire, 1684.

with the above. The terriers then go on to give the positions of all the lands in individual furlongs, proving that the scheme did cover the whole subdivided field-system.

Regular tenurial cycles of 32 and 19 lands are 'hidden' in field books of c1660 and 1433 respectively, for Hardingstone and Muscott. These have already been discussed in some detail (HALL 1980, and HALL 1982b), and further developments will be given below. The regular arrangements as described in these four parishes are unusual.

The parishes of Great Billing, Hardingstone, and Ecton reveal more structure than just a regular solskifte type arrangement of the individual lands making up the yardlands. The groups of 20, 10, and 13 lands at Billing were called hides, as already noted. In this case no information is available about these hides, regarding their purpose, or the date when they were first set out. The subdivisions do not relate to any manorial holdings since Domesday (1086), as the parish was always a single manor, from at least that date until enclosure in 1778. In all probability the hides date from before the Conquest.

More interesting, are the smaller hide division of Ecton and Hardingstone. Ecton had 11 hides each of which had a name (Table 9.1). A map of 1703 shows the positions of these hides throughout most of the parish (N.R.O. 4 and partly published by GÖRANSSON, 1961;) like the component lands, the hides lie in a regular sequence. The map does not indicate the number of lands in each hide, but various terriers make it clear that one rood, when multiplied throughout the hides and furlongs, is equivalent to one yardland. Thus grants of 5 yardlands appear in the furlongs as groups of 5 roods, and $1\frac{1}{2}$ yardlands as $1\frac{1}{2}$ roods (split lengthways) etc (e.g. N.R.O. 5 + 6). A few of the terriers also specify the measure of '10 roods to the hide' (N.R.O. 6 + 7). Thus there are 10 x 11 or 110 lands in the complete sequence. Such a structure is confirmed by the enclosure documents which state that there are 103 yardlands (N.R.O.8), - a few having been lost over the centuries.

Hardingstone had two independent field systems, the East End and the West End fields. This arrangement appears to date from before the Conquest also, in that it does not relate to any manorial holding since. In the East End fields there was a fixed regular cycle of 8 hides each divided into 4 lands, so making 32 lands in all. These hides, like those of Ecton also had names (Table 9:1), which appear to be those of early medieval holders of the hides.

In both cases a date of origin not later than the 13th century is proposed. For Hardingstone there is no contemporary dating, only the forms of names themselves. The Ecton hide names are dated as far back as 1464 (N.R.O.9). Two of them abbots and maries (Marys) were monastic holdings, belonging respectively to the abbot of Warden Abbey, Bedfordshire, and (probably) St Mary Delapre, Northampton. Halls hide was the demesne, Parsons hide the glebe, and the remaining names are probably those of medieval owners. Members of the Laurence family are witnesses to Ecton charters in 1357 and 1360 (N.R.O. 10), and the Billing family are witnesses in 1397 and 1464 (N.R.O. 11).

Table 9:1

ECTON HIDES	HARDINGSTONE HIDES
Marys	Swinsets
Corrall	Doylies
Lawrence	Barrole
Stannymans	Follits
Dickins	Bouden
Abbits	Sickarsbies
Hall	Poles
Billings	Caudwell
Badgers	
Parsons	
Holton	

The hides have an exact parallel with the Danish bolskifte system. Since this system is unknown in Germany Holland or Belgium (Ulsig pers.common 1981), it seems reasonable to look to the 10th century for an explanation of its appearance only in Denmark and England, when the political ties between the two countries were close. Further evidence to support this early date comes from the large number of Scandinavian elements in the furlong names, particularly the use of the Danish dale instead of the English slade for a 'valley' (HALL 1978b). An early date for the Hardingstone field system has already been suggested because the bolskifte cycles are intimately related to the structure of the furlongs and in places predate the final arrangement. There are strong reasons for believing that there have been no significant changes in Midland furlong patterns since at least the 12th century. Fig 9:3 shows the Ecton bolskifte system.

Peasant holdings and the Domesday assessment

It is with great trepidation and wariness that one enters the theme of Domesday assessments, the number and size of virgates (or yardlands), and the areas of holdings and parishes. Nevertheless the detailed parish studies are suggesting there are relationships that are worth pursuing as tools for the structural analysis of field systems.

It has long been known that the size of Domesday hides or carucates, whilst bearing a fixed ratio to some parish areas, does vary a great deal. Similarly the small peasant holdings,

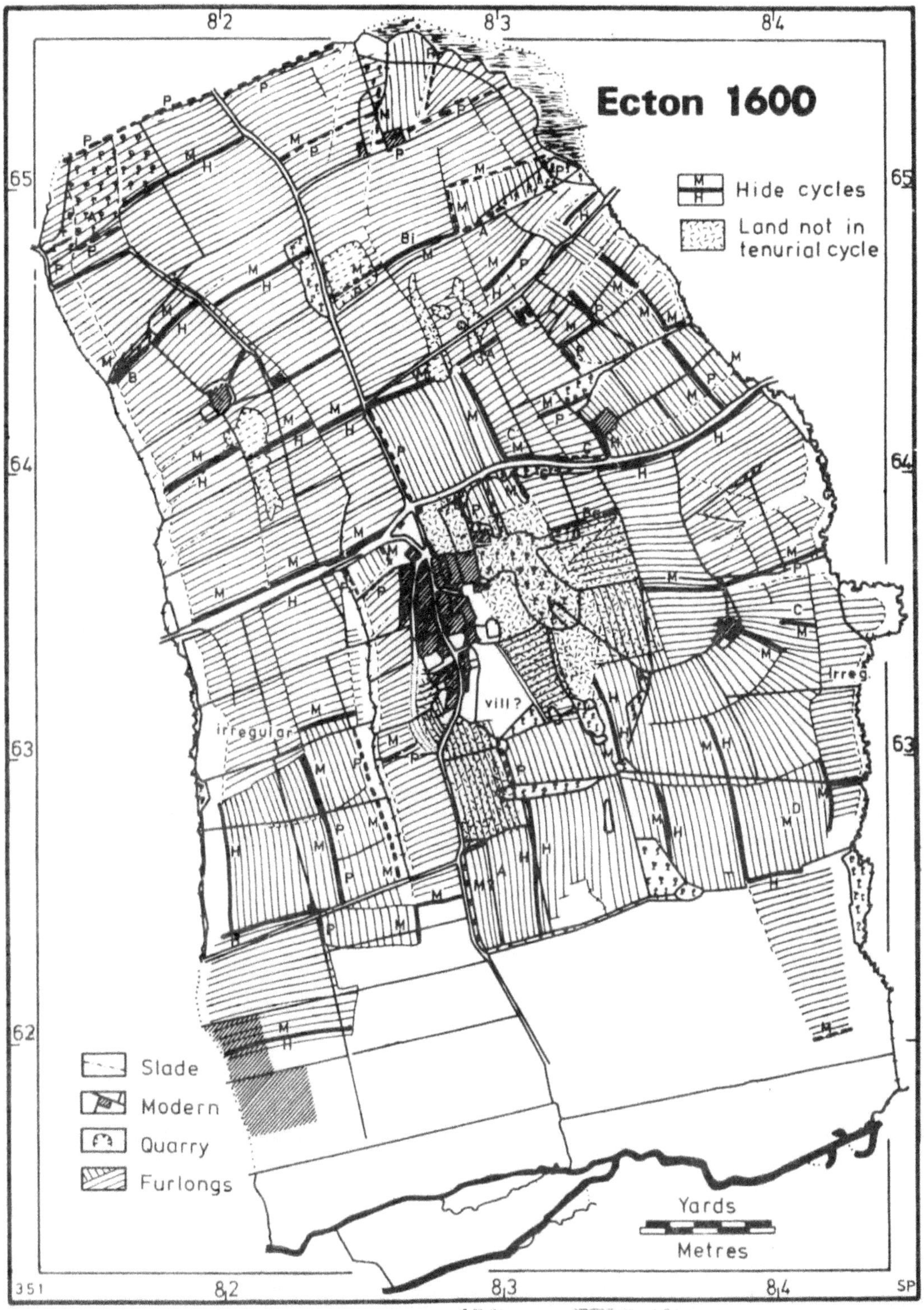

Fig. 9:3 Ecton, Northamptonshire, medieval tenurial cycles (hides).

Note. A map of 1703 shows part of the cycles (NRO map 2115) The complete furlong system was surveyed on the ground, and the cycles completed by studying the furlongs named in various terriers; particularly those for Maries Hide 1619-1721 (NRO E(S) 1054, 1037, 47, 17, 384, 17 and 10), and Parsons Hide 1587-1663 (NRO E(S)106, 626b, 503, and 541). Since there is no terrier for Holtons Hide (the last in the sequence) Parsons Hide (P) is marked on the plan. In all likelihood Holtons Hide was also present. The reconstructed hide cycles (i.e. those not shown on the 1703 map) are represented by broken lines. The terrier evidence, in detail, dates from about 1600, but charters show that the arrangement was in operation by the end of the 13th century. If Holtons Hide be considered the beginning of the cycle then Ecton shows a regular solskifte system, with cycles running north to south and west to east.

called the small virgate, or oxgang, or yardland, likewise vary in acreage.

Detailed parish analyses in Northamptonshire are revealing many useful relationships. First considering the relation between the numbers of yardlands in a whole township and the Domesday assessment. Collected figures for various estates are shown in Table 9:2. The comparative unit has been chosen as the hide. It will be seen that there is a fairly consistent ratio of 1:10-1:12 between many hides and yardlands, even down to details of the demesne in the case of Great Billing. Here the Domesday ploughlands total 8, with 2 in demesne (MORRIS 1979, 57-1): in 1612 the demesne was 12 yardlands out of 48. (BRIDGES 1781, 1405).

These figures suggest that there maybe a consistent relationship between the number of yardlands (small virgates) and the hidation assessments. This is all the more convincing when it is remembered that a 'land' or 'ridge' on the ground, when multiplied throughout all the furlongs represents a yardland. Surely at some stage, when a parish was being assessed, the surveyor merely asked how many yardlands comprised a given estate and then calculated the hidage using a factor which was generally 10 or 12?. The Domesday fractional hide holdings silently support this. Almost all fractional values are multiples of $^{1}/_{10th}$ or $^{1}/_{12th}$ of a hide, in other words these fractions have a physical reality on the ground. Examples are $\frac{1}{3}$, $\frac{2}{3}$, $^{1}/_{5}$, $^{2}/_{5}$, $^{4}/_{5}$, $^{5}/_{6}$, $^{1}/_{12}$, and $^{11}/_{12}$ (MORRIS 1979). A few other fractions exist that do not fit into this scheme, such as $\frac{1}{8}$, $^{1}/_{9}$, $^{1}/_{16}$, of a hide. The implication is that in these unusual cases the hide was rated at 8, 9, and 16 virgates. Earls Barton (Table 9:2), at 20 yardlands to the hide, is probably a 1:10 reckoning doubled. Confirmation that the argument presented above is not just arithmetical manipulation, comes from the village of Clopton. Here a remarkable analytical history was written in c.1255 (now only surviving as a transcript in the Bodleian Library) where it is stated that the hide of one of the fees consisted of 6 virgates, but that these were twice the size of usual virgates. Therefore again we have a ratio of 1:12, and this time stated categorically in an early source.

The approach does not solve all the problems of the variability of the size of Domesday hides, but it gives a tool to help understand 11th-century estates using later quantitative evidence. For instance, Brockhall and Muscott were assessed, in 1086, at 3 (large) virgates only (Baker 1822, 113), and this estate is later identifiable as part of the Leicester Fee. Another estate later called the Warwick Fee, is not mentioned in 1086, but is quantified in 1295 as 7 small virgates (Baker 1822, 119). A field book of 1433 reveals a tenurial cycle of 19, i.e. 19 yardlands. This looks like 12+7, the 7 of the Warwick Fee and 12 others, i.e. one hide. In the early 16th century Brockhall is shown to be 25 yardlands, or

Table 9:2

Relationship between Domesday hides and the number of small virgates or yardlands

Place	Hides in 1086	Number of yardlands	Date	Ratio	Reference
Cold Ashby	5.75	60	c1230	10.4	Bridges I 552
Earls Barton	4.0	80	1760	20	Enclosure (N.R.O.)
Great Addington	3.5	42	1820	12	Enclosure (N.R.O.)
Great Billing	4.0	48	1612	12	Bridges I 405
Hardingstone	7.0	79.5	1765	11.3	Hall 1980 122
Teeton	2.0	25	1600	12.5	N.R.O. LT85
Wollaston	8.0	82	1788	c10.3	Hall 1977 206
		Meadow divisions			
Ashby St Ledgers	4.0	40	c1715	10.0	N.R.O. ASL 1232
Welton	4.0	40	13	10.0	P.R.O. Rentals and Surveys Roll 528

about 2 hides. Perhaps then the Domesday entry should have been 3 hides for the Leicester Fee?.

A strikingly variable factor is the size of the yardland itself. Statistics are presented in Table 9:3 for various Northamptonshire parishes. The figures represent a contemporary statement of the acreage of the arable in a yardland. Yardland areas cannot be calculated using modern parish sizes and knowing the number of yardlands, because of the variable amount of pasture and meadow etc. It can be seen that the yardland can be anything from 15 to 39 acres. It is notable that the smaller values occur in townships laying on good soil, and likely to have been settled and completely given over to agriculture at an early date. In contrast large values are characteristic of parishes with poorer soil, that were not likely to be totally ploughed until the early medieval period.

The interpretaton is that, at sometime, a fixed number of yardlands was taken as the size of the taxation assessment (probably before Domesday). If at that time not all the parish was under arable cultivation, then two possibilities were available when the arable was increased. Either there could be an increase in the number of yardlands of fixed area, or there could be an increase in the area of the yardlands already existing, without a change in their number. The latter process seems to have been chosen.

Table 9:3
Yardland sizes

Place	Yardland size	Date	Reference	Location
East Haddon	29	1752	NRO Enclosure	Upland
Ecton	14	1609	NRO E(S)83&84	Nene valley
Great Addington	28	18th	RNO A165	Nene valley & clay
Hardingstone	20	14th	Serjeantson 1909 p.13	Nene valley
Irchester	20	1318	Kerr 1925 8	Nene valley
Teeton	22	1600	NRO LT87	Upland, good soil
Thornby	32	1752	NRO glebe	Upland, poor soil
Wollaston	25	1788	Hall 1977 206	Nene valley & clay

Preliminary figures suggest that even larger acreages are likely for many upland and woodland parishes, e.g. West Haddon 39 acres.

If this be accepted, then the size of the yardland is a valuable tool for interpreting the early landscape. By using say 18-20 acres as a value typical of an opened-up countryside, and multiplying by the yardland number, then a value is obtained for the area cultivated, and any significant deviation from the total area (less the area of meadow) should represent the area of woodland or waste at the time of fixing the assessment, probably late in the Saxon period.

Great Addington may be taken as an example. The total number of yardlands is 42, and the area of arable in each yardland about 28 acres (N.R.O. 12). Thus the total arable area is 1175 acres, which is consistent with the parish area of 1220 acres, the amount of meadow being very small. If the size of the ancient yardland is taken as 20 acres, then 8x42 acres (28-20 x 42) would be the area not cultivated when the field system was set out, i.e. 336 acres. To the west of the parish lies high clay soil, which is likely to have been the last part to be cultivated. Three of the furlongs here have wold (wood) names; they comprise 103 acres. Next to them is breach furlong which occupies 52 acres, and indicates 'newly broken' ground. The calculations from the yardland statistics, and the furlong microtopography, are in agreement that there was still some waste at the time of the taxation assessment.

It can be seen from the above examples that techniques of reconstructing open field patterns can be used to produce a series of plans for those many parishes that have none. Analysis of furlong names, and of the tenurial arrangements described in fieldbooks, throw much light on the complex way that estates were laid out at an early date. In spite of this, a fairly simple structural organization is evident in parish field systems.

REFERENCES

BAKER G., 1822 The History and Antiquities of the County of Northampton, vol. 1.

BODLEIAN LIBRARY, OXFORD MS. Top. Northants. c19 folio 29.

BRIDGES J., 1791 The History and Antiquities of Northamptonshire

GÖRANSSON S., 1961 'Regular open-field pattern in England, and the Scandinavian Solskifte', Geografiska Annaler, 43, 80-104.

HALL D.N., 1977 Wollaston, portrait of a village

HALL D.N., 1978a 'Elm- a field survey', Proc. Cambridge Antiquarian Soc., 68, 21-46.

HALL D.N., 1978b 'Great Billing Parish survey' Northants Archaeology, 13, 161-167.

HALL D.N., 1978c Medieval Village Research Group, Twenty-sixth Annual Report, 20.

HALL D.N., 1980 'Hardingstone parish survey 1972' Northants Archaeology, 15, 119-132.

HALL D.N., 1981 in Rowley, R.T., (ed) The Origin of Open Field Agriculture, 22-38.

HALL D.N., 1982a Medieval Fields. (Shire Publications, 1982).

HALL D.N., 1982b in Proceedings of the Medieval Archaeology and Medieval Ecology sessions at the Sixteenth International Congress on Medieval Studies, Kalamazoo, Mich. U.S.A., forthcoming.

HARVEY M., 1980 'Regular field and tenurial arrangements in Holderness Yorkshire', J. Historical Geography, 6, 3-16.

HENRY HUNTINGTON LIBRARY, SAN MARINO CALIFORNIA Bridgewater Collection, E 1311/19; microfilm in N.R.O.

KERR W.J.B., 1925 Higham Ferrers and its Ducal and Royal Castle and Park.

LEICESTER MUSEUM ARCHIVES DEPARTMENT ID 50/xii/28 (Register of William Wulston).

MORRIS, J., 1979 <u>Domesday Book</u>, <u>Northamptonshire.</u>

NORTHAMPTONSHIRE RECORD OFFICE (N.R.O.)

1 E(S)155, charter
2 31P/21
3 Great Billing glebe terrier, 1684
4 Map 2115 (1703)
5 E(S)626
6 E(S)84
7 E(S)83
8 X1071
9 E(S)1, & 79
10 E(S) 156 & 164
11 E(S)179 & 185
12 A165

SERJEANTSON R.M., 1909 <u>A History of Delapre Abbey, Northampton.</u>

ACKNOWLEDGEMENTS

I should like to thank P.I.King, Chief Archivist, Northamptonshire Record Office, and his Staff for making available many hundreds of documents, and for permission to publish Figure 2.

CHAPTER 10

RURAL SETTLEMENT IN MEDIEVAL IRELAND: THE EXAMPLE OF THE ROYAL MANORS OF NEWCASTLE LYONS AND ESKER IN SOUTH COUNTY DUBLIN

A Simms
(University College, Dublin)

INTRODUCTION

The Anglo-Norman invasion of Ireland in the late 12th century was followed by the introduction of a formalised feudal system of land-holding which resulted in the creation of manorial estates. There are indications that already under the Gaelic order a type of dependant peasant cultivation had developed, particularly on the big monastic estates. But, the full manorialisation of estate-management came with the Anglo-Normans.

The institution of the Anglo-Norman manor, imported from England, was simultaneously an economic, social and juridical organisation. The land was usually divided into the lord's home farm (demesne land) and the land let out to tenants. The difficult question is what settlement types in Ireland developed within the framework of the manor? Graham (1975) introduced the term 'manorial village', which he defined as a settlement without borough status but always containing a church and generally, but not necessarily, a castle and a mill. It is important to keep in mind that a manor and a village are two completely different concepts, one describing a system of economic, social and juridical organisation and the other a settlement. From a number of detailed case-studies it would appear, that many of the 'manorial villages' were mere manorial centres, from where the manor was administered. The manor and the parish normally constituted one and the same organisational unit in Anglo-Norman Ireland (OTWAY-RUTHVEN, 1968, 119).

Thanks to research by historians (MILLS, 1980; OTWAY-RUTHVEN, 1951, 1965) we have some information on aspects of land-tenure in Anglo-Norman Ireland, but as yet we have no detailed case-studies as to the lay-out of settlements. Glasscock (1971) was the first to draw attention to field evidence of medieval settlement and he also produced the first distribution maps of deserted medieval settlements in Ireland. Most of these sites are undocumented and they can only be identified from the recording of a medieval church and castle site (usually a motte) on the first Ordnance Survey Maps from the 1840s and from the Civil Survey compiled in the 1650s. On this basis and taking into account documentary material as far as available, Graham (1975) published a map showing the distribution and hierarchy of Norman settlement for County Meath on the East coast. He identified 98 'manorial villages' in this county alone. It is likely that there were similar densities for all counties settled by the Normans. A limited number of other regional case-studies have been carried out. Leister (1976) in a book dealt with the problem of the Irish open-field system on the basis of case-studies from Tipperary. Simms (1977) attempted to reconstruct the spatial organisation of an Anglo-Norman settlement by using the example of Duleek in County Meath. Frame (1981) in his recent book on colonial Ireland reconstructed the spatial pattern of the baronial manor of Thurles, and Empey (1982) used the case-study of Knocktopher in County Kilkenny in order to explain the pattern of Anglo-Norman settlement in the context of a baronial manor.

It is the aim of this paper to provide evidence for the form and function of settlements in the framework of two royal manors in South County Dublin. The evidence for Newcastle Lyons is mainly taken from entries in the Calendar of Documents relating to Ireland (Sweetman, ed., 1875), from 18th century estate-maps and relic features on the ground. Unfortunately we do not have a manorial extent for Newcastle Lyons and therefore the manorial extent for nearby Esker, included in the Court Books of Esker and Crumlin, 1592-1600, (Curtis, ed., 1930) will be used in order to illustrate the spatial organisation of a manor in the 16th century.

NEWCASTLE LYONS:

Medieval estate-management and agricultural production

The village of Newcastle Lyons is located on fertile grey-brown podzolic soils approximately 17 km southwest of Dublin at the foothills of the Dublin Mountains. From the medieval documents we know that Newcastle Lyons was a royal manor, that is to say that it became a demesne manor in the hands of the English king. The king also kept six other blocks of land as royal manors in South County Dublin. Those were Esker, Saggart and Crumlin for Anglo-Irish tenants and O Kelly, O Brun, and Othech for Irish tenants (see Fig.10:1). Newcastle Lyons is first mentioned in the Calendar of Documents relating to Ireland in 1215:... taken into the king's hand for the improvement of the manor of Newcastle.(1) The manor came to constitute the Parish of Newcastle. According to the Civil Survey, (2) compiled in the seventeenth century, the parish consisted of seven townlands covering all in all approximately 2000 acres. The Down Survey map, the contemporary cartographical representation of the Civil Survey, shows the location of these townlands belonging to the parish that made up the royal manor (see Fig. 10:2). Townlands are a feature peculiar to Ireland. They are the smallest territorial subdivisions.(3)

In 1228 we find an indirect confirmation of the royal manor at Newcastle, because the king conferred a charter on 'the mother church of his manor Newcastle de Leuan, an

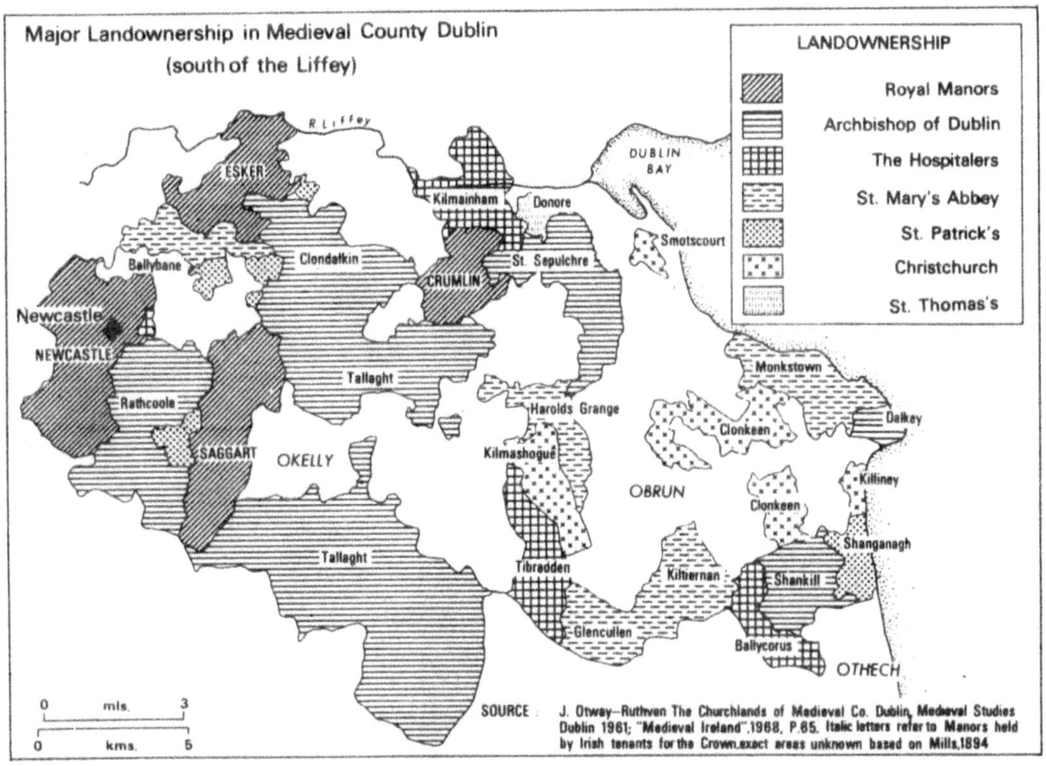

Fig. 10:1 Major landownership in Medieval County Dublin. (source: based on Otway-Ruthven, 1961).

<u>Note.</u> By the time the Civil Survey was compiled (1654-56, vol. VII, ed. R.C. Simington, 1945) the acreage of the royal manors was as follows:

Newcastle Lyons	2000 acres	Esker	1500 acres
Saggart	2000 acres	Crumlin	1200 acres

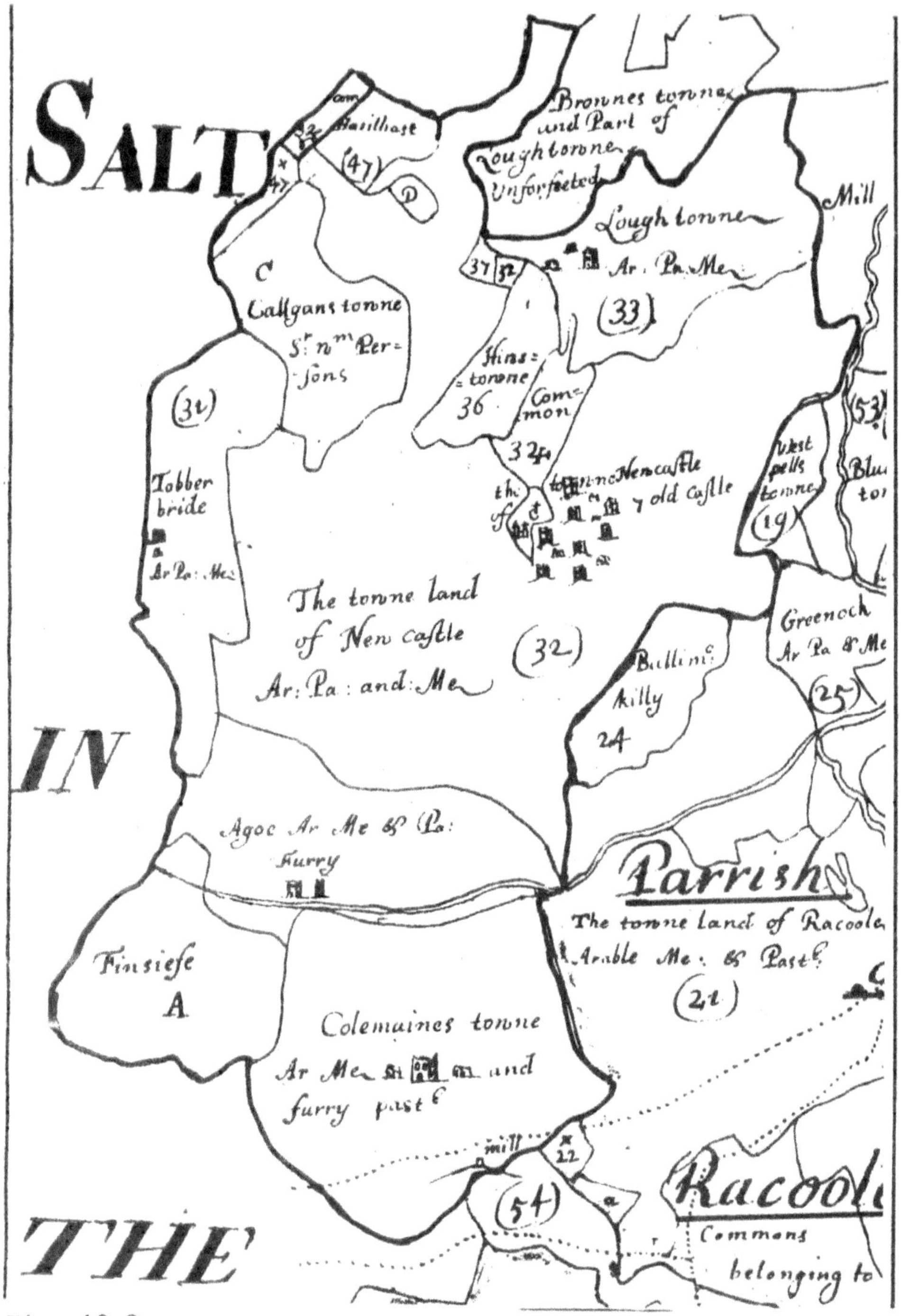

Fig. 10:2

Scale: 160 perches to one inch

The Parish of Newcastle Lyons with constituent townlands in the 1650s. Extract from Down Survey Map, National Library of Ireland, Ms. 714. According to the Civil Survey, 1654-56, (vol. VII, ed. R.C. Simington, 1945) the townlands of the Parish of Newcastle had the following acreages:-

Newcastle	1056-2-10	Loghstowne	219-2-00
(Commons)	38-1-30	Colemanstowne	423-0-00
(Commons)	3-0-00	Agoe	248-0-00
(Commons)	9-2-00	Hinshestowne	53-0-00
(Gleab)	8-2-00	Colganstowne	312-0-00
Tobberbrid	173-0-00	Husilhast	43-0-00

(spellings according to the Civil Survey)

endowment of 55s a year.' (C.D.R.I.,1875,241)

There is documentary evidence that all through the 13th century there was a problem of estate-management in Newcastle Lyons. At first the manor was held under the crown by middlemen. But in 1234/35 the king wrote to his Justiciary in Ireland: '... as more profit would accrue by converting to farm the Irish demesne lands than by retaining them, the King demands the Justiciary to let to farm... at as high a rent as he can, the manor of Newcastle ...' (C.D.R.I. 1875, 335). Twenty-five years later, in 1260, the matter was still not resolved, as the following entry shows:' the tenants of (the Lord) Edward's manor of Newcastle de Leuan having prayed that they should not be oppressed by farmers, Edward notifies to his Justiciary of Ireland that he grants to the tenants that from Michaelmas ensuing they hold this manor for the farm at which it was wont to be held, or for so much as another would give for it'. (C.D.R.I. 1875, 335). These measures resulted in a considerable increase of farm rent by the end of the 13th century (OTWAY-RUTHVEN, 1968, 122).

According to the crown receipts for A.D. 1235 the king received the following income from his Manor of Newcastle Lyons:

£4-00-00 of the mill
£2-64-00 of pleas and perquisites there
£30-00-00 of corn sold there
£14-38-00 of wool, cheese, sheep, skins, oxen, hides
£69-14-00 of rent of the Manor of New Castle of Leuan (MILLS, 1894, 172)

The evidence points towards a mixed arable stock rearing economy on the manor. Obviously, the highest cash income came from grain. Jaeger (1978) has recently suggested in an article on land-use in medieval Ireland that around 1300 A.D. was the time when in Ireland arable land as percentage of the total agricultural land was most important.

From the beginning of the 14th century onwards the productivity of the manor of Newcastle must have been greatly reduced, because of the incursions of the Irish, who came down from the mountains and because of the impact of the Black Death. There is direct evidence for the reduction of the population on the Newcastle Lyons manor from a 14th century report: 'The men of Colmanstown in the royal manor of Newcastle Lyons, Co.Dublin, represented that it had recently been burnt by the Irish, some of the tenants were killed and sixteen were dead on the late pestilence there, so that only three remained' (OTWAY-RUTHVEN, 1968, 324).

From 'rural-borough' to 'parliamentary borough'

The fact that the population in the border area was under continuous threat of attack from Irish raiders might well have been the reason, why Newcastle Lyons, along with a considerable number of other manorial villages in Ireland, was given the legal status of a borough as an incentive for people to settle and stay. Glasscock (1970) introduced the term 'rural borough' for these settlements, which although they were given the legal status of a town, remained villages in functional terms. The first reference to burgesses in Newcastle Lyons appears in an entrance in the Statute Rolls for the late 14th century:' ... that burgesses and commons of the said town have proposed to quit the said town and so it is like to be waste unless it be now remedied. Whereupon the premises considered ... it is enacted and established authority ... that the said town be free from all manner of coign and livery.' (MORRISSEY, 1939, 301). Because of its location near Irish held territory fortifications became important. In 1294 we hear of expenses incurred by protecting the land of Newcastle Lyons against the Irish in the mountains, referred to as 'felons and rebels during this time' (C.D.R.I., 1881, 83). In the late 14th century Newcastle Lyons is granted a special subvention ... 'Provided that they well and securely make their ditches,' (MORRISEY, 1939, 301). The first documentary evidence for a castle in Newcastle Lyons is to be found in a will made in 1475 (Berry's Register of Wills., quoted in BALL, 1905). On the mid-17th century Down Survey Map (Fig.10:2) seven castles are shown and for the same period the Civil Survey (SIMINGTON, ed, 1945) records a population of 205 people. In 1608 Newcastle Lyons had received a licence to hold a weekly market and two fairs each year. In 1613 it received a charter by whch it was erected into a parliamentary borough, which sent two members to the Irish Parliament. The settlement was completely destroyed during the rebellion of 1641. That is why according to the Census of Ireland in c.1659 there were only 115 people in Newcastle Lyons, of whom 94 had Irish names and 21 had English names (PENDER, 1939,p.379).

The village plan: evidence from 18th century estate-maps

The earliest pre-Ordnance Survey map of Newcastle Lyons which John Andrews brought to my attention is the 1722 estate map of George Aylmer (Public Record Office, Dublin, Ms 2989). This map provides essentially the same information as the estate map of Michael Aylmer surveyed by Pat Roe in 1765. (National Library, Dublin, Ms 16 G 5,). This later map has been published in full, but at a greatly reduced scale, by Fred Aalen (1970). A section of this map showing the settlement and some of the fields is reproduced on Fig.10:3. It is of clearer quality than the map dated 1722. The houses are incompletey mapped, because only George Aylmer's property is shown. But it is obvious that the main axis of the settlement is the road. The houses at the head of narrow tofts were aligned in two rows along the road.

The word 'toft' sometimes describes the actual house-site, but it is also used to mean the enclosed land around the

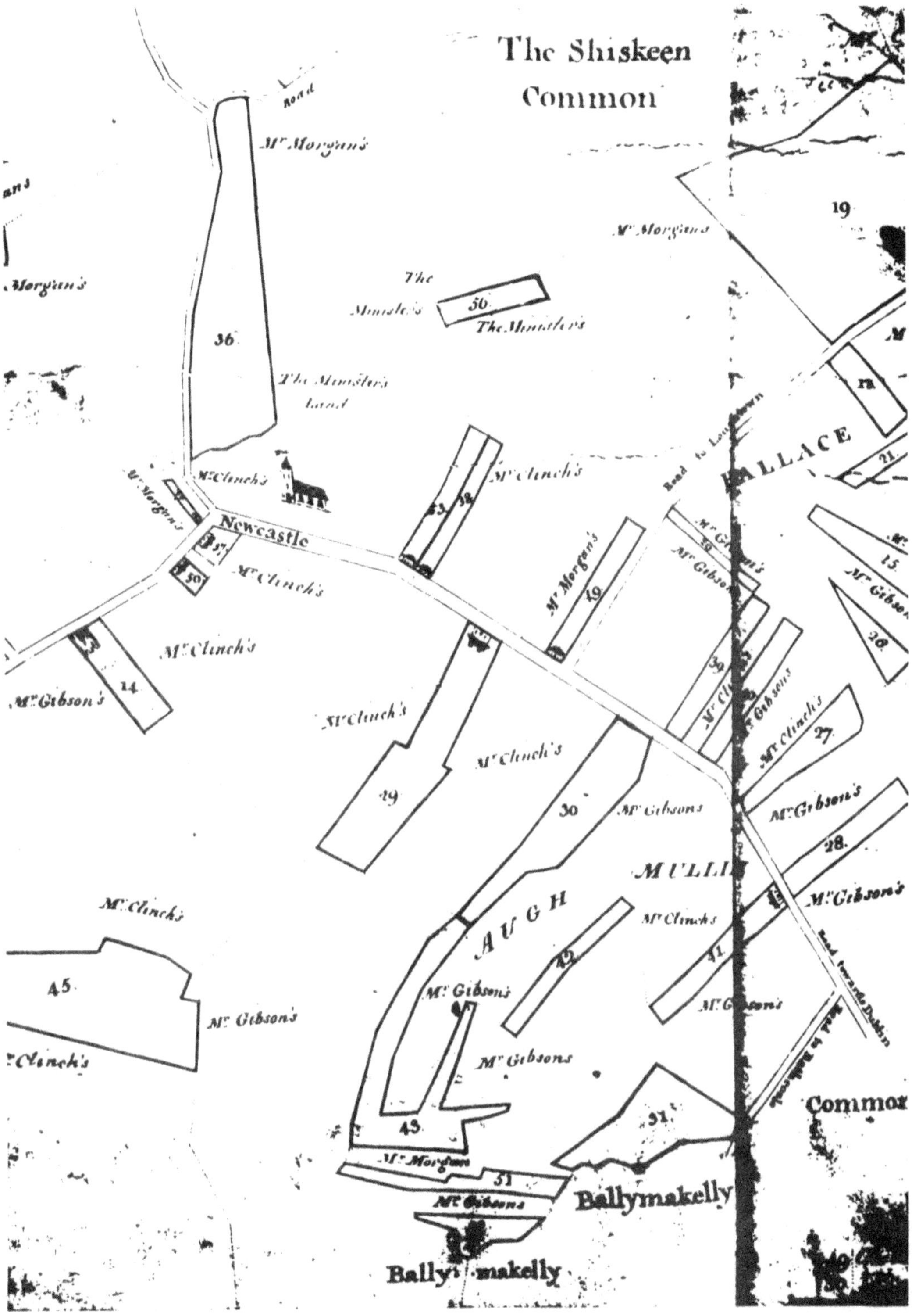

Scale: 20 perches to one inch

Fig. 10:3 Plan of Newcastle Lyons on a section of the estate of Michael Aylmer surveyed by Pat Roe in 1765 (National Library, Dublin, Ms. 16 G 50).

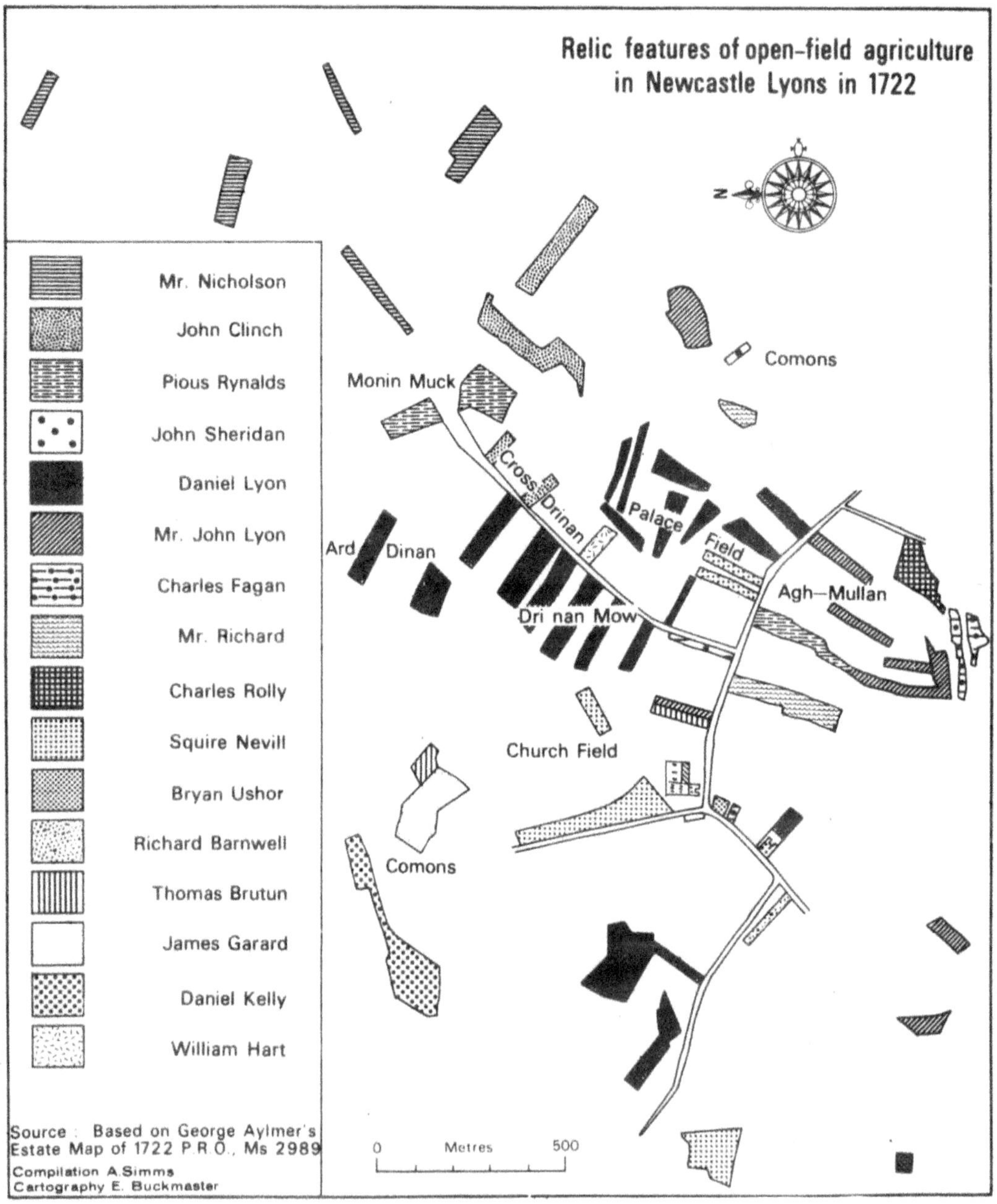

Fig. 10:4 Relic features of open-field agriculture in Newcastle Lyons in 1722

farmhouse, one narrow side towards the common fields and the other towards the road (ROBERTS,1976,296). On Aylmer's map of 1765 there were also tofts adjoining the road without houses. The houses shown on the map are single storey houses with a thatched roof and a central chimney. On the road curving to the southwest a tower-house is clearly marked (plot 14). The drawing of the church does not accurately reflect the battlement-type of church-tower.

On Aylmer's estate map of 1765 there were thirteen tenants listed of whom only five held more than 10 acres and the biggest holding consisted of 32.2.35 acres, distributed in fifteen separate strips. (Fig.10.4). The strips are located in bigger fields which carry names of their own, as for instance Pallace Field and Aughmullin. The ownership of unenclosed strips in a number of large fields appears to be clear evidence for an open-field system, possibly a three-field system. There is evidence from medieval manors on churchland in County Dublin for the division of land into three nearly equal parts, implying the use of the three-field system. For instance in Swords, on the Archbishop's manor, there were 220 acres each in three different fields under wheat, oats and fallow. (Alen's Register, MACNEILL ed., 1950,175). The map also shows common grazing grounds to the east, north and west of the settlement, which are an important part of the open-field agricultural system. (4)

The large number of Irish field names on Aylmer's map as for instance Augh Mullin (Fig.10:3) raise the question whether the names predate the setting-up of the royal manor by Anglo-Norman tenants or whether the influence of the Irish tenants was so strong that they imposed the field names? This raises the difficult question, whether some kind of open-field system in agriculturally favourable areas of Ireland was introduced by the Normans or whether some kind of open-field system was already practised under the Gaelic order before the coming of the Anglo-Normans? We know from the Brehon Laws that at the time of the Norman invasion the Irish population had developed a system of cultivation of fields which was subject to strict rules (CHARLES EDWARDS, 1978).

Relic features on the ground

How does the documentary evidence link with the surviving relic features on the ground? The western end of present-day Newcastle Lyons constitutes a shrunken medieval village consisting of a small flat topped motte, a medieval parish church, two tower-houses, one in the grounds of the glebe-house and the other some 300m SSW along the road, a thatched long-house with old foundation walls, the 18th century glebe-house, a couple of substantial early 19th century farm-houses and a modern bungalow. The eastern end of the village with the 19th century Catholic church, around which had developed a small service centre, has been transformed by the

Plate 10:1 Newcastle Lyons (Fairey Surveys of Ireland, Dublin. 1:10,000: October, 1971).

1 motte, 5m high, 34m diameter (Healy, 1974)
2 15th century church
3, 4 15th century tower-houses
5 granite cross (presumably Early Christian)
6 St. Finian's Well
7 site of rath (ploughed under)

recent addition of a modern housing-estate (see Plate 10.1).

The medieval church (Church of Ireland) with a residential tower attached to it, and the two separate tower-houses date from the 15th century. This church was predated by an earlier medieval church mentioned in 1228 (see above), which in turn must have been preceded by an earlier Christian church, probably on the same site. The fact that the church is dedicated to St Finian, the existence of an ancient granite cross in the church-yard, and the nearby St. Finian's well, are all suggestive of an Early Christian settlement site. The air-photograph also shows relic features of two raths (enclosures built of earthen banks, which served as homesteads from the Iron Age to the medieval period). Along the road frontage in the field opposite the church stone-walls are concealed under the grass indicating house-platforms.

The basic plan of the medieval village must have been determined by the alignment of the motte, the church, the tower-houses and the thatched houses of the peasants along the road forming a simple two-row village. The recently carried out measurement of the concentration of the phosphates on the site produced data of phosphate enrichment which supported the location of known components of the medieval village (5).

A comparison of Aylmer's Estate Map of Newcastle Lyons in 1765 with the air-photo shows that many of the toft boundaries and open-field strips of that date have been preserved in the enclosure boundaries.

The social structure and the spatial lay-out of the royal manor of Esker in 1593 (the documentary evidence).

As there is no manorial extent available for Newcastle Lyons which could give us information on the social structure of the royal manor and its spatial lay-out we are going to look at the manorial extent for the nearby royal manor of Esker (fig.10:1) added at the end of the Court Book Of Esker and Crumlin, 1592-1600 (CURTIS, ed., 1930). At the beginning of the 13th century the royal manor of Esker was farmed by the sheriff and governed by the royal bailiff. In the 13th and 14th century it belonged to a branch of the Bermingham family. By the 16th century the land became divided and rent was paid directly by a number of tenants. At some time in the 14th century the king had started to appoint senechals who would hold courts and collect the rents. This continued right into the 19th century (CURTIS, 1930,137).

We know from 14th century extents referring to the manor of Cloncurry in Kildare (O'LOAN, 1961), Lisronagh in Tipperary (CURTIS,1935-7) and Duleek in County Meath (BROOKS, 1953) that on the manors the different levels in the hierarchy of dependancy were between:(1) the 'freeholders' or free tenants, (2) the 'farmers' or customary tenants, who held

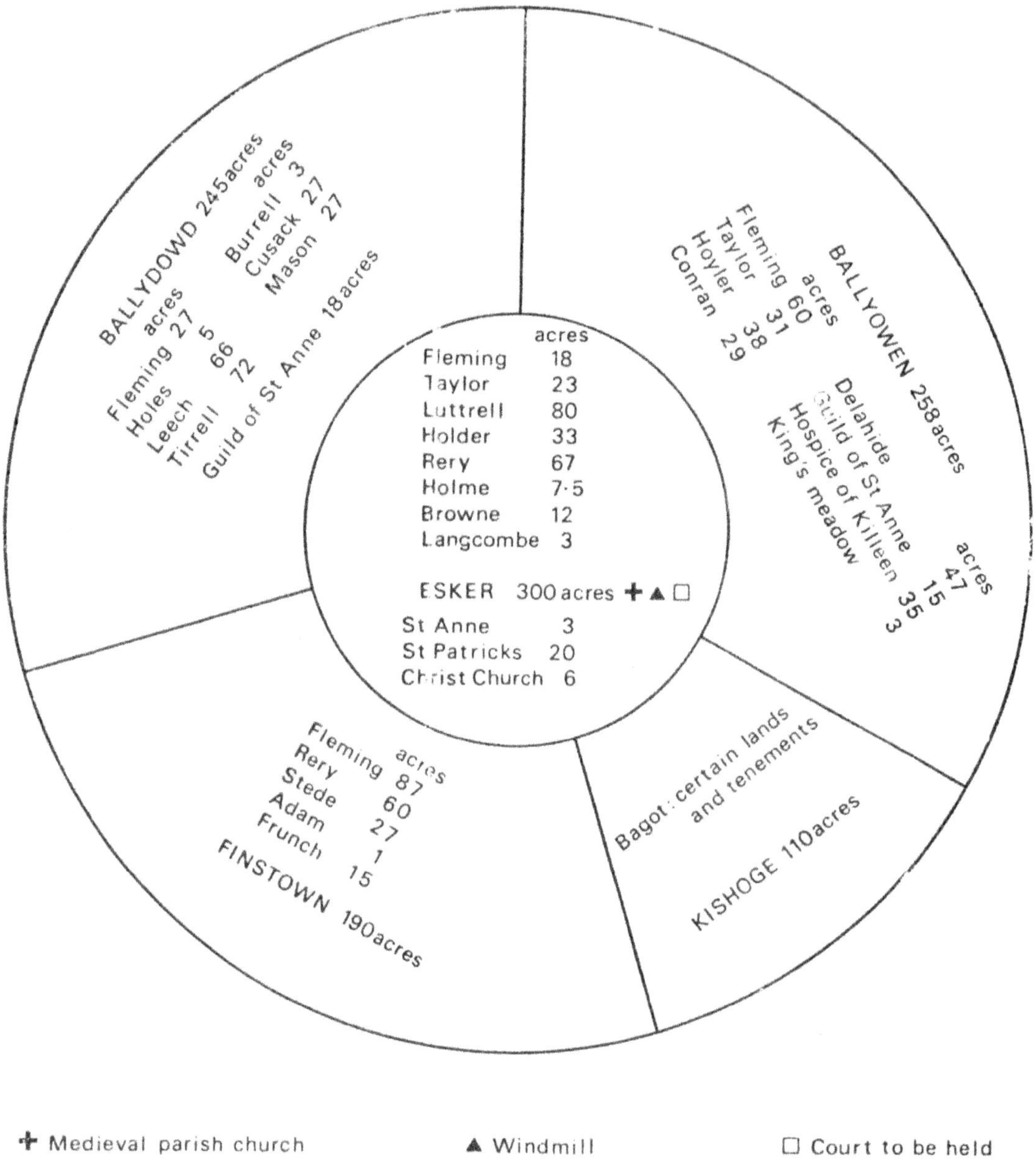

Fig. 10:5 Spatial organisation of landholding on the royal manor of Esker in County Dublin in the 16th century.
(Source: 'Extenta terrarum de Esker', added at the end of the 'Court Book of Esker and Crumlin, 1592 - 1600', ed. E. Curtis, JRSAI, vol. 59, 1929; vol. 60, 1930).

All tenants named are holders in socage, that is to say they were free tenants.

land by custom of the manor and (3) the 'cottiers' or poor tenants, who owned labour services and (4) a special category, the *betaghs* or native tenants, who held land in common and owed labour services.

The categories of tenants mentioned in the Court book as existing at Esker and Crumlin at the end of the 16th century are free tenants or socage holders, a type of free but non military tenure known to English law, *fermors* or farmers, who were the customary tenants, and cottars or poor tenants. All the tenants named in the 'Extent of the Lands and Possessions of Esker' are holders in socage. The total area in the hands of these socage tenants was 800 acres (CURTIS 1950,145) remarks that since the Parish of Esker contains 2500 acres it is possible that the *Extenta* is an imperfect survey limited to the free tenants and omitting the customary tenants and cotters. *Betaghs* are not mentioned at all. In fact, the *Extenta de Esker* ends with the following statement;' And there are no serfs (*nativi*) nor any other profits or commodities pertaining to the said manor'. It is likely that on the royal manors *betaghs* were at an early stage absorbed into the class of customary tenants. By the 16th century servile tenure was rapidly coming to an end and it was formally abolished in 1605.

The *Extenta terrarum de Esker* gives some idea about the spatial organization of the manor as a whole. The attempt has been made to reconstruct it in the form of a diagram (Fig.10:5). There was only one wasted and ruined messuage worth nothing in demesne lands (that is to say in the king's hands) and the total rent paid for Esker including the dependant townlands (*villata*) only amounted to £35 11s 6d, while the perquisites of the Court in ordinary years were worth only £1 6s 8d. The names of the chief tenants are Anglo-Norman or English and there are five church institutions holding land. The organizational centre of the manor was, without any doubt, Esker itself, where the mother church of the manor stood, where the manor court was held and where there was a mill. When the royal manor was first set up there was most likely some kind of fortification, a motte, followed by a castle. It is not obvious from the *Extenta* which tenants lived in Esker itself, but from the dispersed pattern of landownership, showing that some of the major tenants as for instance the Flemings owned bigger chunks of land in the townlands surrounding Esker than in Esker itself, it appears that these tenants might have lived in hamlet-type settlements in the townlands where they owned most land and for which castles are recorded as for instance for Ballyowen and Finstown. In fact, according to Pender (1939,379) the population of Esker was in the mid 17th century 21 people, while Ballydowde had 26, Ballowen 30 and Finstown 18.

It is likely that among those 21 people in Esker were customary tenants. It appears that the settlement pattern of the manor of Esker consisted at the end of the 16th century of an organizational centre and some substantial residences of the stronger tenants and their dependants, who were located in townlands of their own. This model corresponds to that devised for the manor of Colpe in County Meath which was part of the Llanthony lands.(BROOKS, ed.,1953 and SIMMS,1979, 285). The main difference between the two sites is that in County Meath land ownership in the late 14th century was much more consolidated than in South County Dublin so much nearer to the capital. It also appears that land changed hands more frequently in Esker than in Colpe, because none of the land-holders mentioned in the Civil Survey of the 1650s coincides with the tenants listed in the Extenta de Esker from 1593.

CONCLUSION

The manor and the village community have been described as the most powerful institutions of medieval England (POSTAN,1975). No doubt, the Anglo-Normans transferred the manorial system of landholding into their colonial territories in Ireland. But they superimposed it on an Irish base. This Irish base exerts its influence through the continuity of the subdivisions of the land. It is, as Leister (1976) pointed out, most likely that the Anglo-Normans took over bounded property. That is to say, that they continued to use existing land divisions. The subdivisions of land named in the manorial documentation vilae or vills can largely be identified with modern townlands. The origin of townlands can most likely be explained by the subdivision of land for the purpose of farming and grazing in the pre-Norman period. These townlands were mapped by the Ordnance Survey in the 1830s. It is in my opinion the existence of the townland system inherited from the pre-Norman period which militated against the formation of centralised manors in the tradition of the English demesne farming. In Ireland the settlement pattern of the manors is more dispersed. On the basis of his studies in Ulster McNeill (1980, 84) came to the conclusion that 'manors were fiscal centres, places where tenants went to pay rent or deal with minor crimes and administration. They were not villages.' The settlement process must have been as follows: The Anglo-Norman lord received a tract of land as a result of infeudation. He then established a centre for his home-farm and the administration of his estate as a whole. But then, as the example of Esker seems to bear out and as we know from other examples (for instance the De Courcy lands in Ulster), the land was shared out to be held by various tenures in vills (townlands), which were fiscally and administratively dependant on the manor. Perhaps most 'manorial villages' on Graham's (1975) map of County Meath are more manorial centres? There is, of course, no doubt that in Anglo-Norman Ireland there were also nucleated villages with the arable laid out in open-fields and commons as at for instance Newcastle Lyons and other villages in County Dublin, particularly those owned by the Archbishop of Dublin, e.g.Swords, Tallaght and

Rathcoole, which indeed remained villages right into the 20th century. But in all these cases it seems that non-agricultural activities were absorbed into the village and caused growth. Thus the 15th century tower-houses in Newcastle Lyons may have belonged to merchants. Such villages, with non-agricultural activities, were usually granted the title of 'rural borough'. In spite of the nucleated character of these villages, with their arable laid out according to the open-field system, it has been shown that in comparison to the English Midland system the physical extent of the open-fields is smaller. (HOLLWEY, 1980). No doubt, the subdivision of land into townlands each with its own ownership pattern had a limiting influence on the evolution of the system.

The manorial system lasted in Ireland well into the 19th century when the old feudal manor courts were replaced by the authority of the reformed corporations and the new county councils (CURTIS, 1930, 138). The actual changeover from tenantry to owner-occupier must have occurred at the time of the dissolution of churchland, later by the transfer of landownership in the post-Cromwellian period and, last but not least, through the enclosure movement.

Another factor which contributed to the dissolution of the medieval landscape was that as a consequence of the reformation in the 16th century the early ecclesiastical network of central places lost in importance. Jones Hughes (1982, 103) recently pointed out that the planting of the reformed church in these old parish centres appears to have acted more as a repellent than as a magnet to the further development of the site as a central place. He went so far as to say:'the typical parish centre in the tillage lands remained an anxious huddle of mainly colonial elements planted in a hostile or indifferent countryside'. Many of the old parish centres shrunk or were totally deserted. The shrinkage occurred in Newcastle Lyons in the late 17th and 18th century, while from the early 19th century onwards the settlement experienced revival after the building of the Catholic Church. Today it has become a dormitory town of Dublin with 2402 inhabitants at the last census, in 1979. Until recently, when it was taken over by suburban growth, Esker consisted only of a ruined medieval church adjacent to a couple of large farm-houses.

In conclusion, it must be stressed that the pattern which predated the dissolution of the medieval landscape in Ireland was varied, even within limited areas. There were villages in an economic and settlement sense, but there were also manorial centres, which never became fully developed villages. In both instances the pre-Norman settlement pattern had a powerful influence both on the location and the further development of the settlement, because the townland system, which the Anglo-Normans most likely took over from the previous Irish inhabitants, set the framework for all future developments of settlement and field patterns.

REFERENCES

AALEN, F.H.A. 1970 — 'The Origin of Enclosures in Eastern Ireland', Irish Geographical Studies, ed. by N.Stephens and R.E. Glasscock, 209-233, Belfast (1970)

BALL, F.E. 1905 — A History of County Dublin, Part III (1905)

BERESFORD, M.W. and J.G. HURST 1976 — 'Wharram Percy: a Case Study in Microtopography' Medieval Settlement, ed. by P.H.Sawyer (1976)

BREWER J.S. and BULLEN, W. eds.1871 — The Book of Howth: Calendar of the Carew Manuscript (London, 1871)

BROOKS, E.ST.J. ed.1953 — The Irish Cartularies of Llanthony Prima and Secunda, Irish Manuscript Commission (Dublin, 1953).

BUCHANAN, R.H. 1973 — 'Field Systems of Ireland' A.R.H.Baker and R.A.Butlin, (eds) Studies of Field Systems in the British Isles (Cambridge 1973, 580 - 618)

CHARLES-EDWARDS, L.M — Ancient Irish Laws and Early Irish Agriculture, (1978) Lecture to the Dublin Historic Settlement Group 1978 (see also Ph.D Thesis on this topic by the same author).

CURTIS, E. ed.1929 & 30 — 'The Court Books of Crumlin and Esker, 1592 - 1600, in: Jnl.R.Soc. Antiq.Ir., 59(1929), 45-64, and 128-148, and 60(1930), 35-51 and 137-149.

D'ALTON, J. 1838 — The History of County Dublin (1st ed. Cork, 1838, reprint).

DIX, E.R.M.C. 1838 — 'The Lesser Castles of the County Dublin', The Irish Builder, (1898) 72-85

DODGSHON, R.A. 1980 — The Origin of British Field Systems (London, 1980)

EMPEY, C.A. 1982 — 'Medieval Knocktopher:A Study in Manorial Settlement', Old Kilkenny Review, 2,Nr.4 (1982) 329-342

FRAME, R. 1981 — Colonial Ireland, 1169-1369 (Helicon History of Ireland, Dublin, 1981).

GLASSCOCK, R.E. 1970 — 'Moated sites and deserted boroughs and villages: Two neglected aspects of Anglo-Norman settlement in Ireland', Irish Geographical Studies ed. by N.Stephens and R.E.Glasscock, (Belfast, 1970),62-177

1971 — 'The Study of Deserted Medieval Settlements in Ireland', Beresford, M.W. and Hurst J.G. (London 1971) 279 - 291.

GRAHAM, B.J. 1975 — 'Anglo Norman Settlement in County Meath', Proc. Royal Irish Academy 75c (1975), 223-248.

1980 — Medieval Irish Settlement:A Review (Historical Geography Research Series, Nr.3, May 1980)

HARBISON, P. — Guide to the National Monuments of Ireland (new edition, Dublin 1975)

HARVEY, J. 1940 — Dublin, a Study in Environment (1948).

HEALY, P. 1975 — Report on Monuments and Sites of Archaeological Interest in County Dublin. Commissioned by An Foras Forbatha, Conservation and Amenity Advisory Section (1974,1975).

HOLLWEY.M. 1980 — Medieval Field Systems and Rural Settlement in South-West County Dublin. B.A. Thesis, Trinity College, Dublin.

JAEGER, H. 1978 — 'Landund Forstwirtschaftliche Bodennutzung im Mittel-alterlichen Irland', Wirtschaftsgeographische Studien,, Heft 5,3,(Jahrg. 1978)

JOYCE, W. St.J. 1912 — The Neighbourhood of Dublin (Dublin, 1912)

LEISTER, I 1976 — Peasant openfield farming and its territorial organisation in County Tipperary (Marburger Geographische Schriften, 69, 1976)

LEWIS, S. 1970 — A Topographical Dictionary of Ireland. Historical and Statistical Descriptions, II, London (1837) reissued by Kennikat Press, 1970

MACCONVILLE, I. 1980 The Origin and Functioning of the Medieval Royal Manors of South County Dublin, B.A.Thesis, Department of Geography, University College Dublin (1980)

MCNEILL, C. (ed) 1950 Liber Niger Alani 1172 -1534. The Calander of Archbishop Alen's Register, Dublin 1950

MCNEILL, T.E. 1980 Anglo-Norman Ulster: The History and Archaeology of an Irish Barony 1177 - 1400 (Edinburgh, 1980)

MILLS, J. 1890-91 'Tenants and agriculture near Dublin in the Fourteenth Century', Jnl. R. Soc. Antiq. Ir., 21 (1890-91) 54 - 3

MILLS, 1894 'The Norman settlement in Leinster: the cantreds near Dublin', Jnl. R. Soc. Antiq. Ir., 24(1894) 160-175

MILLS, J.(ed) 1905 etc Calendar of Justiciary Rolls Ireland
1295 - 1303 (Dublin, 1905)
1305 - 1307 (Dublin, 1914)
1308 - 1314 (McGriffith ed., 1950)

O'MEARA, REV.C.P. 1903-5 'Notes on Newcastle-Lyons', Kildare Archaeological Journal 4 (1903-1905), 63.

MORRISSEY,J.F. ed,1939 Statute Rolls Ireland Edward IV, Part II (Dublin, 1939)

O'LOAN,J. 1961 'The manor of Cloncurry, Co. Kildare and the feudal system of land-tenure in Ireland' Department of Agriculture Ireland Journal LVII (1961), 14-36

OTWAY-RUTHVEN,J. 1951 'The Organization of Anglo-Irish Agriculture in the Middle Ages', Jnl.R. Soc. Antiq. Ir. 81 (1951) 1-13

1961 'The Churchlands of Medieval County Dublin', in: Medieval Studies, ed. by Watt, Moral and Martin (Dublin 1961)

1965 'The Character of Norman Settlement in Ireland', Irish Historical Studies, 5(1965), 75-84

1968 A History of Medieval Ireland (London, 1968)

PENDER, S. 1939 — A census of Ireland, circa 1659 (Dublin Stationery Office 1939)

POSTAN, M.M. 1975 — The Medieval Economy and Society (Pelican Books, 1975).

REEVES, W. 1862 — 'The Townland Distribution of Ireland', Proceedings of the Royal Irish Academy, 7(1862), 473-490.

ROBERTS, B.K. 1982 — Rural Settlement: An Historical Perspective (Historical Geography Research Series, Nr. 9, April, 1982)

SIMMS, A. 1979 — 'Settlement Patterns and Medieval Colonisation in Ireland, the example of Duleek in Co.Meath' Flatrès, P.(ed) Paysages Rureaux Européens (Rennes, 1979), 159-177.

SIMINGTON, R.C. ed, 1945 — Civil Survey, County Dublin VII, 1654-1656, Irish Manuscript Commisssion (Dublin, 1945).

SWEETMAN, H.S. (ed) — Calendar of Documents Relating to Ireland 1171-1251, (London, 1875)
1252 - 1284 (London 1877)
1285 - 1292 (Dublin 1879)
1293 - 1301 (Dublin 1881)
1302 - 1307 (London 1886)

Pre-Ordnance Survey Maps

A Survey of Parcels of Land being part of the Lands of New Castle in the County of Dublin. Belonging to the Hon.Coll. George Aylmer of Lyons they contain 105.3.05 of Plantation Measure (taken in May 1722 by Henry Roe) Public Record Office, Manuscript 2989.

A map of part of the Lands of New Castle in the County of Dublin. Michael Aylmer Esqu. as surveyed by Pat Roe in 1765, copied by John Roe in 1801
National Library, 16 G 50

Down Survey - Set of Baronial & Parish Maps of County Dublin, National Library, Manuscript 714.

NOTES

1. Calendar of Documents relating to Ireland (abbreviated C.D.R.I.), 1171-1251 1, ed by H.S.Sweetman (London 1875), 88

2. The Civil Survey, 1654-56 VII, ed. R.C. Simington (Dublin,

1945). It is a statistical account on county basis of landownership in Ireland, compiled after the Cromwellian confiscation of Ireland in preparation of the transfer of land from Catholic to Protestant owners.

3. Townlands were first mapped by the Ordnance Survey in the 1830s as the basis for a new land evaluation to a scale of 6 inches: 1 mile. Ireland was subdivided into c. 60 000 townlands. The average size of townlands is 326 acres (REEVES, 1862). The word townland is an English loan word. The Irish equivalent is 'baile'. The origin of townlands is likely to date back into the pre-Norman period, but no definite evidence has been established so far.

4. The problems of open-field agriculture are discussed in the recently published book by T.Rowley, The Origin of Open-Field Agriculture (London, 1981).

5. The potential contribution of soil-phosphate analysis to the study of historic settlements is discussed, by using the example of Newcastle Lyons, in a paper accepted for publication in the Proceedings of the Royal Irish Academy, Dublin, with the title: 'The Medieval Settlement of Newcastle Lyons: the interdisciplinary approach', by K.J.Edwards, F.W.Hamon and A. Simms.

CHAPTER 11

NUCLEAL SETTLEMENT AND ITS TENURIAL RELATIONSHIPS: SOME MORPHOLOGICAL IMPLICATIONS

G R J Jones
(University of Leeds)

Attachment to land is proudly proclaimed in the opening line of the Welsh national anthem, for this can be loosely translated as 'The old land of my fathers is dear to me'. Strictly speaking 'land' in this context should be construed as the territory of the Welsh people and, in any case, the national anthem is of relatively recent origin. Nevertheless these words epitomize my purpose in this paper which is to demonstrate that an old-established attachment to land provides a key to the understanding of early stages in the evolution of settlement in Wales. Since this is an international gathering I wish to suggest that an unravelling of the institutions associated with hereditary land in Wales may also contribute towards an understanding of early stages in the evolution of settlement in the lands of your fathers too; for, as Professor Binchy suggested several decades ago, the narrow circle of kindred comprised by the male descendants of a common great-grandfather, 'originally constituted the primary legal, social and economic unit among all the "Indo-European" peoples' (BINCHY, 1943, 212-27)

There can be little doubt that this group was the narrow circle of the kindred in 'Celtic' society, for some of the functions assigned to it in Irish and in Welsh law are identical. The descendants or a common great-grandfather would have constituted a four-generation inheritance group or agnatic lineage, extending vertically backwards as far as the great-grandfather and horizontally outwards as far as the second cousin (CHARLES-EDWARDS, 1971 304-7; CHARLES-EDWARDS, 1972, 3-33). In Ireland, this was originally the land-holding group for the 'kin land'. Before the seventh century, rights over this hereditary land were shared equally among the adult male members of the four-generation group and, with the passage of each generation, re-allocation occurred. Initially the group was responsible for providing each adult male member, a normal freeman, with a large standard holding; but with the growth of population, the standard holding of some four-generation lineages would have been reduced in size.

The normal tenure taken for granted in the medieval lawbooks of Wales was tir gwelyog (gwely land or hereditary land). The essential feature of gwely land was that the right to it passed to sons in equal shares (JENKINS, 1967, 220-47); but the rules of succession were complicated by the provision for sharing a dead man's holding three times - among brothers, among first cousins and among second cousins, that

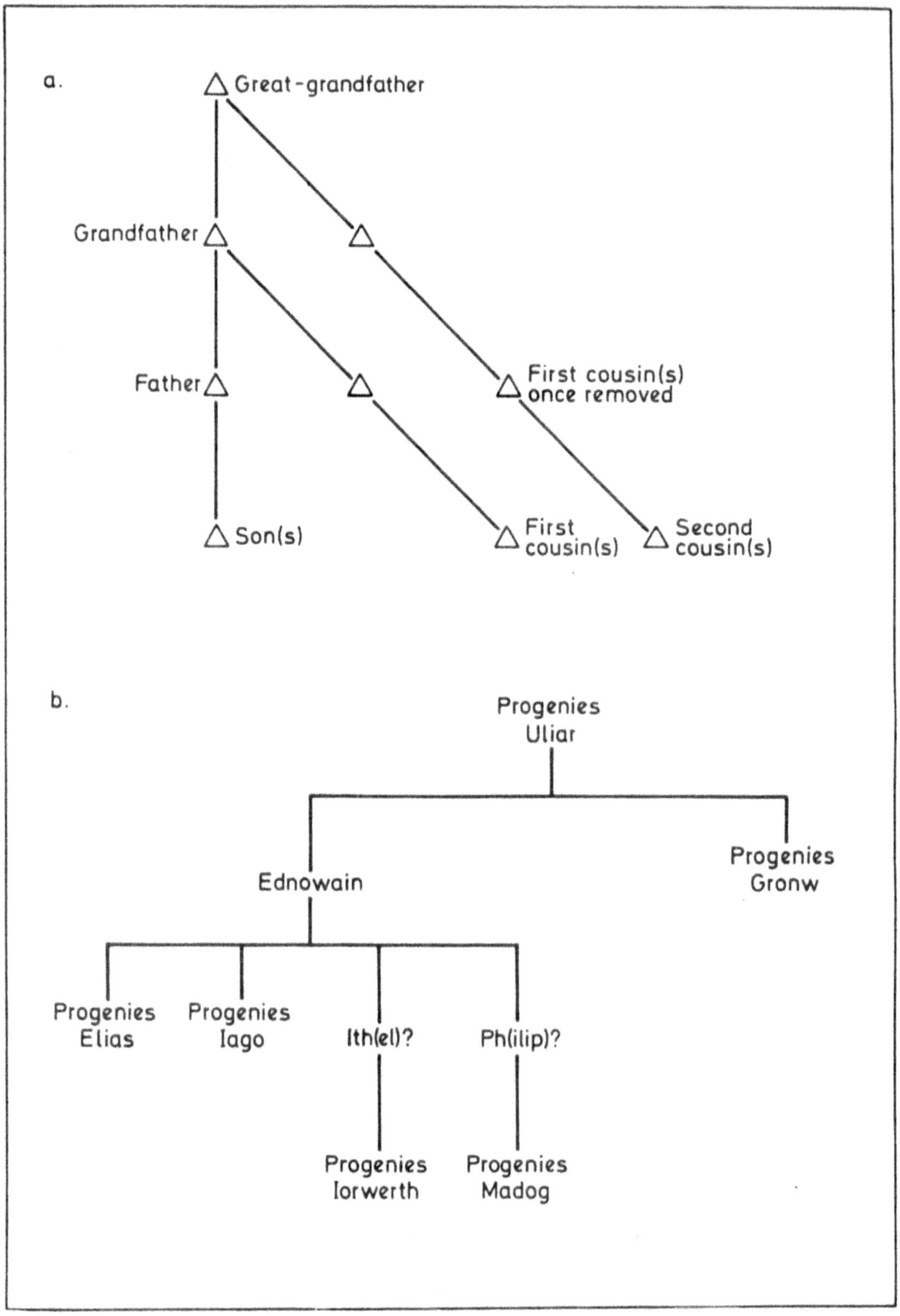

Fig. 11:1 An idealised structure of a four-generation inheritance group and the actual structure of an expanding kindred.

is the sons, the grandsons and the great-grandsons of the dead man (Fig.11:1). If some first cousins or second cousins wanted it they could compel their kinsmen to join in a resharing of their great-grandfather's holding. There was no general obligation, however, to re-open an arrangement for the sharing of land made in the previous generation; and sharing ended with second cousins: 'from then on only common land (cytir) is shared' (JENKINS, 1963, 36, 149-50; WILIAM 1960, 54). Another law-text when dealing with the sharing of patrimony refers to the land of the four-generation group as the joint land (tir cyd) on which 'beyond second cousins no one is to preserve the share (ran) of another' (OWEN, 1841, I, 760-1). This suggests that the four-generation agnatic lineage of Wales, like that of Ireland, was responsible for providing each of its adult male members with a standard holding, at first probably similar to the shareland (rhandir) of 312 'acres', said in some law-texts to contain 12 'acres' for buildings and 300 'acres' for arable, pasture and fuel wood (RICHARDS, 1954,75; EMANUEL, 1967,136,226). These 'acres' were small, so that the 300 'acres' of land would have ranged in statute measure from as little as about 32 acres to 67½ acres.

Possession matured into proprietorship so that men held full prescriptive right to their land if their forebears had held it without interruption for four generations. As each new generation came to inherit they would claim by virtue of descent from a direct male ancestor no more remote than great-grandfather. With each generation therefore the four-generation agnatic lineage was split. After about AD 1100, however, and in some localities possibly long before, sharing by the four-generation kin came to the displaced by equal division among brothers as the normal inheritance custom. Thereafter a kindred would no longer necessarily split with each generation but only as circumstances demanded. Normally the new type of kindred would have expanded with each successive generation into a large group of relatives known in Welsh as a gwely - or in Latin as a lectus - a term originally meaning a 'bed' and serving perhaps as a reminder of the conjugal union from which the enlarged kindred had sprung. But it is also possible that gwely (bed) later came to mean the same as the four-generation inheritance group basic to the older system of land holding (JENKINS AND OWEN, 1980, 199;JONES,1972, 369). After the English conquest of Gwynedd in 1282-3 the term gwely (lectus) was certainly also used of the permanent stake in the soil, later the 'resting-place', of an expanding kindred (progenies). Such was the case at Llanelwy, or St. Asaph, where a rental of c1390 records that the Progenies Uliar occupied the Lectur Uliar which contained a specified area of land (National Library of Wales:SA/MB/22, fols. 6-19; Jones, 1968, II,66-8).

Progenies Uliar was recorded in the same rental as comprising five lesser lineages named after five of Uliar's descendants. These lesser lineages, however, were far from being co-eval. Thus, one, Progenies Gronw was named after a son of Uliar; two, Progenies Elias and Progenies Iago, were named after two grandsons of Uliar; and a further two, Progenies Iorwerth and Progenies Madog were named after two great-grandsons of Uliar. Since there were no lineages further removed in descent from Uliar than the two named after his great-grandsons, it is evident that at Llanelwy the old rule whereby the four-generation kindred was split with every generation had been supplanted (Fig.11:1b).

On the hereditary land of an expanding kindred or lineage each brother on inheriting his share of his patrimony acquired his own house and croft or homestead (tyddyn) as well as parcels or strips of 'scattered land' (tir gwasgar) in one or more open-field arable sharelands and some parcels of meadowland (EMANUEL, 1967, 132-3, 237; OWEN, 1821, II, 688-9; RICHARDS, 1954, 77-9). He also obtained proportional appurtenant rights over common land (cytir), which included wood, pasture, part of which was suitable for temporary cultivation, meadowland and often bogland used as turbary. Different rules, however, applied to one kind of land, tir corddlan, which may be translated as nucleal land (JONES, 1972 340-9). This was not to be shared as homesteads but, rather, as gardens; and if there were buildings thereon they were to be shared as cells or chambers. The gardens of nucleal land appear to have been arranged in a radial fashion around some kind of nucleus; those which radiated from the enclosure which contained a church were long narrow strips said to be 'a legal acre' in length with their ends to the graveyard (JENKINS, 1963, 38, 155-6; OWEN, 1841, I, 140-1, 180-1; WILIAM, 1960, 44,58). Such gardens were regularly manured so that they could be cultivated, year in year out, like an infield; and probably, although separated only by turf balks, they were encompassed by a strong fence so that they could not be entered by beasts.

In the lawbooks which deal with nucleal land, the rules about sharing it are preceded immediately by the statements that 'a mill, a weir and an orchard are called the three ornaments of a kindred (cenedl); and these three are not to be shared nor removed but their produce shared between those who may have a right to it'. The office of 'head of kindred' (pencenedl) was still in being in the thirteenth century and persisted into the early fourteenth century (EDWARDS, 1940, 132; REES, 1975, 84). It is not surprising therefore that a book of case-law of the thirteenth century, probably incorporating contemporary rather than archaic material, should cite these same three ornaments of a kindred and in precisely the same order (JENKINS, 1973,10). Nevertheless, when sharing and re-sharing of ordinary hereditary land among the members of a four-generation inheritance group came to be supplanted under normal circumstances by equal division among brothers per stirpes, as in Progenies Uliar (Fig 11:1b), the same changed rule appears to have been applied to the ornaments of a kindred. Thus, in one post-Conquest law-text these ornaments, listed in the order 'an orchard, a mill and a weir' but called buildings, are said to be 'in common among brothers' (OWEN, 1841, II, 688-9). The same law-text adds that 'all lands are to be shared except these lands: a bog; and an oak wood; and a quarry'. Earlier, in a Latin law-text of the mid-thirteenth century the buildings said to be common among brothers are listed as a church, a fishery and a mill; again a law-text of the later thirteenth century lists these buildings in Latin, and in Welsh, as 'a church, a mill, and a fishery, that is a weir' (EMMANUEL, 1967, 231, 289). A later law-text records that every habitation, presumably every outlying habitation, 'ought to have two footpaths, one to its church, and one to its watering-place' (OWEN, 1841, II, 270-1). It is likely therefore that similar small nucleal settlements developed near wells. Just as the freemen of an expanding kindred could have under-tenants who held small parcels of nucleal land so too the members of a four-generation inheritance group are likely to have had under-tenants. Hence the relatively large area of 12 'acres' devoted to 'buildings' in the shareland (rhandir). In this event the small nucleal settlements containing the ornaments of a four-generation kindred are likely to have been very old-established.

That such features as the ornaments of a kindred, or small parcels of nucleal land in nucleal settlements, were no mere figments of the imagination of medieval lawyers is clearly revealed by fourteenth-century surveys of the estate of the Bishop of St Asaph at Llanelwy. This estate comprised the four medieval townships of Brynpolyn, Gwerneigron, Gwernglefryd and Talar. Besides 210 acres of arable demesne land for the sustenance of the Bishop's palace, the Bishop's garden, the cottage and garden of one brewer, and the holdings of eight bond tenants who performed labour services at Llanelwy, the four townships contained the lands of seven lecti, including those of Lectus Uliar. The rental of c. 1390 is incomplete but, in the sections which survive, five lecti were surveyed in full; and to judge from the area of these five lecti and of the Bishop's demesne lands the two remaining lecti were only small units. For the five surveyed lecti some 770 constituent parcels were recorded. In many cases the locations of these parcels and other features were indicated with a degree of precision in the rental; and very frequently their names have been preserved in later sources, principally maps (Clwyd Record Office, Hawarden: MS L.20/E; QS.DE.14, 18, 65; Public Record Office: IR 29/50/37, SC 6 1143/2; University College of North Wales: Mostyn MS 275, 2708, 2709, 2729, 8554). As a result it has been possible to plot the locations of some 300 features, mainly parcels of land, recorded in the rental (Figs. 11:2,3 and 4).

In the five surveyed lecti some parcels were still held by right of inheritance. Most of the remaining parcels had been

Fig. 11:2 Some significant features named in four town-ships of St. Asaph parish.

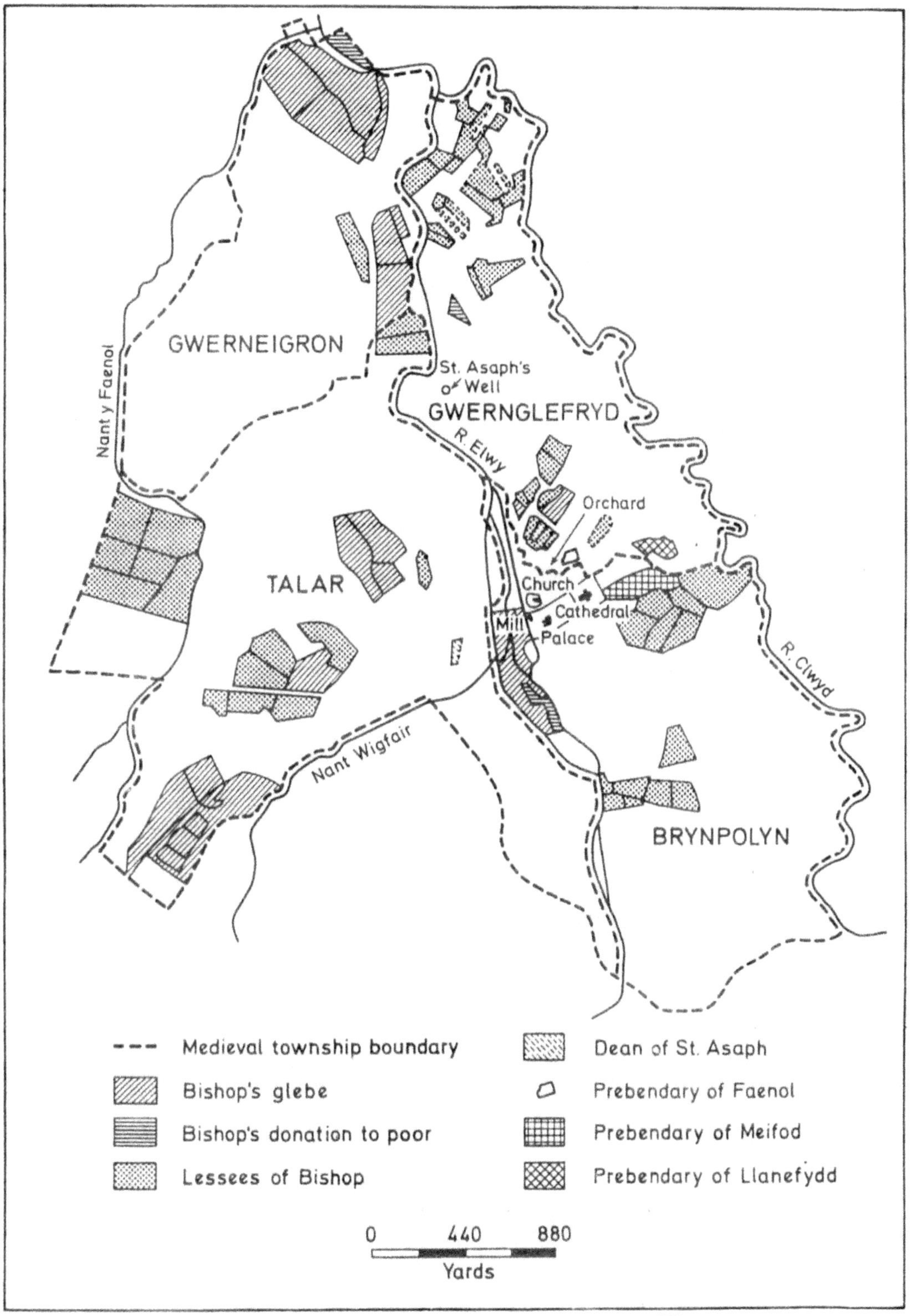

Fig. 11:3 The disposition of Church lands in four townships of St. Asaph parish in 1845.

gaged by means of Welsh vifgage (prid) to would-be purchasers (SMITH, 1976, pp. 537-50); and a small number of parcels were described as acquired lands. On the map showing the sites of located parcels (Fig.11:4) a distinction is drawn between hereditary parcels and the other gaged or acquired parcels. In addition some indication is given of their size. The largest parcels usually consisted of gaged or acquired lands gained by those intent on incorporating unenclosed and intermixed parcels of arable land (tir) or meadow (dol) and common pasture (cytir) or wood (coed) into enclosed fields, thus creating more compact holdings. In the rental, land was measured in customary acres, each apparently containing 10,240 statute square yards (2.12 statute acres). The largest parcel was one of 21 customary acres in Lectus Kedivor, and included part of Y Maes Mawr (The Big Open Field) in Brynpolyn (Fig.11:2). The smallest contained only $1\frac{1}{2}$ customary rods (96 statute square yards) in Lectus Segynnabiaid and was said to be 'within the vill of Llanelwy near the cemetery of St Asaph', in other words near the parish church. Close by was a parcel of 6 customary rods in Lectus Aldrid 'between the manor (ie the palace) of the Lord Bishop and the house of Cynan the Precentor'. Since this house was undoubtedly sited on the small radial parcel which was held by the Prebendary of Faenol (Fig.11:3), this parcel must have been located in Llanelwy, the loosely nucleated settlement later called St. Asaph (Edwards, 1801, I, pp. 188-98). Altogether within this 'vill of Llanelwy' there were on a total area of about 6 customary acres some 38 parcels. These, belonging to all five surveyed lecti, are obviously too numerous to be shown on Figure 4.

Among the many parcels within the vill of Llanelwy was Gardd y Perllan (Garden of the Orchard), part of Lectus Uliar, where Gwenllian and Tangwistl, 'daughters of the Orchard', held 24 customary rods of hereditary land. In addition they held by gage (prid) 54 customary rods of the land of the daughter of the Smith 'within the Garden of the Orchard'. Apparently therefore gardens could contain smaller parcels held by different proprietors, as befits the statements in the law-texts about nucleal land.

Meredudd ap Madog ap Cynwrig held by hereditary right in Lectus Segynnabiaid 18 customary rods 'within the vill of Llanelwy 'with one house', and a second parcel of 32 customary rods between the Bishop's manor and the house of Cynan the Precentor. Meredudd also held by hereditary right larger outlying parcels. Thus in Brynpolyn township in one place with one house and in two parcels annexed to his house' he had $8\frac{7}{8}$ customary acres; near his house on the river bank he had $4\frac{1}{4}$ customary acres and at Dolau Elwy (Elwy Meadows) 2 acres 72 rods; and near the mill in this vicinity he had various parcels containing 4 acres 7 rods. Meredudd's fairly substantial holding of 21 acres 9 rods, by customary measure, therefore comprised tiny parcels at the core of Llanelwy and

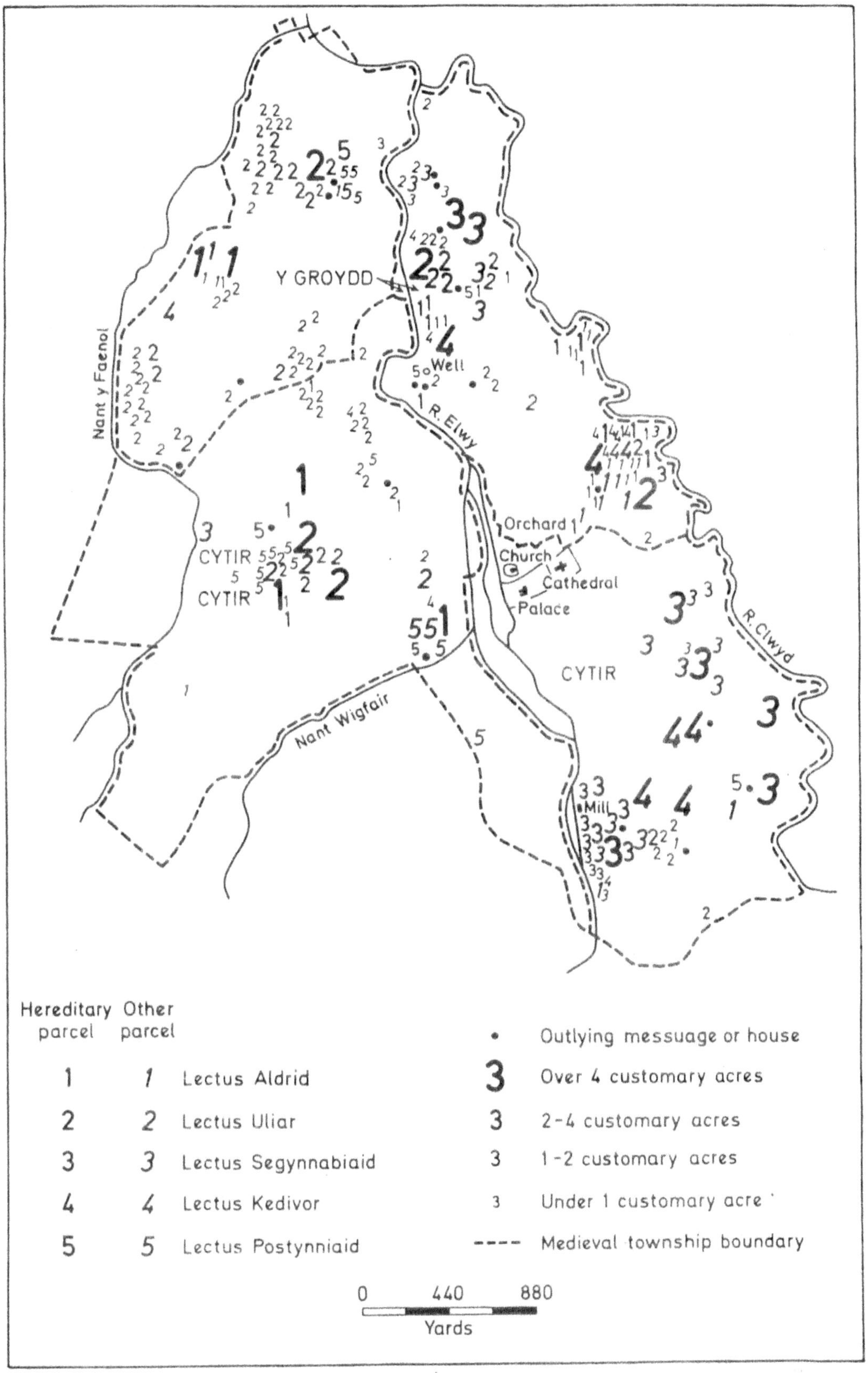

Fig. 11:4 The disposition of parcels of land in five lecti of the Llanelwy estate of the bishopric of St. Asaph, c. 1390.

much larger outlying parcels (Fig.11:4). As the rental indicates, a typical substantial proprietor's homestead or messuage (tyddyn) would be located near the outlying parcels; and the same appears to have been true of Cae'r Tyddyn (Field of the Homestead) in the township of Gwernglefryd and of Hendre (Old Settlement) in Brynpolyn, although neither can be identified in the rental. Thus the messuage of Gruffydd ap Robyn in Lectus Aldrid was in the township of Talar near Cae Coch (Red Field), but in addition he had 48 customary rods 'in his place in Llanelwy with his houses and buildings' Since there were numbers of houses and buildings on some of the core parcels within Llanelwy it is likely that they consisted of nucleal land (tir corddlan) including orchards as well as gardens, and were occupied by under-tenants. Certainly it was recorded in the rental that in five of the seven lecti there were some progenies which should have tenants. Nucleal land was not confined, however, to Llanelwy proper for, about half a mile to the north, there were ancient radial gardens at the Well of St. Asaph; and it was recorded that Progenies Ior-(werth) Velyn (of the Mill?) in Lectus Segynnabiaid ought to have tenants.

By c 1390 most of these nucleal lands or 'ornaments' appear to have been divided between the heirs of the lecti. There was however one exception, an area of 3 acres 50 rods (7 statute acres) on Y Groydd (The Pebbly Strands) said to lie on both sides of the River Elwy in Gwernglefryd (JONES, 1981, 218). These pebbly strands and the adjoining banks were probably used as the site of a weir. A part of Lectus Uliar, this portion of Y Groydd was probably a late surviving ornament of a kindred. It was in the hereditary tenure of a group of six heirs, namely two cousins, an uncle and three others who, at best, were second cousins. These could perhaps have been the members of a four-generation inheritance group, for such a group still served functions within the wider kindred during the later Middle Ages (DAVIES, 1978, 558-62); and a clear pointer to its former role in succession to land at Llanelwy is given by the reference to parts of both Lectur Uliar and Lectus Aldrid as Tir y Cefnder (Land of the Cousin), and to part of Lectus Aldrid as Tir Mab y Cefnder (Land of the Son of the Cousin), in other words, land of the second cousin. Nevertheless this group of heirs holding part of Y Groydd was probably made up of more distantly related heirs of the expanding kindred, the Progenies Uliar, which occupied Lectus Uliar.

There were lands belonging to all the five surveyed lecti 'within the vill of Llanelwy', but the same was not true of lands adjoining the outlying homesteads in the four townships. Usually the lands near any outlying homestead belonged to fewer of the lecti; but even so, outside the typical home croft, the hereditary arable land appears to have consisted originally of intermixed parcels of open field held by different heirs.

The pattern which had emerged on the Llanelwy estate by 1390 was clearly old-established. The cathedral was a late intrusion, for the plan to establish an episcopal seat was evidently still in an embryo state in 1125 and the new see was only an established fact by 1143 (HARRIS, 1956, 5-15). Earlier, however, Lanuuile (ie. Llanelwy) was named in Domesday Book and, since the new see of St. Asaph was Norman inspired, it is likely that the small area of land recorded at Llanelwy in 1086 became part of the Bishop's demesne (FARLEY, 1783, I, fol. 269). Although the church indicated by the prefix llan in LLanelwy was not recorded in Domesday Book it was already in being by the ninth century and probably long before (MILLER, 1977, 45-6; MILLER, 1979, 84). The name of Lectus Segynnabiaid itself appears to be derived from that of the segynnab (secundus abbas) or prior of a clas (CHARLES-EDWARDS, 1971a, 180-90), a clerical community of canons under an abbot or bishop-abbot. Further cnfirmation is provided by the statement in the rental of c. 1390 that some members of Progenies Gronw ap Uliar, a lineage which ought to have tenants, were canons who did not owe the same tenurial obligations as their kinsmen . It is probable that the parish church of Llanelwy was at first a monastic church whose territorial endowments were later transferred to the bishopric of St. Asaph. Before this took place, however, members of the monastic church community had evidently established lecti on lands formerly occupied by four-generation groups whose members lived in outlying homesteads but also possessed 'ornaments' in the Elwy valley. Among them no doubt were nucleal lands around the ancient church, near to, but below, the more elevated site probably occupied by the Roman fort of Varis to the east of that used over a millennium later for the cathedral (GUILBERT, 1980).

The kind of pattern portrayed for the medieval estate of Llanelwy, and the necessarily tentative account of its antecedents, may well provide a key to the further elucidation of recent findings on Saxon settlement patterns as revealed by systematic field-walking over parish areas. As a result it has been clearly demonstrated, for some districts at least, that early and middle Saxon occupation was of a dispersed nature 'more like that of the Iron Age or Roman period than the later medieval' and accordingly points to 'an initial continuity of settlement patterns ... from the Roman period' (HALL AND MARTIN, 1978, 6). Thus, within the Northamptonshire parish of Brixworth no less than nineteen dispersed Saxon domestic settlement sites have been identified in an area containing three located pagan Saxon cemeteries. Yet, to judge from the absence of wheel-made Saxo-Norman pottery, all the dispersed sites were deserted before the middle of the ninth century and presumably there was some concerted action to establish the medieval village of Brixworth. The core of this village adjoined the remarkably early Saxon church, on or near the site of a major Roman structure, a site later to become that of a presumed pre-Danish monastery (EVERSON, 1977, 55-9; HALL, 1977, 129-30). Given these dispositions, some important questions arise even in this well-investigated parish. Why was one particular site selected for the nucleated settlement which apparently supplanted the dispersed early and middle Saxon sites? Was it because the church site was already the ornament of a kindred? Nor was Brixworth the only Northamptonshire example for in Great Doddington parish, to take one of many, a dispersed settlement pattern persisted into the middle Saxon period, though here the process of primary nucleation in the village was seemingly complete by the ninth century (FOARD, 1978, 370)

The questions which can be posed for Northamptonshire examples can be as appropriately raised for many other parts of England, but nowhere perhaps with more justification than the parish of Goodmanham in East Yorkshire. Goodmanham, according to Bede writing c. 730, was the site of a pagan temple containing idols (COLGRAVE and MYNORS, 1969, 178-87). Here the high-priest Coifi, after he had abjured heathenism c. 626, committed an act of desecration by casting his spear into the temple, before it and its enclosures were set on fire. By 1086, according to Domesday Book, Goodmanham was in part an outlying dependency of a royal multiple estate with its caput at Weighton some two miles to the south-west (FARLEY, 1783, I,fols. 248, 302, 306, 322); but among the other outlying dependencies was Sancton, earlier the site of a major pagan cremation cemetery nearly four miles south-south-east of Goodmanham. Yet Goodmanham, together with a number of other satellites, including Londesborough, earlier the site of a major pagan inhumation cemetery nearly four miles to the north-north-west, was also in 1086 a dependency of a church estate. The caput of this estate, whose possession by the Archbishop of York clearly stemmed from a royal grant, was at Everingham, nearly six miles to the west of Goodmanham. The former site of a hall at Goodmanham (Fig.11:5) has been equated with that of the royal court where, as Bede reports, King Edwin of Northumbria, after his conversion to Christianity, discussed with his wise men the relative merits of this new faith and heathenism (FAULL, 1974, 8-12). Thus Coifi's use of the King's stallion to ride to the Goodmanham temple has been viewed solely as a symbolic gesture to break the taboo which forbade the high-priest from riding anything but a mare. Yet, as Dr David Wilson of the British Museum has suggested to me, it is equally reasonable to regard Coifi's equestrian act as serving in part the practical purpose of rapidly completing a longer journey to the pagan temple from a more distant royal court; and this could have been from Weighton, Everingham or even York, the royal capital where Edwin was soon afterwards baptised. In other words it is feasible to envisage Goodmanham as a large-scale variant of that kind of settlement pattern represented on a smaller scale at Llanelwy by the relationships between outlying homesteads and the ornaments of kindreds near the church. In this event the pagan sanctuary

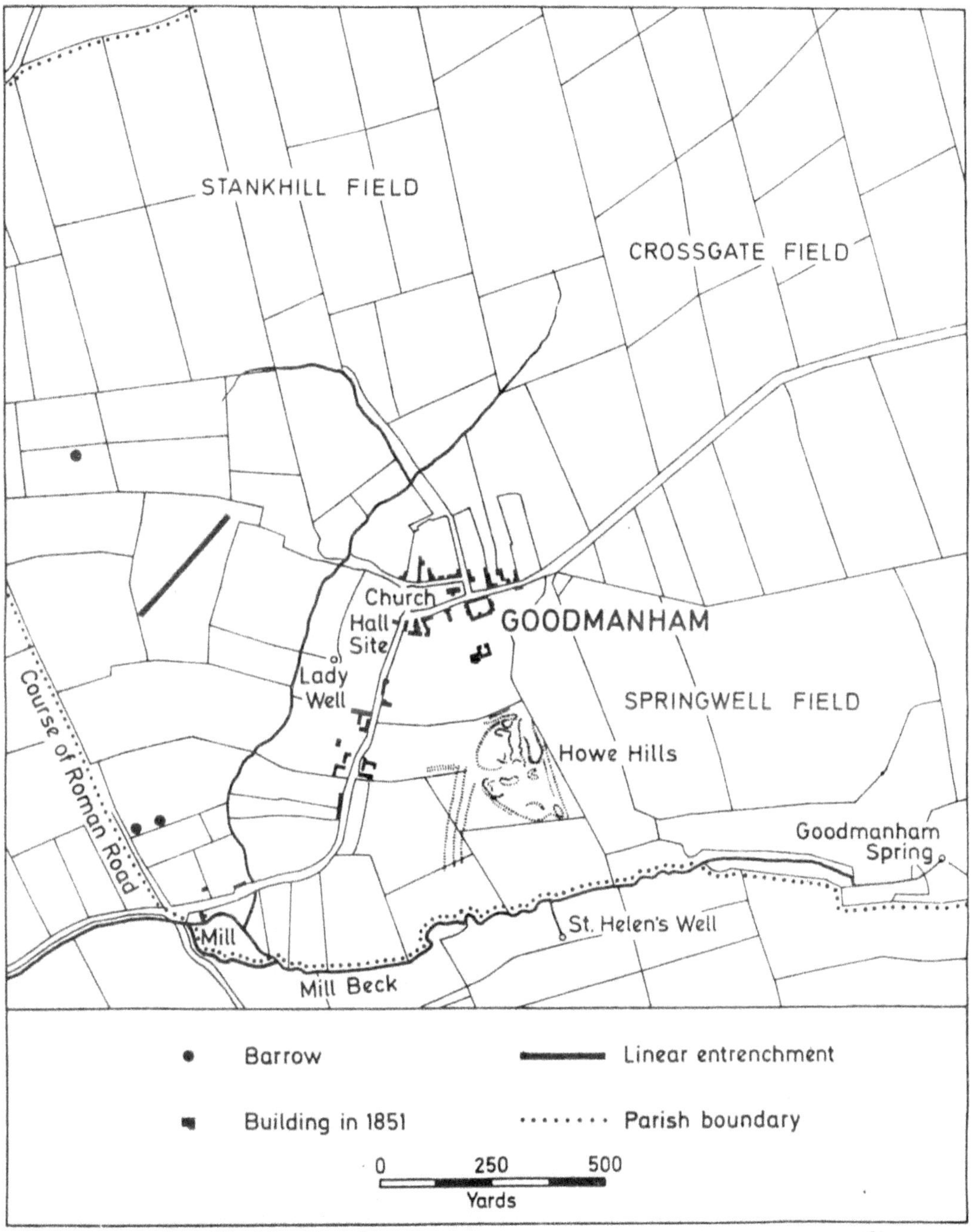

Fig. 11:5 Some features of settlement in the vicinity of Goodmanham.

at Goodmanham could have been the ornament of a kindred, possibly a royal kindred. Certainly by 1086 the 16 carucates, or ploughland-based units of assessment to geld, at Goodmanham formed part of a number of important estates including not only that of the King and the Archbishop but also those of three important Norman tenants-in-chief (FARLEY, 1783, I, fols. 298, 302, 306, 322). Nevertheless, a precursor of Domesday Book, a geld roll produced before 1086 (HARVEY, 1973, 753-73), records that there was a further half-carucate here and, significantly, that 'no one has it' (FARLEY, 1973, I, fol. 381b). Could this have been a late reference to the ornament of a kindred on the site of the pagan temple?

The current interpretation that the site of the destroyed pagan temple was used for Goodmanham church is buttressed by Pope Gregory's general advice, cited by Bede, that missionaries to Britain should use the sites and buildings of pagan temples for Christian purposes (FAULL, 1974, 10; RAMM, 1979). If the half-carucate does refer to the area of the temple, this, however, could not have been accommodated on the site of the church and the adjoining sub-rectangular green (Fig. 11:5). On the other hand the half-carucate could have been accommodated on Howe Hills, a site interpreted by early antiquaries as that of the pagan temple; and if the lynchets on the western side of Howe Hills are included the earthworks cover an area approaching 30 acres, that is half a 60-acre carucate. Early investigators found nothing here save chalk pits which were already old in the eighteenth century. Yet, curiously enough, Howe Hills stood within, though near, the western limit of Springwell Field, one of the three large common-fields of Goodmanham which survived until 1777 (University of York, Borthwick Institute: PR Good. 18 Dep.); and it will be recalled that according to a late Welsh law-text a quarry was specifically excluded from the lands to be shared.

The pre-Domesday geld roll refers to the Goodmanham land of one Norman tenant-in-chief in two places, once in association with the King's dependency and one in association with the Archbishop's dependency and the half-carucate. Possibly therefore there were, as Mr Herman Ramm has suggested, two medieval settlement nuclei at Goodmanham. If the King's dependency was near the hall and the Archbishop's at the church, then Howe Hills were in the angle between them (Fig.11:5). Moreover, Howe Hills were midway between Goodmanham church and St. Helen's well in the neighbouring parish of Weighton, a well linked with Weighton Hall (University of York, Borthwick Institute, TA 378/5). Dedications to St. Helen appear to commemorate the mother of Constantine, the Roman Emperor who at York in the early fourth century adopted Christianity for the Roman Empire; and, in Yorkshire, Helen dedications occur frequently near Roman or sub-Roman sites.

Near the Roman road, in the area between the two Iron Age barrows and the mill south-west of Goodmanham village, evidence was found in the nineteenth century of a possible small Saxon cemetery together with Roman burials, and recently Roman pottery has also been discovered in one of the barrows. North-west of the village, crop marks on aerial photographs provide evidence of another barrow and a linear entrenchment probably of the Iron Age, while in Crossgate Field and Springwell Field they indicate the presence of Romano-British field systems. There is therefore considerable testimony for the presence of widely scattered pre-Roman and Roman settlement in this district, part of Deira, a kingdom incorporated into Northumbria, but whose name was of British origin. It is not improbable therefore that the pagan temple at Goodmanham, whatever its precise site, was established in relationship to the scattered and nucleal settlements of a pre-existing Romano-British Christian community, only to be supplanted in turn by a Christian church.

Alike in western and in eastern Britain, sacred sites, whether pagan or Christian, appear to be associated with the ornaments of a kindred and thus with nucleal lands. Shrines associated with similar tenurial institutions occur even in the easternmost territories occupied by the 'Indo-European' peoples (WEST AND BUHLER, 1884, I, 185, 196-7; II, 730-45, 784-6, 830-3, 848-9). Under these circumstances the kind of ancient attachment to nucleal land in nucleal settlements considered in this paper should have far-reaching implications for the understanding of early stages in the evolution of settlement over a large part of the Old World.

BIBLIOGRAPHY & SOURCES

(a) Published:

BINCHY, D.A. (1943) 'The Linguistic and Historical Value of the Irish Law Tracts', Proceedings of the British Academy, XXIX, pp. 195-227.

CHARLES-EDWARDS, T.M. (1971) A comparison of Old Irish with Medieval Welsh Land Law, Unpublished D.Phil thesis, University of Oxford.

CHARLES-EDWARDS, T.M. (1971a) 'The Heir-Apparent in Irish and Welsh Law', Celtica, IX, pp. 180-90.

CHARLES-EDWARDS, T.M. (1972) 'Kinship, status and the origins of the hide', Past and Present, LVI, pp. 3-33.

COLGRAVE, B and MYNORS, R.A.B. (1969) (ed). Bede's Ecclesiastical History of the English People, Oxford

DAVIES, R.R. (1978) Lordship and Society in the March of Wales, 1282-1400, Oxford

EDWARDS, E. (1801) Browne Willis' Survey of St. Asaph, London

EDWARDS, J.G. (1940) (Ed) Littere Wallie, Cardiff

EMANUEL, H.D. (1967) The Latin Texts of the Welsh Laws, Cardiff

EVERSON, P. (1977) 'Excavations in the Vicarage Garden at Brixworth, Journal of the British Archaeological Association, CXXX, pp. 55-122.

FARLEY, A. (1973) (Ed) Domesday Book, London, I-II.

FAULL, M.L. (1974) 'Roman and Anglian Settlement Patterns in Yorkshire', Northern History, IX, pp. 1-25

FOARD, G. (1978) 'Systematic fieldwalking and the investigation of Saxon settlement in Northamptonshire', World Archaeology, IX, pp. 357-74

GUILBERT, G. (1980) Unpublished papers, Clwyd-Powys Archaeological Trust, Welshpool.

HALL, D.N. (1977) 'Excavations in Brixworth Churchyard, 1971', Journal of the British Archaeological Association CXXX pp. 123-32.

HALL, D.N. and MARTIN, P.W. (1979) 'Brixworth, Northamptonshire - An Intensive (1979) Field Survey', Journal of the British Archaeological Association, CXXXII, pp. 1-6.

HARRIS, S.M. (1956) 'Liturgical Commemorations of Welsh Saints: II, St. Asaf', Journal of the Historical Society of the Church of Wales VI, 5-24.

HARVEY, S.P.J. (1971) 'Domesday Book and its predecessors', English History Review, LXXXVI, 753-73.

JENKINS, D. (1963) (Ed) Llyfr Colan, Cardiff.

JENKINS, D. (1967) 'A Lawyer looks at Welsh Land Law', Transactions of the Honourable Society of Cymmrodorion, pp. 220-48.

JENKINS D. (1973) (Ed). Damweiniau Colan, Aberystwyth.

JENKINS D. and OWEN, M.E. (1980) (eds.) The Welsh Law of Women, Cardiff.

JONES, G.R.J. (1972) 'Post-Roman Wales' in Finberg, H.P.R. (Ed) The Agrarian History of England and Wales, I(i), pp. 281-382.

JONES, O.E. (1968) Llyfr Coch Asaph, Unpublished M.A. Thesis, University of Wales, I-II.

MILLER, M. (1977) 'Date-guessing and Dyfed', Studia Celtica, XII, pp. 33-61.

MILLER, M. (1979) The Saints of Gwynedd, London.

OWEN, A. (1841) (Ed) Ancient Laws and Institutes of Wales, London, I-II.

RAMM, H.G. (1979) Unpublished papers, Royal Commission on Historical Monuments, York.

REES, W. (1975) (Ed) Calendar of Ancient Petitions Relating to Wales, Cardiff.

RICHARDS, M. (1954) (Ed) The Laws of Hywel Dda (The Book of Blegywryd), Liverpool.

SMITH, L.B. (1976) 'The Gage and the Land Market in Late Medieval Wales', The Economic History Review, XXIX, pp. 537-50

WEST, R. and BUHLER, J.G. (1884) A Digest of Hindu Law of Inheritance, Partition and Adoption, Bombay, I-II.

WILLIAM, A.R. (1960) (Ed) Llyfr Iorwerth, Cardiff.

(b) Unpublished:

Clwyd Record Office, Hawarden: MS L.20/E, QS.DE.14, 18, 65.
National Library of Wales: SA/MB/22.
Public Record Office: IR 29/50/37, SC 6 1143/2.
University College of North Wales: Mostyn MS, 275, 2708, 2709, 2729, 8554.
University of York, Borthwick Institute: PR Good. 18 Dep., TA 378/5.

CHAPTER 12

FEUDAL WOODLAND COLONIZATION AS A STRATEGY OF THE CAROLINGIAN EMPIRE IN THE CONQUEST OF SAXONY; RECONSTRUCTION OF THE SPATIAL PATTERNS OF EXPANSION AND COLONIST SETTLEMENT MORPHOLOGY IN THE LEINE-WESER-REGION.

H-J. Nitz
(University of Göttingen)

Spatial expansion at the cost of weaker neighbours is one of the early attibutes of imperial states. The Frankonian Empire of the early Middle Ages behaved just the same as its predecessor on the continent, the Roman Empire or its contemporaries, and so did the other early Germanic states, provided they did not themselves fall victim to the expansion of stronger neighbours.

THE GEOPOLITICAL SITUATION OF CONQUEST

The Frankonian Empire was the most successful medieval state on the continent. An early phase of expansion occurred under the Merovingian kings of the 5th and 6th centuries, extending from the core region along the Rhine to the south and east, as far as the Saale river in central Germany and beyond the Danube in the south. A second phase of expansion to the south-east and south-west was initiated by the need of defence, against the Awars (Awaren) in the region which later on became Hungary and the Moors expanding from Spain. The most successful expansion was carried out to the north-east by Charlemagne during the last three decades of the 8th century. The strategic reason for this expansion seems obvious if one studies the map of the political situation by about the middle of the 8th century (Fig 12:1). The boundary of the Frankonian Empire to the north-east had the shape of a wide open angle and here the territory of the Saxon tribal confederation of the Westfalians, the Engern, and the Ostfalians pushed like a wedge into the core of the Carolingian territory. Aachen, the main residence of Charlemage, was located only 100 km from the border.

Under the Merovingian kings there seems to have existed some kind of agreement with the Saxons, indeed even cooperation, as when the Thuringian kingdom was conquered in the year 531. The trend of further expansion of the Saxons to the west with attacks and drives close to and even across the Rhine seems to have been the main motivation for Charlemagne to decide to conquer the Saxon territory and, thereby, remove the danger to his state. Another impetus for the king was propagation of Christianity. Charles and the church worked in close cooperation. Attempts by missionaries in Saxony to that date had not been very successful, but the political conquest of Saxony actually resulted in her complete conversion to

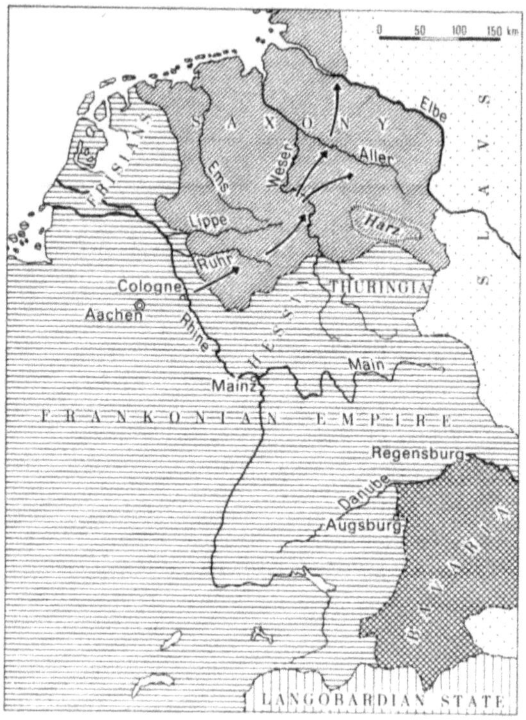

Fig. 12:1 The geopolitical situation of the Frankonian conquest of Saxony. Arrows indicate the main movements of Frankonian campaigns. (Map based on 'Völker und Kulturen. Ein Kartenwerk zur Geschichte'. G. Westermann Verlag, 1971, p. 28, I).

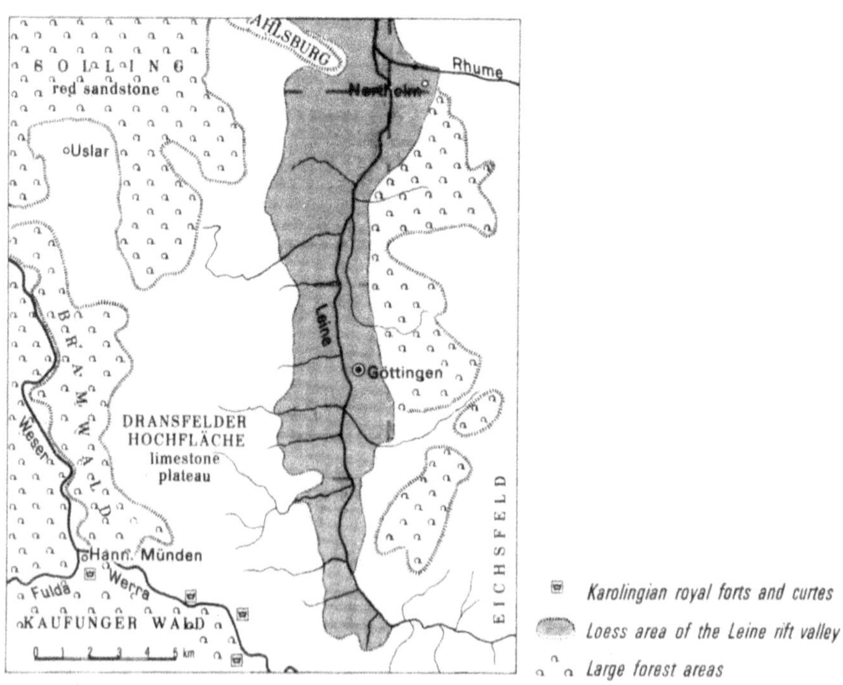

Fig. 12:2 The Leine-Weser-region.

Christianity.

I do not intend to discuss the progress of the conquest in detail but to concentrate on a section of the southernmost part of Saxony between the Harz mountains and the Weser River (Fig. 12:1). In terms of warfare nothing dramatic happened in this region. The military expeditions by the Franks and the main battles were executed in the central regions of Saxony. So, the south was actually cut off from the rest of Saxony. The fact that nothing was reported about any military activity in this region suggests that it had been conquered rather quickly and without any difficulties.

FRANKONIAN COLONIZATION STRATEGIES FOR THE INTEGRATION OF CONQUERED TERRITORIES.

In general, three strategies for the integration of conquered territories seem to have been applied by the Frankonian state authorities:
First, the royal authorities confiscated parts of the occupied territory, especially strategic regions, in order to control the conquered country. Large numbers of Frankish farmer-soldiers were settled on this confiscated land, from which the earlier inhabitants had been expelled. These new settlements were laid out in a planned pattern. The author has published several studies on this so-called Frankonian State colonization (NITZ 1967, 1963, 1971) But, in contrast to the centrally-organized Roman Empire, the Frankonian state was never able to establish an all-embracing administration and military system. Therefore, it had to develop the feudal system in which the king depended on the assistance of a hierarchy of landlords - noblemen and ecclesiastical institutions - who in return were granted fiefs. This was the framework for the second, the king handed over confiscated areas to high-ranking nobles who had rendered their feudal services to the empire as officials, as counts for instance. This impirial nobility - Reichsadel in German - with close commitment to the king and state, was provided with the economic means as well as strong political motivation to integrate the conquered region into the empire and to 'Frankonisize' them. Not only were they given settled areas but even more frequently large tracts of woodland, which, by ways of colonization, could be transformed into new cores of purely Frankish settlements from which the Frankonization of the remaining region could be carried out more easily.

Thirdly, in order to promote Christianity, bishops and monasteries of the Frankish empire were entrusted with certain districts to build up mission bishoprics and a system of mission churches. The Church was endowed with confiscated settlements as well as with forest areas for colonization to give them an econimic basis for their religious activities. Beyond this purpose these institutions of the State Church (Reichskirche) had also to contribute annual dues to the king and render military assistance by their vassals.

In the southern part of Saxony bordered by the upper Weser river in the west and by the Harz mountains in the east (Fig.12:1) the second and the third strategy seem to have been applied. R. Wenskus (1972) has identified three families of the impirial nobility and the bishop of Mainz from South-west Germany as the major forces of military occupation followed by peaceful colonization of southern Saxony. Two of the families were indeed of Saxon descent but, at an earlier stage, had joined the Franks and served as royal functionaries in the adjoining regions of Thuringia and Hesse. The expansion started from a base line along the upper Leine and Werra rivers where they had the control of sections of the former frontier line with royal forts and curtis-settlements as centres (Fig 12:2). From these bases they started the conquest and the consequent woodland colonization. Not infrequently - so it seems - Saxon feudal families who had surrendered to the Frankish king were left in their positions, though their property was reduced by confiscation. These people energetically joined in the woodland colonization to retain their status.

The area under closer study (Fig.12:2) is situated east of the upper Weser river. The Leine river flows in the eastern part of the area as the axis of a small rift valley, about 3 to 4 km wide and covered with fertile loess. It is today completely settled and cultivated. To the west of the rift valley stretches a limestone plateau called Dransfelder Hochflache, which is about 100 to 200 metres higher and 10 km wide, with medium quality loam soils, well-settled, with some surviving patches of woodland. To the west and north-west it is framed by densely forested red sandstone hills and plateaus with rather poor soils and only a few settlements. On the opposite side of the rift valley the limestone forms a narrow plateau with a rather steep cuesta to the east, mostly forested, which further to the east is followed by the undulating hilly country of the Eichsfeld, where the lower beds of red sandstone are covered by loess.

For the study of some spatial aspects of the Carolingian woodland colonization process five steps of analysis have been undertaken:-

1) Identification of the Carolingian colonization areas and of the earlier Saxon settlement areas by the application of the place-name method

   In Figure 12:3 three groups of place-name suffixes have been identified:

a) The main medieval suffix-types have been marked by a symbol each:

+ -hausen ('houses')	- -rode ('woodland clearing')
F -feld ('field')	B -buren (old term for 'houses')

b) Such suffix-types which, in place-name research, have been identified as older than the former group have been put into one group (large black circles). They include a larger variety of suffixes each of which has but a few representatives; among them are -tun (corresponding to the English -ton), -ingen and -heim, -stedt, old types of stream-names as -aha, -a,-ara, and so-called 'dark names' which cannot be explained.

c) Such place-names which are either younger than the first group or which have been in use throughout both periods are marked by dots. Also isolated -rode-villages for which a late origin can be proved, have been included into this group.

From the compactness of the area of -hausen-names it can be concluded that it was settled during a continuous and comparatively short colonization period. In the donation registers of the Benedictine monastery of Corvey the -hausen-names are dominant during the 9th and 10th centuries, with a peak in the third quarter of the 9th century (FIESEL 1972, 101). For the personal names as the first element of the place-name it can be proved from various monastic registers of Carolingian Germany that they were in use during the 8th and 9th centuries. The compact area of -rode and ingerode-suffixes in the south-east cannot be explained as a later extension of colonization because they are situated on loess with even more favourable settlement conditions than the limestone plateau. The suffix -rode was already in use in the neighbouring northern Hessia and Thuringia by the end of the 8th century. About half of the personal names in the -rode place-names were already outmoded by the 11th century (FLECHSIG, 1965). This means that the colonization of the -rode-villages must have taken place by the 10th century at the latest.

The Leine rift valley, in place-name terms, clearly comes out as the early-settled core region. This is also proved by the distribution of pre-Christian cemeteries of the 7th and early 8th centuries (Fig.12:3) with 10 out of 11 sites situated close to the villages of the older suffix-types (LAST 1970). From this we gain a terminus post quem for the -hausen-villages which, in their majority, cannot have been founded before the late 8th century.

2) Dating the beginning of the colonization process - The contribution of soil science

A second method of relative and absolute dating of the colonization process and its rather sudden beginning is offered

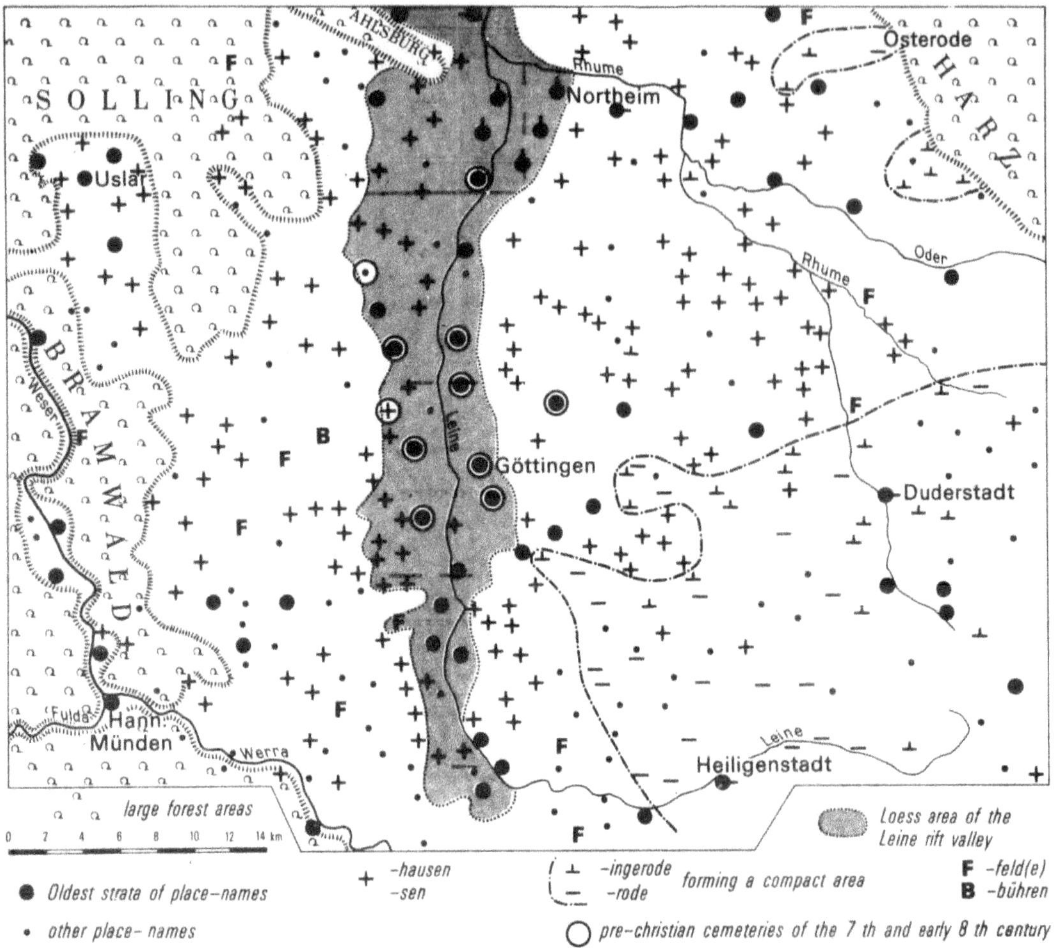

Fig. 12:3 Place-names of the Harz-Weser-region.

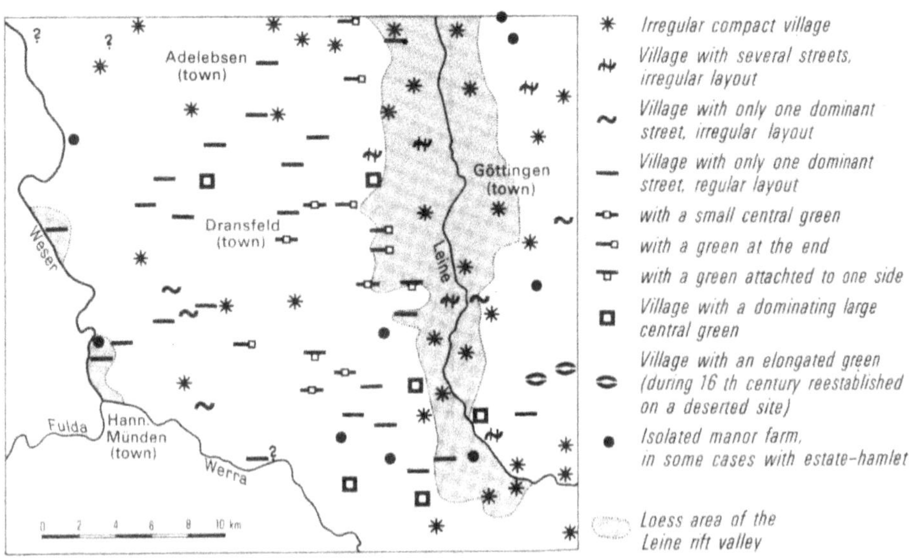

Fig. 12:4 Village plan types of the Leine-Weser-region.

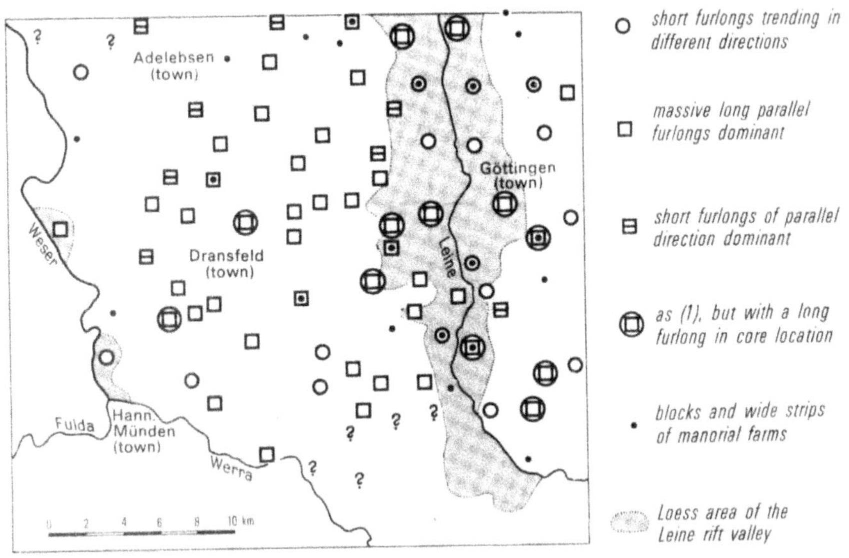

Fig. 12:5 Field pattern types of the Leine-Weser-region.

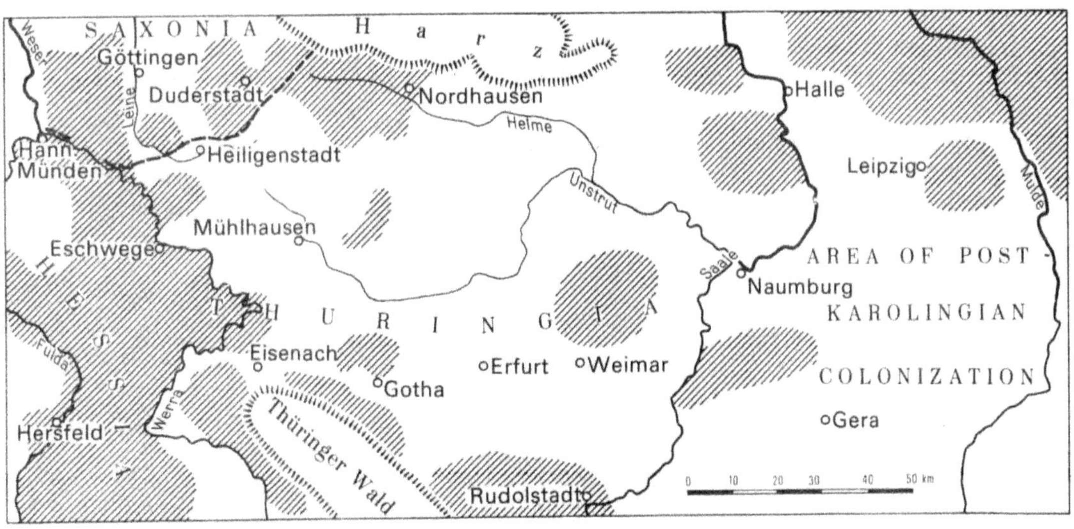

Fig. 12:7 Areas of street villages in south Saxonia, northeast Hesse and Thuringia (based on M. Born, 1970 and W. Emmerich, 1968).

by analysing the alluvial sediments resulting from soil erosion due to extensive forest clearing and cultivation. Two soil scientists of Göttingen University, Meyer and Wildhagen, found three strata of alluvium:

The first stratum: Iron Age alluvium on the fans of the tributary valleys of the Leine river. As proved by pollen analysis this sedimentation ended at around 600 A.D.

The second stratum is an older medieval stratum, while the third stratum a later medieval stratum. For our problem the first medieval stratum - 60 cm - is important. With a sharp boundary it is superimposed on the Iron Age stratum of the lateral fans, and for the first time the alluvium covers the whole Leine flood plain indicating the large volume of sediments. The sedimentation starts with loess rich in humus, then pure loess follows, and finally, reddish brown loamy alluvium. Its origin has been traced as the upper Leine catchment area of red sandstone in the border region of northern Thuringia and southern Saxony. The interpretation of the first medieval stratum by Meyer and Wiledhagen (1972) reads as follows: a very sudden loading of the Leine river with masses of sediments which can only be the result of large scale clearing of forests in the catchment area of the upper Leine river. This interpretation corresponds very well with our consideration of political events and with the interpretation of the place-name map with the compact -rode-area in the south-east.

But where are the alluvial sediments of the contemporary clearance process on the limestone plateau west of the Leine rift valley? They have been traced in the rather wide tributary valleys leading from the plateau to the Leine river. Two radiocarbon dates of the alluvial stratum read 790 ± 28 and 850 ± 24. So, the sedimentation analysis is a clear proof of the dating already derived from historical sources and considerations. The colonization process must have commenced rather abruptly in the last quarter of the 8th century in connection with the Carolingian conquest of south Saxony. This leads to the third step:

3) Historico-geographical analysis of the settlement pattern

What types of village and field patterns are characteristic for the colonization areas? I have studied in detail the limestone plateau and the Leine rift valley. Figure 12:4 shows the village types: The central Leine valley is mainly covered with the Haufendorf-type (irregular compact agglomerations), the limestone plateau and the fringe of the rift valley with street villages, some connected with a village green. The street villages exhibit clear characteristics of regularity which may originate from planning.

The distribution of the types of fields patterns (Figure

12:5) coincides with the distribution of the village types: The limestone plateau street villages are connected with extensive parallel long strip fields (Figure 12:6). In the core areas of the Leine rift valley, on the other hand, irregular and short strip fields prevail - in some cases with a small core area of long strips close to the village.

4) Reconstructing the original pattern of a colonization village

The next step of the study is a closer analysis of one street village and its long-strip fields. I shall try to answer the question: What is the original pattern of a colonization village of the late 8th century?

The street village Lemshausen (Figure 12:6) - its medieval name is Lyamanneshuson - is situated in a valley along a small street running west to east with the fields on two low interfluvial ridges. A cross-country road running from north to south separates a younger addition to the east from the older built-up area in the west. This consists of two rows of tofts: the southern row with longer tofts separated from the fields by a broad back lane. This element is absent in the northern row where the shorter tofts are separated from the fields by a line clearly parallel to the stream. I think these are clear elements of regularity.

A closer view on the village reveals the structure of the two rows of building plots. I start with a hypothesis: if it is a planned medieval village the building plots should have been laid out according to a standard breadth. I first take up the southern row. At the eastern end two plots have an equal breadth of about 36 m. This breadth is repeated on a third plot which was vacant in the 19th century. Two pairs of narrow plots taken together again repeat the breadth of about 36 m. They seem to have been subdivided secondarily. So, there have been 5 adjoining plots of equal breadth.

The largest house-plot belonged to a small manor house. The village chapel is attached to it. The manor plot has a larger breadth of 82 m and is longer than the others projecting farther into the street. Therefore, it is about three times the size of the 5 farm units to the east. At the west end of the row there are again two peasant tofts which are narrower, but longer than the five ones to the east and therefore equal to them in respect of toft area. Summing up: the row in an earlier stage was made up of ten units: 5 to the east, 3 for the manor toft, and 2 units to the west. On the map these units have been marked by stronger lines.

The written medieval sources fit well into the results of the meteorological analysis: in 1432 ten Hufen (mansi) of Lemshausen belonged to the large manorial estate of Plesse castle north of Gottingen, made up of a manorial farm of three

Fig. 12:6 Cadastral map of Lemshausen (sw of Göttingen)
Inset map: reconstruction of original tofts (strong lines).

Hufen and 7 peasant Hufen. The manorial farm can be traced back to a nobleman named Hathamar who, in the year 1030, transferred his property to the bishop of Paderborn who was the owner of the Plesse estate. This leads me to the conclusion that the southern row of 10 units with its regular lay-out can be dated back to the early Middle Ages.

The opposite row again looks rather regular. If we analyse the breadth of the tofts 7 standard units and one double unit can be distinguished, making up 9 units (on the map again marked by stronger lines). 5 of them can be identified as belonging to the medieval estate of the Count of Everstein, including the double unit. One problem remains: is this row of the same age as the southern one?

The field pattern analysis will help us in revealing that the settlement was established in two stages:

The 3-Hufen-manor unit was still intact in 1873. The shaded plots belong to it. On the other hand, the medieval peasant Hufen-units have been fragmented and so far I have not been able to reconstruct them. If the settlement started with a single row and ten Hufen in the fields then in the earliest fields the manor should own a portion of 3 by 10. Indeed, in the furlongs immediately adjoining the southern row which I have marked by a dotted line the proportion of the manor is 3 by 9, 3 by 11, and 3 by 12. The areas beyond in the south and east are very moist with heavy soil and must therefore be recognized as later extensions.

On the nothern side of the valley the quality of fields is equally good. Here we could expect the fields being brought under cultivation when the 9 units of the second row were established. In this northern area the farms of both rows own land. If this was so from the beginning the manor's ratio should be 3 by 19. The actual proportion in the 19th century map fits well into this calculation, being 3 by 19 north of the field lane and 3 by 21 south of it. So, the analysis has revealed two stages in the foundation process. The colonization village was founded as a single-row street village with long parallel strip fields adjacent to it. Its manorial structure with a manor farm and 7 peasant Hufen-holdings dates back to the very beginning.

So far, no definite answer can be given about the dating of the village foundation though Lemshausen must belong to the earliest group of Carolingian villages because it is located on the fringe of the pre-Carolingian settlement area of the rift valley. It is not yet clear whether all the double row villages did start with a single row. This has still to be tested by further studies.

5) Origins of the model

In the regional context the final question is: from where

stems the model for this planned type of street village with long strip fields which was applied by the Carolingian nobility in the conquest of southern Saxony? I have compiled a distribution map of street villages in southern Saxony and the adjoining border regions of the Frankish empire to the south and south-east of north-eastern Hesse and Northern Thuringia. It is quite obvious that the street villages of our region are just an extension of the large street village region of the Frankish territories in the south. Here again, the early settled basins are without street villages which dominate in the lower hill and plateau regions. From historical sources it can be concluded that here colonization was started some decades before the conquest of Saxony, about 750. It was in these border regions of the empire where those families of the imperial nobility acted as officials of the Frankish state and where they had their earliest feudal estates. Hence, there can be no doubt that it was from here that the model of the colonization village was transferred to southern Saxony.

REFERENCES

BORN M., (1972) 'Siedlungsgang und Siedlungsformen in Hessen', Hessisches Jahrbuch für Landesgeschichte, 22 (1972), map 'Dorfformen in Nordhessen'.

EMMERICH, W., (1968) 'Die siedlungsgeschichtlichen Grundlagen', Geschichte Thüringens I (eds.H. Patze/W. Schlesinger) (1968), map.

FIESEL, L., (1972) 'Franken im Ausbau altsächsischen Landes', Niedersächsisches Jahrbuch fur Landesgeschichte, 44 (1972), 74-158.

FLECHSIG, W., (1965) 'Ortsnamen als Quellen fur die Siedlungsgeschichte des Leinetals', Deutsche Königspfalzen, vol.2 (1965),83-113.

LAST, M. and K. BRANDT 'Zur Frühgeschichte Göttingens', Neue Ausgrabungen (1970) und Forschungen in Niedersachsen, 6 (1970), Part I, 191-203 (by Last).

NITZ, H.-J. (1961) 'Regelmässige Langstreifenfluren und fränkische Staatskolonisation, Geographische Rundschau,13 (1961), 350-365.

NITZ, H.-J (1963) 'Siedlungsgeographische Beiträge zum Problem der fränkischen Staatskolonisation im süddeutschen Raum', Zeitschrift für Agrargeschichte und Agrarsoziologie, 11 (1963), 34-62.

NITZ, H.-J (1971) 'Langstreifenfluren zwischen Ems und Saale - Wege und Ergebnisse ihrer Erforschung in den letzten drei Jahrzehnten', Braunschweiger Geographische Studien 3 (1971), 11-34.

WENSKUS, R. (1972) 'Das sudliche Niedersachsen im fruhen Mittelalter', Festschrift für Hermann Heimpel, Vol.III (1972) 348-398.

WILDHAGEN, H. and B. MEYER 'Holozane Bodenentwicklung, Sedimentbildung und (1972) Morphogenese im FluBauenbereich des Göttinger Leinetalgrabens', Göttinger Bodenkundliche Berichte, 21 (1972).

CHAPTER 13

LATE-MEDIEVAL DUTCH SETTLEMENTS IN THE LOWER RHINELAND

W. Krings
(University of Bamberg)

The region I have chosen for my investigation is the Lower Rhine Plain on both sides of the present-day border between the Netherlands and North Rhine-Westphalia. This region has usually been characterized as an area of dispersed settlement (Streusiedlungsgebiet) (1) and it has obviously offered no particular interest to geographers and historians. Their concern was merely to find out, and to explain, the boundary-line separating the northern area of dispersed settlement from the southern pattern of large nucleated villages (Dorfsiedlung) (2). There is, however, plenty of evidence to show that the settlement geography of the Lower Rhineland is much more complicated than it had appeared at first sight, indeed one may say that the region, today, is an unplanned open-air museum of planned settlements, urban as well as rural, for instance, the Roman Colonia Ulpia Traiana founded about AD 100. This is now being partly rebuilt on the basis of deliberate archaeological excavations (3). There are also examples of heathland cultivation dating from the seventeenth, eighteenth and early nineteenth centuries (4), or agricultural colonies established after World War II for refugees from Eastern Germany (5).

The varied pattern of planned and unplanned spontaneous settlements reflects the variation of the natural environment and its pedological and hydrological conditions. In the course of history these conditions have been evaluated according to the particular circumstances, human resources, technical skills, political organization. My study is centered on the late Middle Ages. By the twelfth century the low-lying parts of the northern Rhineland were not yet completely cultivated, and the shallow, ill-drained valley grounds and the fen areas (called Bruche, broeken) still remained exempt from activity. They were sparsely used as common pasture for the livestock. An exception were the lowlands immediately bordering the Rhine River. Although being exposed to the risk of flooding, these had already been filled up by small settlements using natural levées or built upon artificial mounds (called Polle). Beyond this area, the first reclamation activities were due to monasteries or feudal lords, but their impact is difficult to discern within the modern landscape.

The situation changed when political territories of considerable extent came into existence during the thirteenth and fourteenth centuries: the Counties, later on Duchies, of Cleves (Kleve) and Guelders (Geldern). Both these territories attempted to extend their sphere of influence, to develop their economic resources and to give security to their inhabitants

(6). Both were able to profit from an innovation radiating from Flanders and the County of Holland.

This innovation consisted in a special organization and technique of draining low-lying grounds (7). I can only indicate the main points. The procedure went on as follows:
First step: to regulate the water discharge. This normally required the digging of one or more ditches (called weteringen, Leitgraben) crossing the area. Furthermore, it was necessary to lay out dikes and ditches on the periphery intercepting the ground and surface water infiltrating from higher adjacent lands. Second step: to conduct a survey of the drainage area. The surveyor used a standard unit of length of 30 rods called Rijnlandse roeden (8) after the district of Rijnland in Southern Holland (in the environs of the town Leyden). It measures 113 metres or about 123 yards. By that means, the drained area was divided into straight strips of equal width but varying length. The strip plots were call slagen/Schlage or Hufen (9). This system of land allotment was adopted in order to minimize the expenses which each colonist (broeker) had to bear for the clearing of the ditches and other works. According to the terminology proposed by the International Working-group for the Geographical Terminology of the Agricultural Landscape the present-day field structure can be described as a group 'of former wide strips within a pattern of fragmented holding, later on divided into smaller strips' (aufgeteilte Breitstreifengemengeflur) (10). Wide strips are defined as 'strips of more than 40 m width' (11).
Third step: to recruit the colonists and to organize the drainage and reclamation works. This was done by managers (single persons or a group of persons) called Lokatoren. It can be proved that in Cleves and Guelders some of the Lokatoren really had immigrated from Holland or Flanders or had kindred within the Dutch or Flemish knighthood.
Fourth step: to organize the colonists' community in order to guarantee the maintenance of the drainage installations. For this purpose the community was authorized to choose a drainage board consisting of seven so-called heemraden/Heimrate. Considering that the land was cultivated individually, we can classify these communities as co-operative settlements (12). By reason of the four points I have talked about, we can conclude that we are dealing with a regulated and planned colonization and as a result with planned settlement (13).

As the Count assigned the land to the Lokatoren and as he issued a document fixing the conditions and the legal status of the colonists, we should be able to reconstruct the diffusion of the so-called 'Dutch settlement', provided that the documents are preserved. However, even if they are lost, or not to be found within the records, there are other ways to get positive results. The diffusion of the Flemish and Dutch reclamation systems has been studied some years ago by an historian, Franz Petri (14). He collected a lot of details concerning the territories of Cleves and Guelders (15). The results of Petri's research together with my own supplements (16) are shown on Fig. 13:1. Dutch settlements are to be found widely spread over all low-lying parts of the map. They clearly correlate with gley soils, basin clay soils or peat fen districts (17). The distribution of the settlements differs from that of the so-called Hollerland near Breman, where the Dutch settlements cover a large coherent area (18).

For the historical geographer it is interesting to compare the different reclamation areas mapped on Fig.13:1. A first aspect concerns the size of the reclamation schemes. On Fig.13:2 I have assembled several examples. As you see, they vary in form and size. There are small units like No. 2 (Tiller Bruch) or No. 5 (Niederbruch near Xanten, letter B). On the other hand, there are larger districts composed of several units like No. 1 (Kranenburg) or No. 4 (Veen). In these latter cases, we would like to know whether the different units have been established simultaneously or not. The question is not yet sufficiently resolved, for it is difficult to refer the records to an exactly delimitated area. For the case of Veen (No. 4), Petri cites two different documents. The first which has been issued in 1299 proves that the Count of Cleves received payments from Dutch settlers (<u>'van den Hollanderen, die in synem lande wonen</u>') (19). The second document reveals that in 1340 the Count has assented to the reclamation of a peat fen (<u>Torfveen</u>) on condition that it would be executed according to the Dutch system (20). As the soil map shows, only one single peat fen exists in the district in question. It is covered by the units C,D and F. So we are allowed to assume that the units A and B have been reclaimed first by Dutch immigrants, while C and D are the work of later local reclaimers.

I go on to another point of interest. In many cases the whole area of the original scheme has been individually transformed into arable land or artificial pasture. There are, however, other cases where parts of the land continued to be used as common. Figure.13:3 shows the Baaler Bruch situated near the village of Weeze (see Fig.13:1, No. 6). You can recognize the straight central ditch (<u>Leitgraben</u>) which discharged the surplus water into the Maas River together with the lateral dikes. I have reconstructed the original division into strip plots of 30 rods (<u>Rijnlandsche roeden</u>) which was still descernible in an eighteenth century cadastral survey. Until the beginning of our century, only the land along the dikes has been brought into cultivation (21). The rest of the reclamation area, though divided into individual strips, remained common, and was used for grazing sheep.

The apparent differences between the reclamation projects are explained by a series of factors: soil conditions, drainage conditions, local population pressure and social stratification, facilities granted by the Count (or is legal successors), market relations and so on.

Fig. 13:1 Inventory of the identified and supposed Dutch settlements:

1 Arnhemsche Broek near Arnhem
2 Hollanderbroek near Elst
3 Lindenholt and other areas near Beuningen
4 Hollander Broek near Mill
5 Siebengewald
6 Hülmer Heide and Baaler Bruch near Weeze
7 Laarbruch near Weeze
8 Wembscher Bruch near Weeze-Wemb
9 Steinberger Bruch near Uedem
10 Uedemerbruch near Uedem
11 area of Sonsbeck
12 area of Veen
13 Labbecker Bruch near Labbeck
14 Urselwald/Neiderbruch near Xanten
15 Raderbruch near Appeldorn
16 Bernem and Arselaerbroek near Kalkar-Altkalkar
17 Tiller Bruch near Till-Moyland
18 Kranenbruch near Kranenburg
19 Maldensche Broek near Malden
20 Hetter near Vrasselt-Praest
21 area called In Holland near Praest
22 Kattenbruch near Haldern (supposed)
23 Wertherbruch near Werth
24 Die Hufen, southern extension of Wertherbruch (supposed)
25 Ringenberger Bruch near Ringenberg
26 Silvoldsche Slagen near Silvolde (supposed)
27 Azewijnsche Broek (supposed) near Gendringen (supposed)
28 Vinkwijksche Broek near Zeddam (supposed)
29 Kwartier near Didam (supposed)
30 the Liemers district (not studied by the author)

Sources:

F. Petri, op. cit. (1972): 2, 4, 10, 11, 12, 14, 15, 17, 18, 19, 20, 21, 23, 25.

F. Gorissen, Niederrheinischer Städteatlas II, 1 Nimwegen. Kleve 1956: 3.

W. Krings, op. cit. (1976/1980): 5, 6, 7, 8.

J.H. Mooren and F. Nettesheim. op. cit.: 9

F. Weibels, Die Großgrundherrschaft Xanten im Mittelalter. Krefeld 1959: 13.

F. Gorissen, Niederrheinischer Städteatlas I, 2 Kalkar. Kleve 1953: 16.

A.G. Van Dalen, Gelderse Historie in de Liemers. The Hague and Rotterdam 1971: 30.

Topographical and/or place-name evidence: 1, 22, 24, 26, 27, 28, 29.

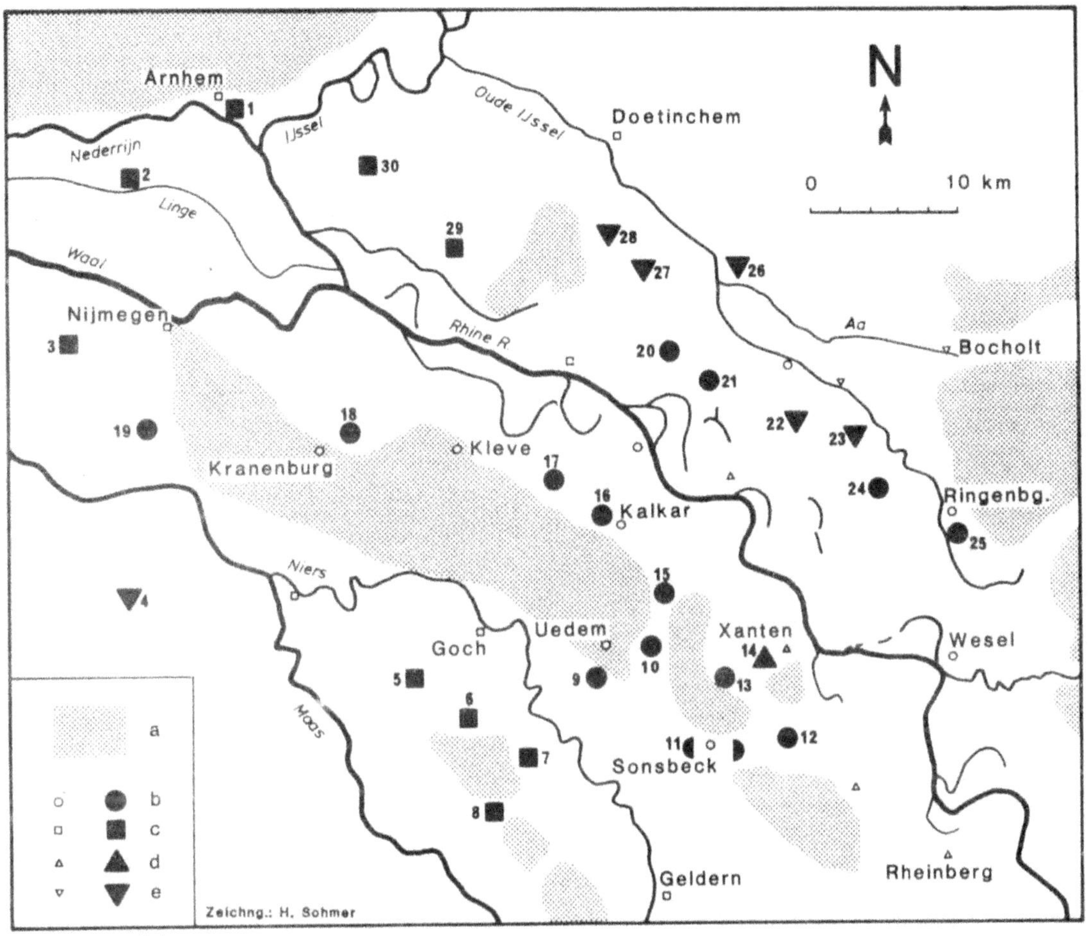

a. high-lying parts (Pre-Pleistocene heights, push-moraines, sandy areas, Main-Terrace, and younger Middle-Terrace).

b. towns (open symbol and Dutch settlements (filled in symbol) of the territory of Cleves.

c. id., territory of Guelders.

d. id., territory of Archbishopric of Cologne.

e. id., other territories.

For the figures 1 - 30 see list p.

Fig. 13:1 Distribution of the identified and supposed Dutch settlements in the Lower Rhineland.

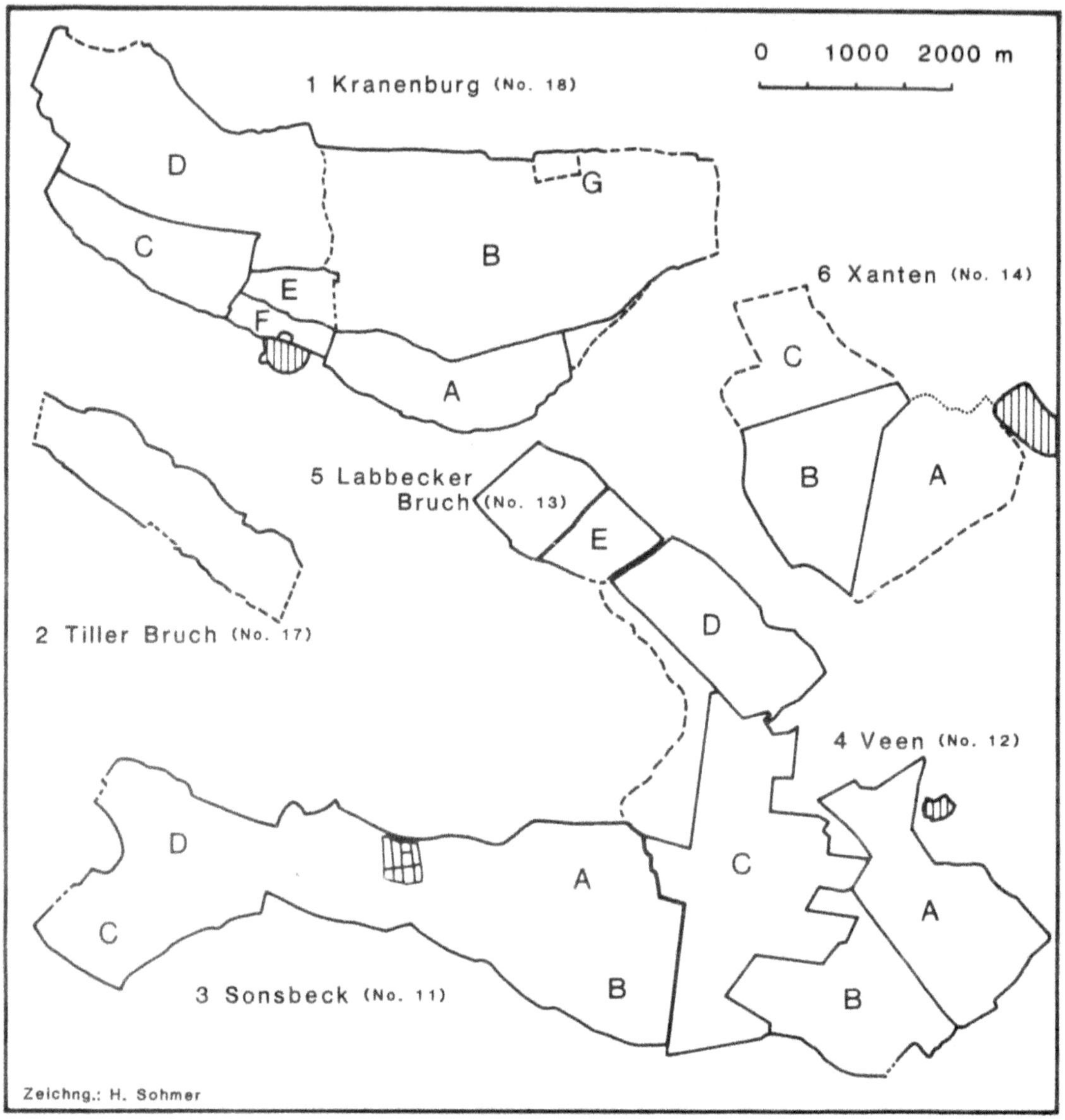

1 A Die kurze Houben (short strip plots)
B Die lange Houben (long strip plots)
C Das leege Bruch (leeg=low-lying)
D Zyfflicher Busch, cleared woodland
E Reyer Bruch
F common of the town of Kranenburg
G Bürgerkämpe, enclosed blocks of younger origin
3 A Hammerbruch
B Stadtveen
C Mengsischbruch
D Breienbruch
4 A, B reclaimed before 1299
C, D, F Torfveen, reclaimed after 1340
E later extension of D (?)
5 A Hochburch
B Niederbruch, reclaimed after 1315
C later extension of B (?)

Fig. 13:2 Size comparison of five Dutch settlement areas. The figures in ( ) refer to the numbering on Fig. 13:1.

Fig. 13:3

Reconstruction of the original Dutch strip division in the Baaier Bruch - Hülmer Heide south of Goch (see Fig. 13:1, No. 6)

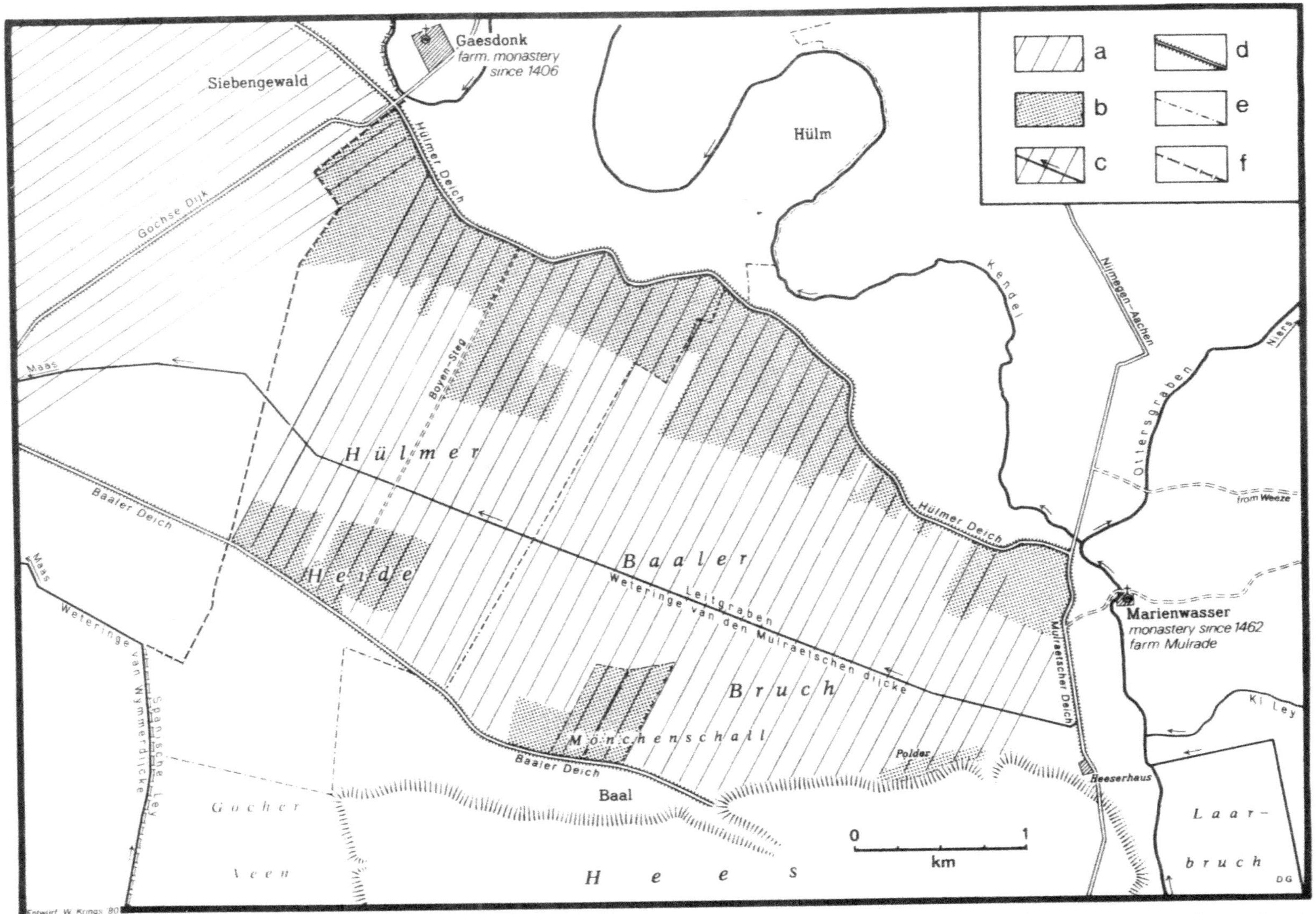

a. long strips of 30 rods (Rijnlandse roeden, 113m) width
b. reclaimed area in 1843 (after map 1:25,000 of the Preussisches Topographisches Bureau, first edition, sheets 4302/4303)
c. main drainage ditch (wetering, Leitgraben)
d. dikes
e. 19th century township boundary
f. state boundary since 1815

West

Plate 13:1 East 15th May, 1979

Plate 13:1 Aerial view of the Baaler Bruch (see Fig. 13:1, No. 6, and Fig. 13:3). Note the winding lane which coincides with a dike (Hülmer Deich). To the right of this dike there are the meanders of the Kendel brooklet, scattered farmsteads and, at the bottom, the single-row village of Oberhelsum. On the left-hand side one sees the small strip plots of the reclaimed fen and the larger rectangular blocks which date back to the beginning of the 20th century when the remaining parts of the fen area were reclaimed by steam ploughing.
Scale 1:12,500
Reproduced by permission of Landesvermessungsamt Nordrhein-Westfalen, September 24th, 1981 (D 538/81). Released by Landesvermessungsamt Nordrhein-Westfalen, No. 26/79, July 25th, 1979.

South

Plate 13:2 North 31st March, 1977

Plate 13:2 Aerial view of Wertherbruch (see Fig. 13:1, No. 23). The characteristic regular strip pattern of the Dutch reclamation scheme is well revealed in this view. From bottom left to top right: the main settlement axis of Wertherbruch with the parish church and a modern, nucleated extension of the village, the so-called outer drainage ditch (Aussenwässerung), another ditch called Vorwässerung, and, close to the motorway (E 36), the border ditch (Klevische Landwehr).
Scale 1:12,000
Reproduced by permission of Landesvermessungsamt Nordrhein-Westfalen, September 24th, 1981 (D538/81). Released by Landesvermessungsamt Nordrhein-Westfalen, No. 11/77, May 3rd, 1977.

In the case of Steinbergen (see Fig. 13:1, No. 9), we know that the settlers deserted their land and that the Count tried to lure them back by reducing the rents (22). Certainly, while the project did not definitely fail, neither did it develop its full form.

A third aspect we must study is the location of dwellings super-imposed on the original strip divisions. The arrangement of the colonists' dwellings often seems to have been guided by pre-existing elements such as dikes or roads, but there was no uniform system. In the case of the Baaler Bruch, a loose row of new dwellings, most of them cottages, has appeared along the northern dike (Hulmer Deich). They constitute a clear contrast to the old farmsteads scattered between the meanders of the Kendel brooklet (see Plate 13:1).In other cases a straight linear pattern is to be found following one of the main drainage ditches (Stadtveen near Sonsbeck, Uedemerbruch, Wertherbruch/see air photograph 2).

In the case of Sonsbek (see Fig. 13:1, No. 11), the reclamation activities led to the foundation of a town. Figure 13:4 shows part of the present ground-plan of the town where it is easy to recognize the original rectangular lay-out subdivided by parallel streets. This ground-plan preserves traces of the division into uniform wide strips and the west-east-extent corresponds with three and a half of these units.

I come to a final question. I have tried to sketch the impact which the Dutch settlements have had on the landscape of the Lower Rhineland. Considerable relic featues of these patterns have survived in modern landscapes. Their further existence, however, is threatened by large-scale improvement of the road network and by land consolidation schemes (23). Therefore, I thnk, we are challenged to contribute to a 'soft' environmental policy respecting the relics of the rural past. This would mean that we have to support the following objectives: (24)

- to identify and record the existing historical patterns,
- to protect important features and to ensure their survival as far as possible,
- to encourage public interest and awareness in historic rural landscapes.

I should like to end by giving you an example to illustrate these aims.

At the beginning of my paper, I have mentioned the reconstruction of a Roman colony. This project is being realized in front of the gates of the medieval town of Xanten. Moreover, the imposing Gothic cathedral destroyed in World War II, as well as several other religious and civil buildings of the town, have been saved by remarkable reconstruction or restoration works. The historic sites and a projected installation for water sports are designated to form a regional

1 unit 1/2 unit
of 30 Rijnlandse roeden

Fig. 13:4 Persistence of the original Dutch strip allotment within the ground-plan of the town of Sonsbeck.
Part of the sheets SONSBECK WEST and SONSBECK of the Deutsche Grundkarte 1:5,000. Reproduced by permission of the Oberkreisdirektor Wesel - Katasteramt, 24. 9. 81 No. K 39

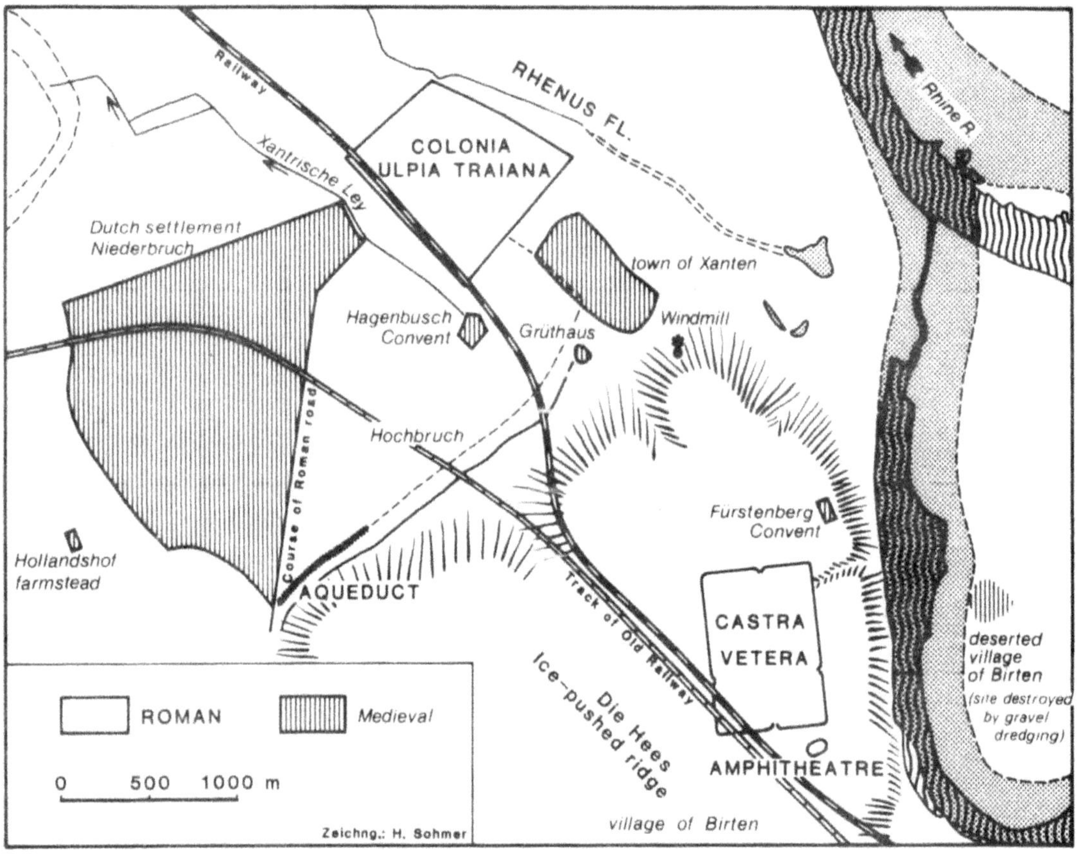

Fig. 13:5 Historic sites of the Xanten area

center for short-time recreation.

In order to promote public interest for the measures initiated and subsidized by the government of North Rhine-Westfalia, Xanten has been chosen for one of five model projects in order to represent the Federal Republic of Germany within the European Architectural Heritage Year 1975 (25). For these reasons the small town merits a wider attention.

There are good guidebooks for the historic sites of Xanten and her surroundings. None of them, however, refers to the legacy of medieval wasteland cultivation. In 1315 the Chapter of the cathedral decided to reclaim a wood district called Urselwald (see Fig. 13:2, No. 6, letter B). It had to be divided into sedecim jurnales secundum mensuram Hollandensem ('sixteen plots according to the Dutch survey system') (26). The reclaimed area which was named Niederbruch is still used for agriculture, while the probably earlier cultivated area, letter A (Hochbruch), has become an industrial zone (see Fig.13:5). I think it is, indeed, very desirable to safeguard the traces of the Roman colonization and the remains of the medieval town, but we also should take care of the persistent patterns of a rural settlement like Niederbruch which remind us of an important chapter of European landscape history.

NOTES & REFERENCES

1. See for instance map Mitteleuropa, Landliche Siedlungs formen 1:10,000,000, in Sydow-Wagners Methodischer Schul-Atlas, compiled by H. Haack and H. Lautensach, Gotha 21. ed. 1942, map No. 23 b.

2. See map Grenze der Dorf - und Hofsiedlung im Rheinlande, in J. Niessen (ed), Geschichtlicher Handatlas der Deutschen Lander am Rhein / Mittel - und Niederrhein. Cologne - Lorrach 1950, 47.

3. The project is called Archaologischer Park Xanten. It is maintained by the Lanschaftsverband Rheinland (Cologne). Information is available at the entrance or by the Regional-museum Xanten, Kurfürstenstrasse 7-9, D 4232 Xanten.

4. See W. Krings, Wertung und Umwertung von Allmenden im Rhein-Maas-Gebiet vom Spatmittelalter bis zur Mitte des 19. Jahrhunderts. Eine historische-sozialgeographische Studie (Maaslandse Monografieen 20). Assen/Amsterdam 1976.

5. See J. Bahr and W. Golte, Die Reichswaldsiedlung nach 25 Jahren. Enstehung und Entwicklung des jungsten Kolonisations-werkes im Reichswald bei Kleve, in Reinishche Heimatpflege N.F. 14, 1977, 251-257.
Ch. Abs, Goch-Nierswalde. Entwicklungsskizze der landwirtschaftlichen Siedlerstellen in einer Siedlung der Nachkriegszeit (The development of a post-war agricultural settlement), in G. Aymans (ed), Niederrheinishche Studien (Arbeiten zur rheinischen Landeskunde 46). Bonn 1980, 137-143.

6. See D. Kastner, Die Territorialpolitik der Grafen von Kleve (Veroff. d. Historischen Vereins fur den Niederrhein 11). Dusseldorf 1972; w. Jappe Alberts, Geschiedenis van Gelderland van de vroegste tijden tot het einde der middeleeuwen. The Hague 1966.

7. The standard work is H. van der Linden, 'De Cope'. Bijdrage tot de rechtsgeschiedenis van openlegging der Hollands-Utrechtse laagvlakte (Bijdragen van het Instituut voor Rechtsgeschiedenis der Rijksuniversiteit te Utrect 1). Assen 1956. - See also A. M. Lambert, The Making of the Dutch Landscape. London and New York 1971, 100-106.

8. S. J. Fockema Andreae, 'De Rijnlandsche Roede, Geschiedenis eener Oud-Nederlandsche Landmaat', in Tijdschrift K.N.A.G. XLIX, 1932,635-657; B. H. Slicher van Bath, The Agrarian History of Western Europe A.D. 500-1850. London 1963, 152.

9. For the term 'slagen' see E. Verwijs and J. Verdam, Middelnederlandsch Woordenboek. The Hague, vol. 7, col. 1189-1190.

10. H.Uhlig (ed), Basic Material for the Terminology of the Agricultural Landscape. Vol. I Types of Field Patterns. Giessen 1967, E 56.

11. Uhlig, op. cit. E 31.

12. H. Uhlig (ed), Basic Material ... Vol. II Rural Settlements. Giessen 1972, 85.

13. See Uhlig, op. cit. (1972), 85, and M. Born, Geographie der landlichen Siedlungen 1 (Teubner Studienbucher Geographie). Stuttgart 1977, 53-55.

14. F. Petri, Entstehung und Verbreitung der niederlandischen Marschenkolonisation in Europa (mit Ausnahme der Ostsiedlung), in Konstanzer Arbeitskreis fur mittelalterliche Geschichte (ed), Vortrage und Forschungen XVIII, Sigmaringen 1974, 695-754.

15. F. Petri, 'Die Hollandersiedlungen am Klevischen Niederrhein und ihr Platz in der Geschichte der niederlandisch-nieder-rheinischen Kulturbeziehungen', in Festschrift M. Zender, Bonn 1972, 1117-1129.

16. See inventory facing Fig. 13:1.

17. For soil maps see Atlas van Nederland. The Hague 1963-, IV. 4 Gelderland and IV. 9 Limburg (1.200000). - Bodenubersichtskarte von Nordrhein-Westfalen 1:300000 (E. Muckenhausen and H. Wortmann), ed. by Amt fur Bodenforschung. Hannover 1953. - Ubersichtskarte von Nordrhein-West -falen 1:100000, B. Bodenkarte, Bl. C 4302 Bocholt. Krefeld 1967.

18 See D. Fliedner, Die Kulturlandschaft der Hamme-Wumme-Niederung. Gestalt und Entwicklung des Siedlungsraumes nordlich von Bremen (Gottinger Geogr. Abh. 55). Gottingen 1970.

19. Petri, op. cit. (1972), 1120.

20. Petri, op. cit. (1972), 1124

21. For further details see W. Krings, 'Persistente Muster in der Agrarlandschaft des Baaler Bruchs, Gemeinde Weeze' (Persistence of late-medieval land-use patterns in a reclaimed fen), in G. Aymans (ed), Niederrheinische Studien (Arbeiten zur rheinischen Landeskunde 64). Bonn 1980, 101-116.

22. J. H. Mooren and F. Nettesheim, 'Heberegister der Einkunfte der Grafschaft Cleve aus dem Ende des 14. oder dem Anfange des 15. Jahrhunderts', in Annalen des Historischen Vereins fur den Niederrhein 31, 1977, 113.

23. See, for example, Krings, op. cit. (1980), 114.

24. After F. Aldsworth, Survey and Conservation Objectives in West Sussex, in P. F. Brandon and R. Millman (eds), Historic Landscapes: Identification, Recording and Management (Occasional Paper of the Department of Geography, Polytechnic of North London), 45-6: See also A. P. de Klerk, 'Historische Geografie en ruilverkaveling. Enkele overwegingen ter bescherming van het Eemnesser kulturrlandschap', in Geografisch Tijdschrift K.N.A.B. 11, 1977, 434-447: F. N. Nagel, 'Konzept zur Erfassung von erhaltenswerten kulturgeographischen Elementen in landlichen Siedlungen', in Berichte zur deutschen Landeskunde 53, 1979, 81-93.

25. See Landeskonservator Rheinland (ed), Xanten, Europaische Beispielstadt (Arbeitsheft 9). Koln 1975: Deutsches Nationalkomitee fur das Europaische Denkmalschutzjahr 1975 (ed), D 5 Xanten. Berlin 1975.

26. cited after Petri, op. cit. (1972), 1124.

CHAPTER 14

MORPHOGENETIC TYPOLOGY OF VILLAGES AND THEIR REGULATION IN POLAND : A CASE STUDY FROM SILESIA AND POMERANIA

H. Szulc
(Polish Academy of Sciences)

In Poland the tradition of the morphogenetic study of villages dates back to the beginning of the twentieth century, but although many regional studies exist sythesising works are few. (1) Many reasons account for this, but two seem to be particularly significant: (1) the absence of group research, involving representatives of various disciplines and (2) a great diversity of source materials as a result of numerous changes in the boundaries of Poland, indeed its partition for over a century.

First geographical-historical studies of rural settlement were undertaken not by geographers but by historians, such as K. Potkanski, O. Balzer, K. Tymieniecki, H. Lowmianski, F. Bujak, K. Dobrowolski. The contribution of archaeologists - mainly J. Kostrzewski - is also important, especially as regards early settlement. Geographers joined in much later. They mainly represented the landscape school of settlement and therefore they broadened the scope of research and of problems dealt with, as well as enlarged the source base by including statistical materials, village historical plans and findings of field research.

Certain attempts at synthetical works made in the post-war period in the form of historical-geographical lexicons of all localities documented in historical sources prior to the 15th century, as well as Poland's historical atlas on the scale 1:250 000 presenting the settlement network, the size of villages, and property relations at the end of the 16th century. However, the lexicons and the atlas have been compiled for a few regions only.(2)

My own research into the morphogensis of villages and their regulation refers mainly to Western and North-Western Poland, i.e. the regions of Silesia (3) and Pomerania (4), and to a lesser degree the Sub-Carpathian region (5). Therefore, all examples quoted in my paper are taken from those areas.

Silesia and Pomerania seem to be of great interest for all those who study the morphogensis of villages, since: (1) source materials i.e. the manuscript plans of Polish villages from the turn of the 18th and 19th century as well as registers are still available; (2) many relics of old settlement are well preserved; (3) settlements of two different ethnical groups, namely the Slavs and the Germans, still exist in the areas where those population groups have intermingled. The Limes sorabicus, i.e. the boundary running along the rivers Elbe and Saale, is clearly seen on the map compiled by K.H. Schroder and G. Schwarz (6).

In my research into the morphogenesis of the Silesian and Pomeranian villages I have selected the turn of the 18th and 19th century, because big changes took place then in the spatial layouts and the socio-economic structure of the villages. They were caused by the regulation of villages after the emancipation of the peasants and the separation of the peasant and manor land. The large-scale contemporary manuscript plans are still available as well as numerous statistical and historical material.

The cartographic material is my basic source, namely the village plans made for Silesia and Pomerania by a body named "the General Commission" at the turn of the 18th and 19th centuries. For many years I have searched archives for village plans. Unfortunately, I have not found many which have survived the hostilities: for example, in Opole Silesia only 20% of all the plans have been preserved. The situation is even worse in Pomerania: there are only about 300 plans for an area covering three voivodships: Szczecin, Koszalin and Slupsk. The plans are set in order, classified and photographed. A card-index, which presents historical data until the 15th century, has been prepared for those villages the plans of which have been preserved. As regards those, however, for which monographs have been compiled, a full historical documentation until the early 19th century has been gathered, as well as source materials, mainly for the 16th and 17th centuries, available in archives.

The aim of my research was the reconstruction of the network of rural settlements for the early 19th century; I have also taken into consideration their forms and socio-economic structure. The main problem with which I was concerned as to distinguish a regular from an irregular village, and to identify the period in which the field and the settlement patterns developed. This means that I had to distringuish a measured regular village from an unmeasured irregular one, in which the mansus was a taxation unit and not a measure.

My research method is the investigation of village models. I have arrived at a conclusion that large numbers of village plans represent the same morphogenetic type, and therefore subsequently it is possible to formulate hypotheses per analogiam as regards changes in the rural settlement of the whole area.

I have selected such villages situated in Opole Silesia and Pomerania which are well-documented (that is their historical plans are available), and which are typical examples of villages differing between themselves in their layouts of fields and sites, their origin, forms of ownership and topographical situation.

MORPHOGENETIC PHASES

As regards the forms of rural settlement in Poland the following five main morphogenetic periods can be distinguished: (1) the early medieval (early feudal), before location of the villages on German law; until the 12th century; (2) the medieval, feudal, when the villages had already been located on German law; until the 15th century; (3) the late-feudal and early-capitalist; until the early 19th century (4) the capitalist; until the outbreak of World War II; (5) the socialist, in People's Poland.

The Medieval Foundations

The knowledge of the forms of the early-medieval villages is fragmentary and based mainly on hypotheses. But a few relic villages have been preserved and have been identified on the plans from the early 19th century. The new research methods (7), mainly based on metrological-genetic and archaeological techniques, have made it possible to identify the sequence of village developments and establish certain elements of their plans.

Settlements with dispersed, non-compact forms appear to have been the most common, but compact villages, with buildings arranged in rows were also present, particularly where site was a conditioning factor, along a river valley or the shore of a lake. Studies of the earliest medieval villages are, however, limited by the fact that almost all Polish work has been concerned with buildings and field systems.

When field strips are wide shaped as squares or rectangles, we can speak of an irregular, block system of fields. This system has been indentified in many medieval villages, the fields of which were regulated as late as the 15th century and which were located in regions subject to Polish law. As a rule such villages were small, consisting of only a few farms, in the form of a small, irregular, street village or a small green village, well-adapted to the topgraphy. Goraszowice (Fig. 14:1) is a good example of such a village (8): the field pattern is irregular and the settlement remained under Polish law until the 15th century, although it was mentioned in the 13th century registers. (9) The investigation of its plan from the early 19th century provides interesting results because the picture of the spatial layout of the former village is quite clearly the block system. The fields can be arranged in two groups, relics of the two former hamlets which belonged to two families.

The genesis, function and distribution of green villages, especially round ones, have been widely discussed both in Poland and abroad (10). The field pattern in such villages is usually either a block or strip system. They are believed to be

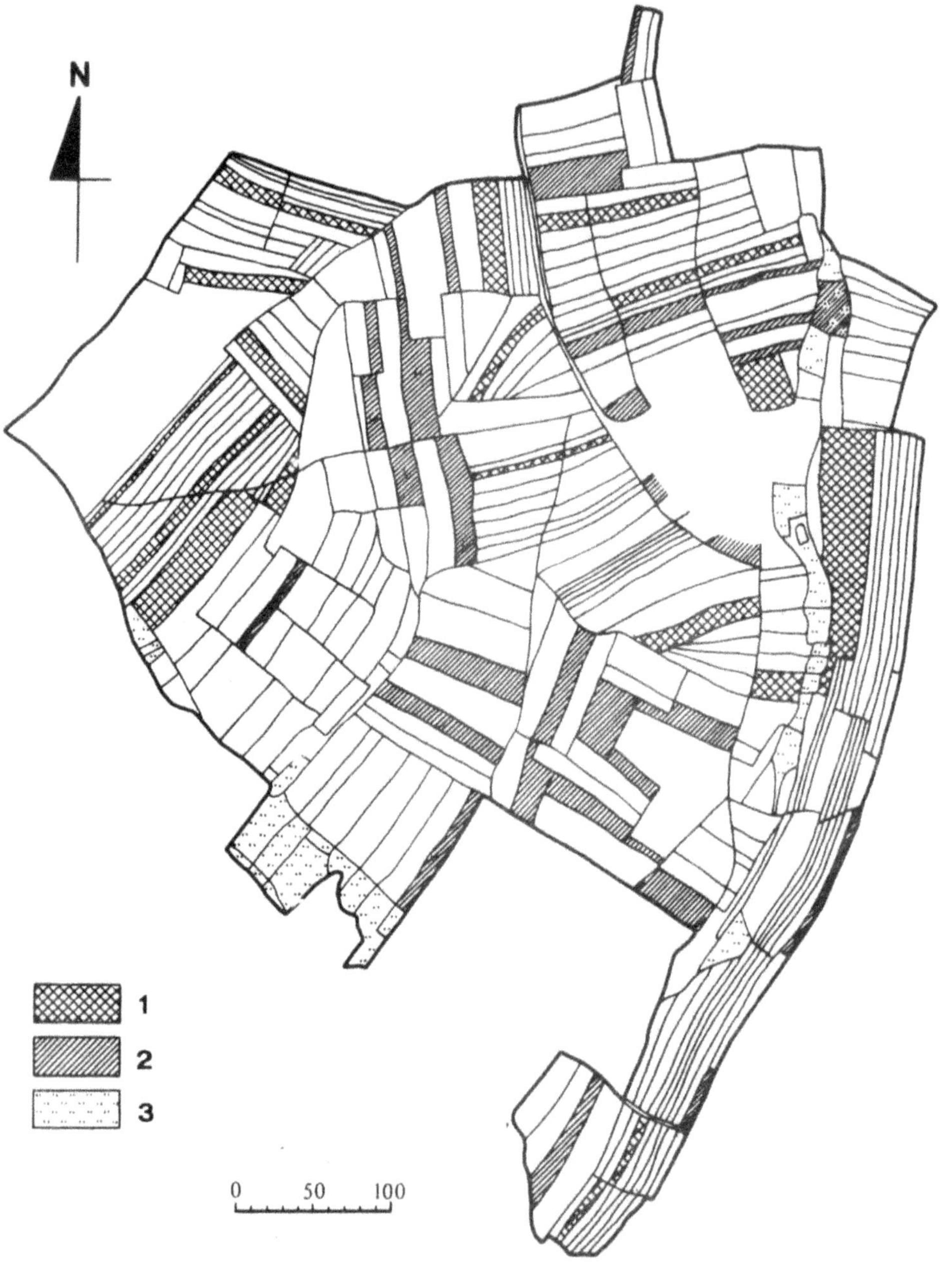

Fig. 14:1 Goraszowice, plan of 1827. The village was mentioned for the first time in 1284. The location on Polish law confirmed in the 15th century. Irregular layout; former blocks subdivided into property strips. Greatly dispersed strips.
1 - property of one peasant; 2 - property of another peasant; 3 - meadows.
The scale is given in Rhine rods.

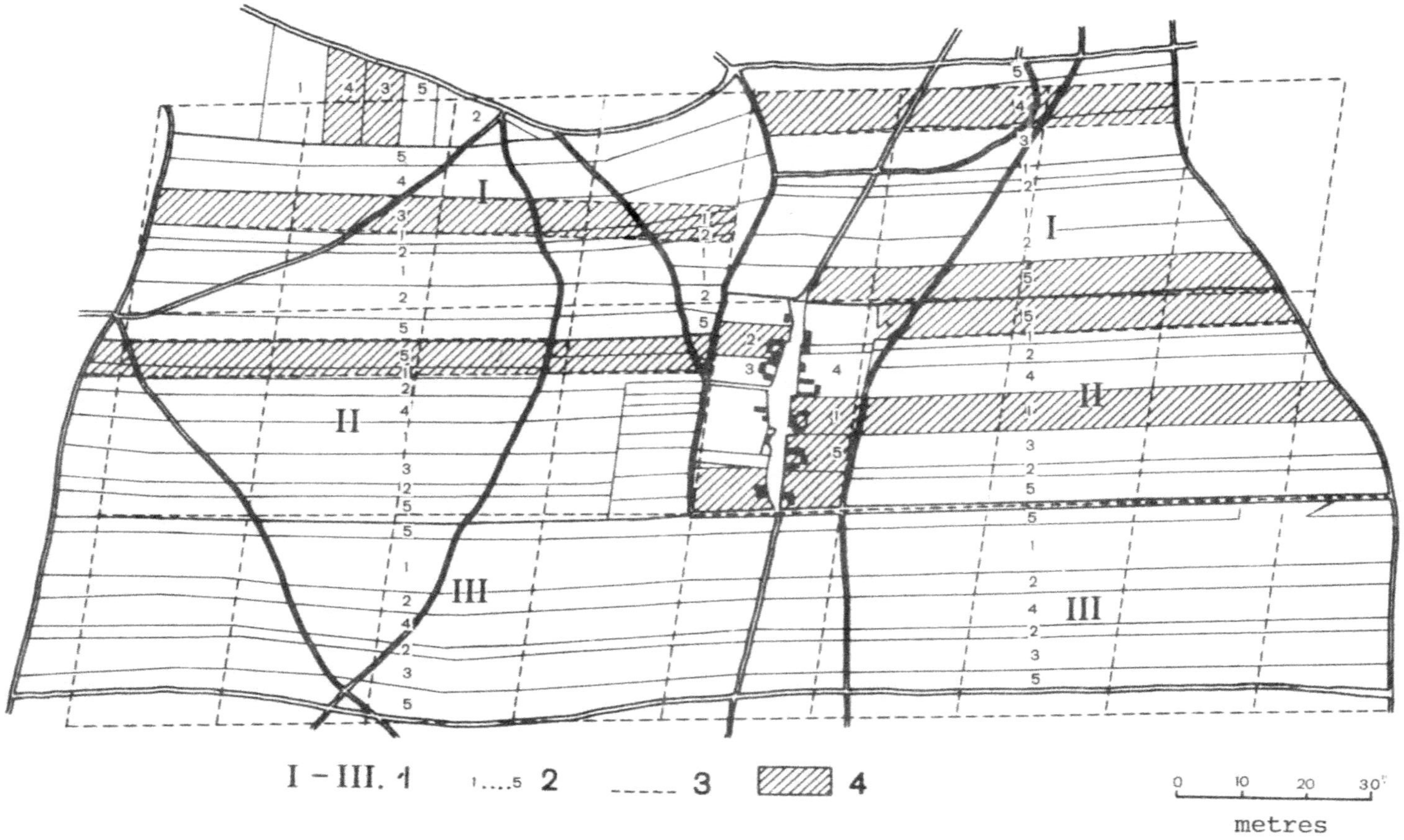

Fig. 14:2 Budziejowice, an attempt at the reconstruction of the principles underlying the planning of the village. The situation on the basis of the 1833 plan:
1 - I-III - fields; 2 - 1-5 - property strips;
3 - the introduction of construction lines in the plan; 4 - farmsteads and fields with the same width.

one of the oldest village types in Poland, relics of the Middle Ages, and are most commonly found in Pomerania (11).

Planted villages are a better known type, settlements which were regulated or even founded anew in the Middle Ages. The metrological-historical methods make it possible to distinguish them from the irregular villages, and to establish the frameworks around which they were constructed, and the measurements which were used to lay out the plans. In some irregular open-field villages the numbers of mansi identified in nineteenth century sources remains the same as those known to have been present before the process of regulation during the Middle Ages. For instance, in the village of Budziejowice (Fig. 14:2), mentioned in 1253, (12) the village plan is based on the principle of the regular, open, three course rotation field system. Its area - according to the acts of 1666 (13), amounted to 7.5 'mansi'. This size is corroborated in the village plan dating from the early 19th century and in the register of 1833. The regular pattern of the village is visible in many elements such as the size and layout of the fields, the situation of the site in relation to the village boundaries, the proportion of the length to the width of the site, recurrent widths of the farmsteads in the site and the widths of the fields.

Relics of such regular medieval villages with a street-green site and an open-field system have been preserved mainly in Silesia, in the southern part of Lubusz Land, partly in Greater Poland, in a large part of Pomerania and between Olsztyn and Bialystok.

Regular medieval villages also took the form of long double rows. These developed as a result of forest clearings, and their fields normally took the form of very long, parallel strips. The 'mansus' in Silesia was usually the Franconian 'mansus' measuring 24 ha. In the plans of certain double-row villages I have managed to reconstuct the number of mansi established at the time of the foundation (14). The relics of double-row villages in Poland are found in the Sudetes, the Carpathians, the southern part of the Lublin plateau, and also southwards of Zielona Gora.

Post-Medieval Developments

The sixteenth century saw the appearance of important changes in the socio-economic structures of village populations. The development of the manor, which then took place, had an effect on the existing settlement network and structure. This process was closely linked with the increase of prices of cereals, following the expansion of the export and internal trade with newly developed towns. Manors developed either in the centre of villages or on their peripheries, becoming an integral part of the village, and thus forming the manorial-peasant village. At the same time, also purely manorial villages were established. Buildings in the manor consisted of typical farm buildings and the residence of the owner; moreover, there was usually also a short street village inhabited by crofters. Manor fields were consolidated and formed large manor blocks, while crofters' fields occupied a small area divided into property strips.

Various types of colonies represent another form of settlement, for example, Dutch colonies founded at the Vistula estuary at the end of the 16th century and in the 17th century, or colonies founded after Poland's partition (at the end of the 18th century) on the initiative of the occupying states, Prussia and Austria. This new colonization wave, called 'Frederician' or 'Josephinian' settlement, overlapped the former settlement networks in the areas occupied by Prussia and Austria, that is in western and southern Poland. As a rule, those colonial foundations grouped together people engaged in the same occupation: farmers, handicraftsmen, foresters, miners, metallurgical workers. The function fulfilled by the colony bore a decisive influence upon its spatial pattern. Colonies of farmers were usually in the form of a very regular street village with a very compact construction. The field system was similar to that in the open-field village with broad, schematic strips, the proportions and size of which were however different than those in medieval villages (e.g. the Frederician colony of Piastowice (Fig. 14:3). (15).

Further transformations in Poland's agricultural landscape throughout the 19th and at the beginning of the 20th century were caused by the progress in agricultural techniques, as well as socio-economic changes. Changes in the village forms and the settlement networks were also different in Poland's separate parts, because of the different economic policy of the occupying states. A detailed discussion of this phenomena is however not possible within the framework of this paper. I will therefore give only one example. In Greater Poland, Pomerania and Silesia agricultural reforms, introduced in the early 19th century, in connection with the emancipation of the peasants, exerted a strong influence on changes which affected the village spatial patterns. The effect of the agrarian reforms was the consolidation of land within the villages as seen on the plan of the village of Ratnowice (Fig. 14:4). The separation of manorial and peasant land also took place.

The processes of reconstruction of the villages intensified in the inter-war period. They mainly consisted of the parcellation of the manors and land-consolidation. However, in Lesser Poland, fields were still dispersed and this patchworkk system was very troublesome. Moreover, the subdivision of farms brought about a greater density of settlement, the formation of hamlets and dispersed settlements.

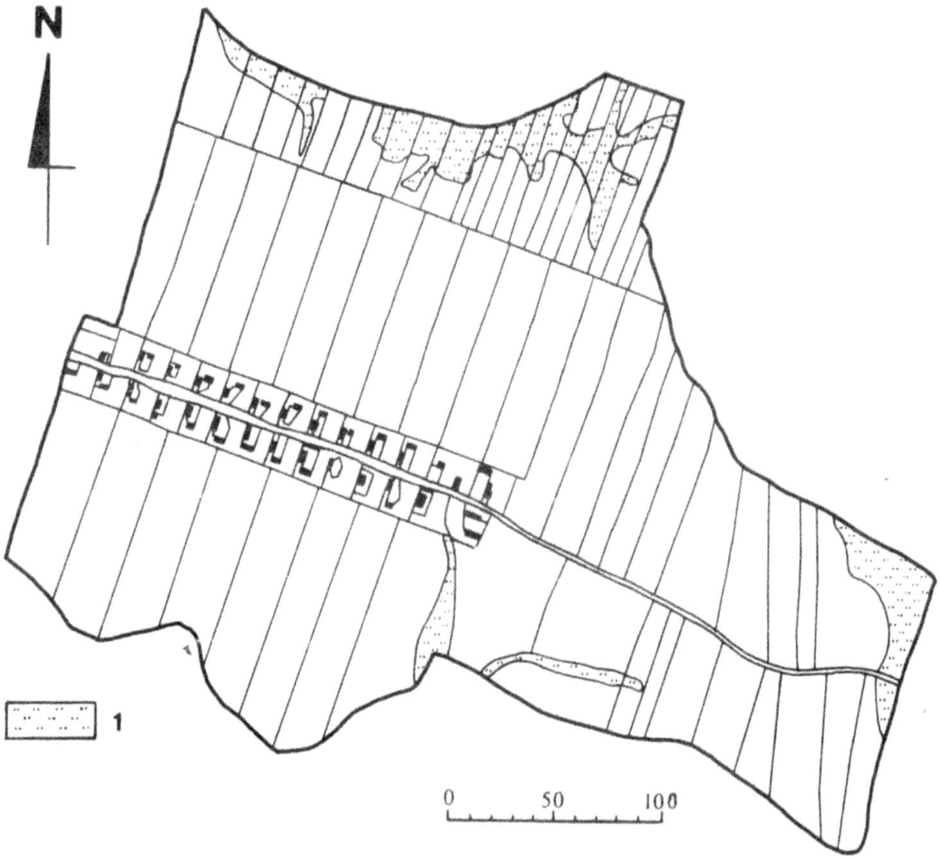

Fig. 14:3 Piastowice, plan of 1851. A Frederician colony. The regular layout of the site and schematic property strips. The scale is given in Rhine rods.

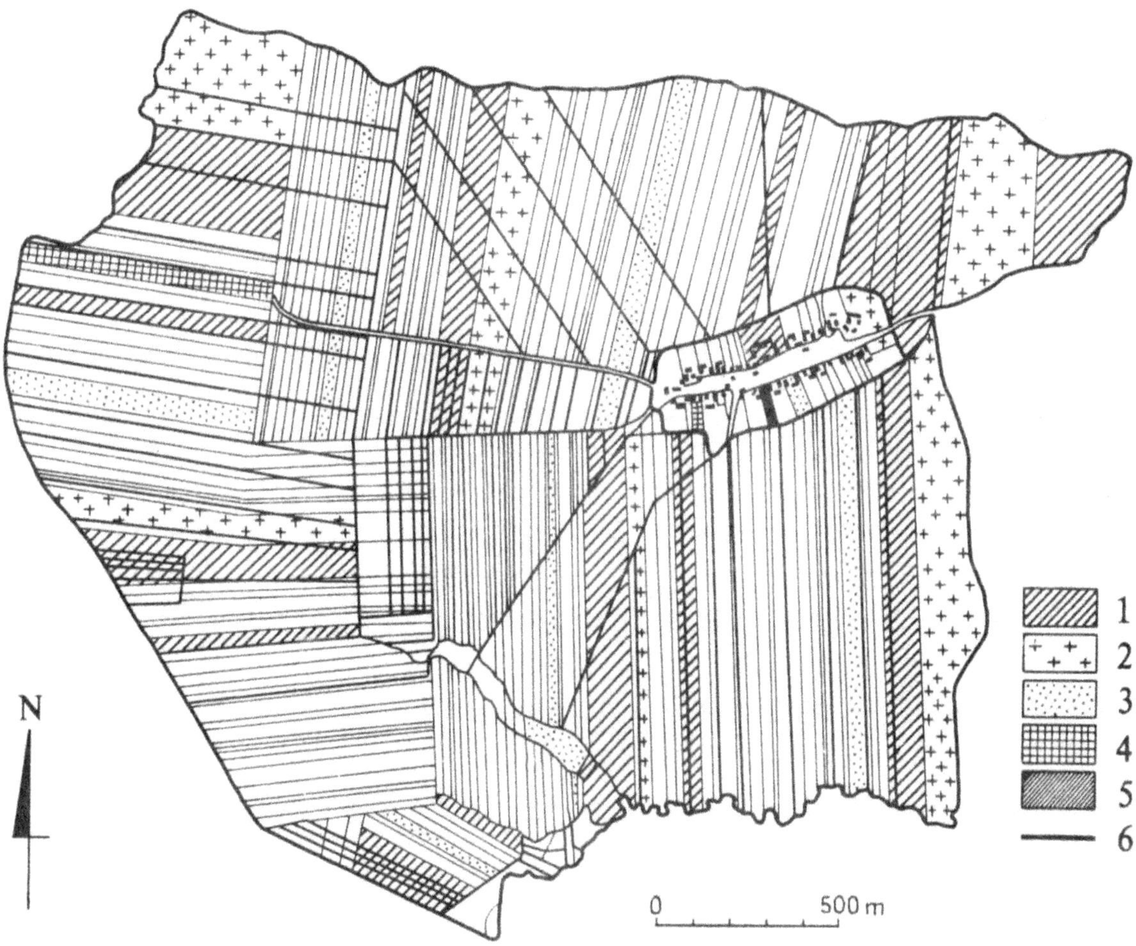

Fig. 14:4 The village of Ratnowice, located on German law in the Middle Ages. Example of a regular, open-field village before and after the regulation in the early 20th century.
1 - 5 - properties of five peasants;
6 - boundary of the strips after the regulation.

The new rural settlement forms in People's Poland have been brought about by changes in the socio-agrarian structure. The spatial changes which took place immediately after 1945 were caused by the parcellation of manor land and large peasant farms, and the establishment of new farms of agricultural farms as well as collective farms.

The scope of changes in the village spatial patterns differed from region to region. The biggest changes affected the northern voivodships (Szczecin, Koszalin, Slupsk, Elblag and Olsztyn): socialized farms (i.e. state and collective farms) account there for from 40 to 60 per cent of the total agricultural land.

It is of interest to investigate to what extent the earlier layouts of sites and field systems have been retained in the present-day villages. Spatial transformations in the villages in which farms are owned exclusively by individual peasants are irrelevant. As a rule the typical phenomenon is the subdivision of land within the site and fields. In double-row villages fields are subdivided vertically, and when the vertical subdivision is prevented by the width of the strips, the fields are cut horizontally and a new line of farmsteads is built parallel to the former one. Thus, the former pattern of a double-row is obliterated, and it becomes transformed into a patchwork system. In the area where sites were greatly dispersed, mainly in the Sub-Carpathian region, the consolidation of land was started after the war.

The greatest spatial changes take place as an effect of changed ownership. For example, the organization of collective farms in villages inhabited by individual farmers bring about changes in the village site. As a rule, the number of farmers in the collective farm is much bigger than the earlier number of farmsteads. Again, this makes it necessary to divide the farm-steads within the site and to construct the second line of farmsteads outside the former site (16).

A majority of state farms are also organized on the basis of the former settlement layouts, which initially are adapted to the state farm needs and in future will be destroyed (17). In the post-war period, besides the various forms of transformations, new settlment forms, new collective and state farms are also developing which are endowed with residential, social and services, as well as production centres. Often they have also recreation grounds, such as parks, sports grounds and playing fields. However, the number of newly established agricultural collective and state farms is not large.

CONCLUSION

It seems that the described study of the morphogensis of rural settlements, situated mainly in Silesia and Pomerania, may be summed up by the following conclusions:

1. Settlements have continued to develop on fertile land since the early Middle Ages.
2. The metrological analysis makes it possible to differentiate the irregular from the regular village, and to establish the period in which field and settlement layouts developed.
3. Not many relics of the early Middle Ages have been preserved either in historical plans or in the present-day villages. The historical plans or field research show only the traces of former settlements but not the former settlement forms. It is only the application of the new methods such as morphogenetic mapping, soil analysis, air photography, the analysis of the village historical metrology, which makes it possible to identify the patterns of the old sites and fields.
4. The most regular layouts have been found in the villages which were the property of princes, landlords and monasteries (predominantly of the Cistercian Order). This form of ownership largely contributed in Silesia and Pomerania to the formation of regular settlement structure throughout the Middle Ages.
5. Transformations in the old forms of villages, taking place in People's Poland, are closely associated with the genetic forms, the rate of transformations, and the current state of husbandry. Studies of the transformation types are interesting from both the cognitive and practical viewpoints. The main object of such studies is to establish in what ways the new layouts of fields and sites overlap the former ones and what extent it is possible to adapt the former layouts of sites and fields to the new forms of mechanized, big-scale and socialized economy.
6. Maps of morphogenetic types of villages and their transformations may well serve to sum up the model studies of the morphogenetic types of villages. I have made an attempt at the preparation of such a map for Opole Silesia (18). I would like, however, to make a map of morphogenetic types for Poland as a whole, and to take part in the all-European discussion on the terminology and models of village morphogentic types. This may eventually lead to the preparation of a map of European morphogenetic types.

REFERENCES

1. In my paper I am making a reference to my article, 'The Development of the Agricultural Landscape of Poland', Geographia Polonica 22 (1972), 85-103.
2. Atlas Historyczny Polski, Polska Akademia Nauk, Instytut Historii, (Polish Academy's of Sciences Historical Institute), Wojewodztwo Plockie okolo 1578 r. (Plock voivodship around 1578), S. Herbst (ed.) (1958); Prusy Krolewskie w drugiej polovwie XVI w. (Royal Prussia in the second half of the 16th century, by M. Biskup (1961); Wojewodztwo Lubelskie w drugiej polowie XVI w (Lublin voivodship in the second half of the 16th century) by S. Wojciechowski (1966); Mazowsze w drugiej polowie XVI w.(Mazovia in the second half of the 16th century W. Palucki (ed.) (1973); Slask w koncu XVIII w. (Silesia at the end of the 18th century) J. Janczak and T. Ladogorski (eds), (1976).
3. H. Szulc, Osiedla podwroclawskie na poczatku XIX w. (Sum. Suburban settlements in the vicinity of Wroclaw at the beginning of the 19th c.), Wroclaw-Warszawa-Krakow 1963, 107 + 4 maps. H. Szulc, 'Typy wsi Slaska Opolskiego na poczatku XIX w. i ich geneza' (Sum.: Types of rural settlements of Opole Silesia at the beginning of the 19th century and their origin), Prace Geogr. IG PAN, 66 (Warszawa 1968), 108 + 2 maps. H. Szulc, 'Studies on the Silesian village in the light of plans from the beginning of the 19th century', Kwart.Hist.Kult.Mat., 16(1968), 4, 621-639.
4. H. Szulc, 'On the Origin and Development of Small Green Villages in the Region of West Pomerania', I Paesaggi Rurali Europei, Atti Del Convegno Internationale Indetto a Perugia dal 7 al 12 Maggio 1973 Dalla, Conference Europeenne Permanente Pour L'Etude Du Paysage Rural (Perugia 1975), 493-506. H. Szulc, 'Regular green villages in West Pomerania', Geographia Polonica 38(1978), 265-270.
5. H. Szulc, 'Charakterystyka osadnictwa wiejskiego powiatu gorlic kiego, (Sum.: Description of rural settlement in Gorlice district), Dokumentacja Geograficzna 2(1974), 11-26.
6. K.H. Schroder and G. Schwarz, 'Die landlichen Siedlungsformen in Mitteleuropa, Grundzuge und Probleme ihrer Entwicklung', Forschungen zur Deutschen Landeskunde, 175(1969).
7. On the subject of new methods applied in historical geography please see my following papers: 'O nowych drogach badan w geografii historycznej' (Sum. About the new methods of research in historical geography), Czas. Geogr. 35(1964); oraz 'Badania geograficzno-historyczne nad osadnictwem wiejskim w Skandynawii' (Sum.: Geographical and historical research on the rural

settlement of Scandinavia), Kwart. Hist. Kult. Mat., 17(1969), 3,473-492.

8. H. Szulc, Typy wsi ... 42 and 45.
9. Liber fundationis, Registrum Nissense.
10. H. Szulc, On the origin and development of small green villages ...
    H. Szulc, Regular green villages ...
11. H. Szulc, Osiedla podwroclawskie ...
12. Codex diplomaticus Silesiae, regestr 839.
13. Visitationsberichte der Diozese Breslau, (ed.) Jugnitz, v.I, (Breslau 1902)
14. H. Szulc, Typy wsi Slaska Opolskiego ... 54, 55, 56.
    H. Szulc, Studies on the Silesian Village ... 627.
15. H. Szulc, Studies on the Silesian Village ... 61.
16. I have described this village in two time cross-sections 1949 and 1955 in my paper: 'Les changements de formes de l'habitat rural en Pologne', Recherches de Geographie Rurale Hommage au Professeur Frans Dussart, (Liège 1979), Figs.9 and 10.
17. I have described the state farm which was set on the area of the former manor in my paper: 'Changes in the layout of Polish Villages with special regard to the last thirty years',Rural transformation in Hungary and Poland. A Polish-Hungarian Academy of Sciences, Budapest (1979), 149-165.
18. H. Szulc, Typy wsi Slaska Opolskiego ... map 'Typology of rural settlements in the Opole Silesia at the beginning of the XIX C'.

CHAPTER 15

BARKNÅRE - SWEDISH HAMLET IN A MARGINAL AREA BUT AT THE CENTRE OF AN INTERDISCIPLINARY RESEARCH PROJECT

B. Windelhed
(University of Stockholm)

The hamlet of Barknåre was used by Sporrong (1983, in this volume) to exemplify the forest settlements in the marginal area of the region around the Malar basin in central Swed-n.

This hamlet, located in Hållnäs parish, is the study area for a present-day Swedish research project (see Fig 15:1). The agrarian landscape in Hållnäs was formed during the Early Medieval Age, the hamlets are small and lack regulations otherwise common in central parts of the region. Production has traditionally been based upon stock-raising, with a modest grain production combined with fishing and forestry.

For studies of the agrarian landscape there are advantages in chosing a marginal area such as Hållnäs. Firstly, we can find here older agrarian patterns well-preserved in the landscape. These patterns are a useful agrarian-historical empirical material because important structures in the Swedish agrarian landscape were created during periods lacking written documents. Secondly, the parish is geologically comparatively young due to the land elevation. This means that we can follow the settlement development from the colonisation phase through diffusion towards more stable phases (these concepts are discussed by Sporrong 1983).

The title of this project, 'The Individual, The Society and the Agrarian Landscape', accounts for those relationships we are studying at Hållnäs. The anlysis of the agrarian landscape will include changes and influence with respect to both individual's and the society's ideas and decisions. The project is interdisciplinary. Archaeologists, historians, ethnologists, human geographers and natural scientists are all working together.

The purpose of my part of the project, the human geography, is to describe and explain the settlement development and land-use at Barknåre. Primarily, the empirical material consists of maps and information from fossilised traces of agrarian activites found in the field. The study has a retrospective view starting with an excellent cadastral map from 1752, rich in details (see Fig 15:2). From this time it is possible to follow the development backwards during a period of 1 000 years 'down' to the Viking Age when, according to present knowledge, the first stable settlement was established in this area.

The aim is to interpret changes in the location of

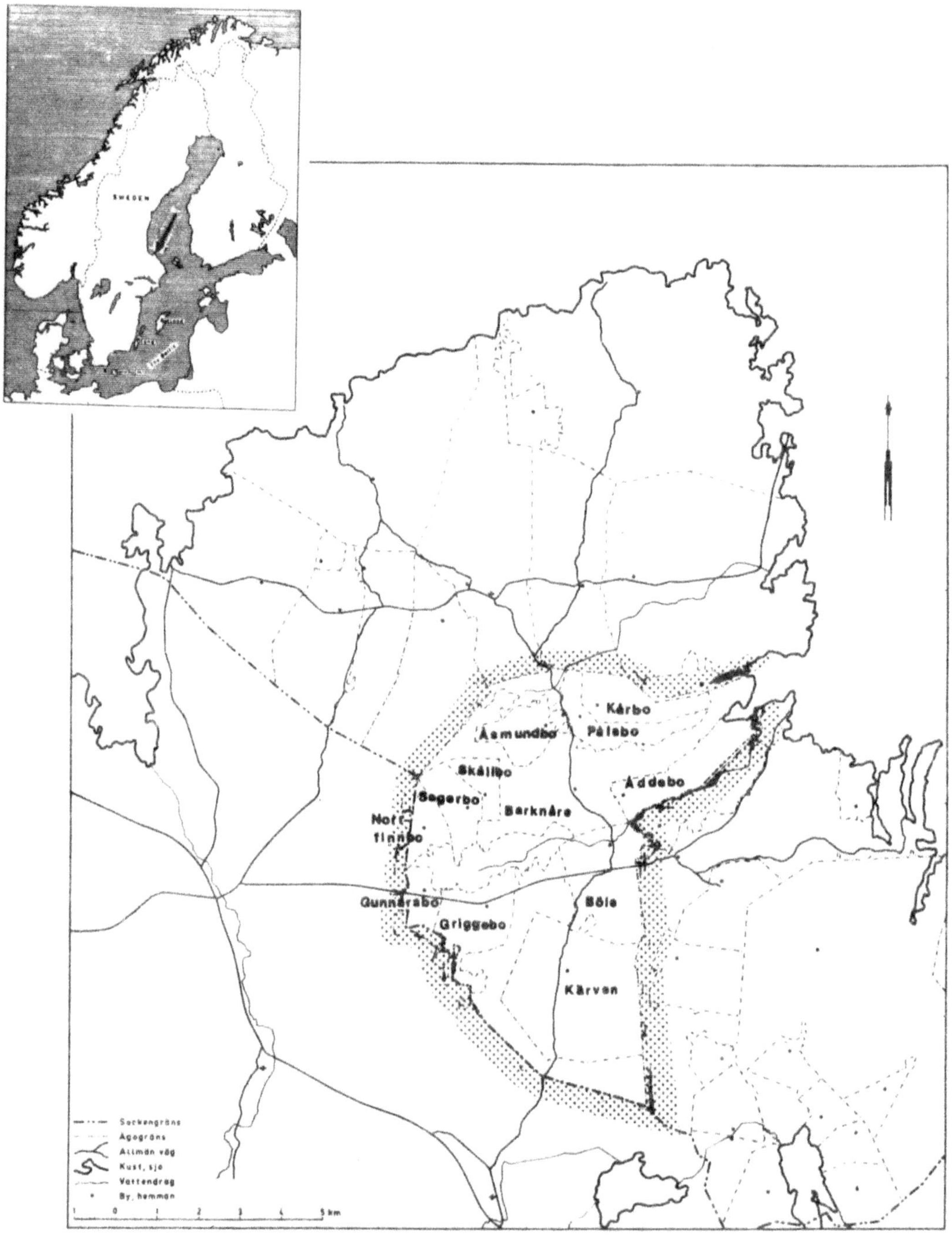

Fig. 15:1 Map showing the study area, Barknåre hamlet in Hållnäs parish in Sweden. The parish is situated along the coast in the northern part of the province of Uppland, about 200 km north of Stockholm.
On this parish map the Barknåre primary unit is marked together with Barknåre hamlet and the <u>boda</u>-units.

settlements and cultivated lands, production and farming techniques. By following the exploitation of resources through time, I hopefully will be able to contribute to one of the general problem complexes within the project, namely the importance of the processes of social change in the creation of the agrarian landscape. Conversely this can be formulated as follows: How can different physical phenomena in the agrarian landscape be explained in relation to the structure and the function of the (local) society during different times?.

To simplify the presentation of problems within the study area, I will here define some of the most important settlement categories according to the map of 1752.

The study area, which covers about 60 km (2), will be called Barknåre primary unit (see Fig.15:1). Located here is Barknåre hamlet, the real hamlet with 8 farms located in the settlement area, along with two large fields and other production lands (see Fig.15:2). Within the hamlet the land was collectively used, but privately owned. The hamlet was surrounded by 12 independent settlement units which in 1752 consisted of usually one and, at a maximum, 5 farms. Because many of them have names like Pålsbo and Kårbo we call them boda-units (Fig.15:1; see BRINK in this volume).

Barknåre hamlet can be seen as the result of an early colonization in the area. Here we find graves from the late Viking Age. The boda-units can be interpreted as diffusion within the territory of the primary unit. The later establishment of the boda-units, in comparison to the hamlet, is supported by their lack of prehistoric graves and by interpretation of the early written documents from the area. It seems that the boda-units were located at the periphery of a mother unit, covering the area we call the Barknåre primary unit.

According to evidence from place-names Brink (1979 and 1982, in this volume) has, for areas north of the Mälar basin region, discussed the development of settlement units with boda-names in an agrarian system. To extend the resource area for the permanent farm, periodic settlement forms were used, either for arable farming (so called SW. bodland) or for cattle breeding (so called SW fäbodar). In time the areas used extensively for arable farming became independent farms. A corresponding development has also been proposed for the Barknåre primary unit (DAHLBACK 1974), which will be used as a general hypothesis in my study.

FIELD WORK

The following section will show how the field work has been done, in order to estimate when the different settlement units were established and also when and how the land use

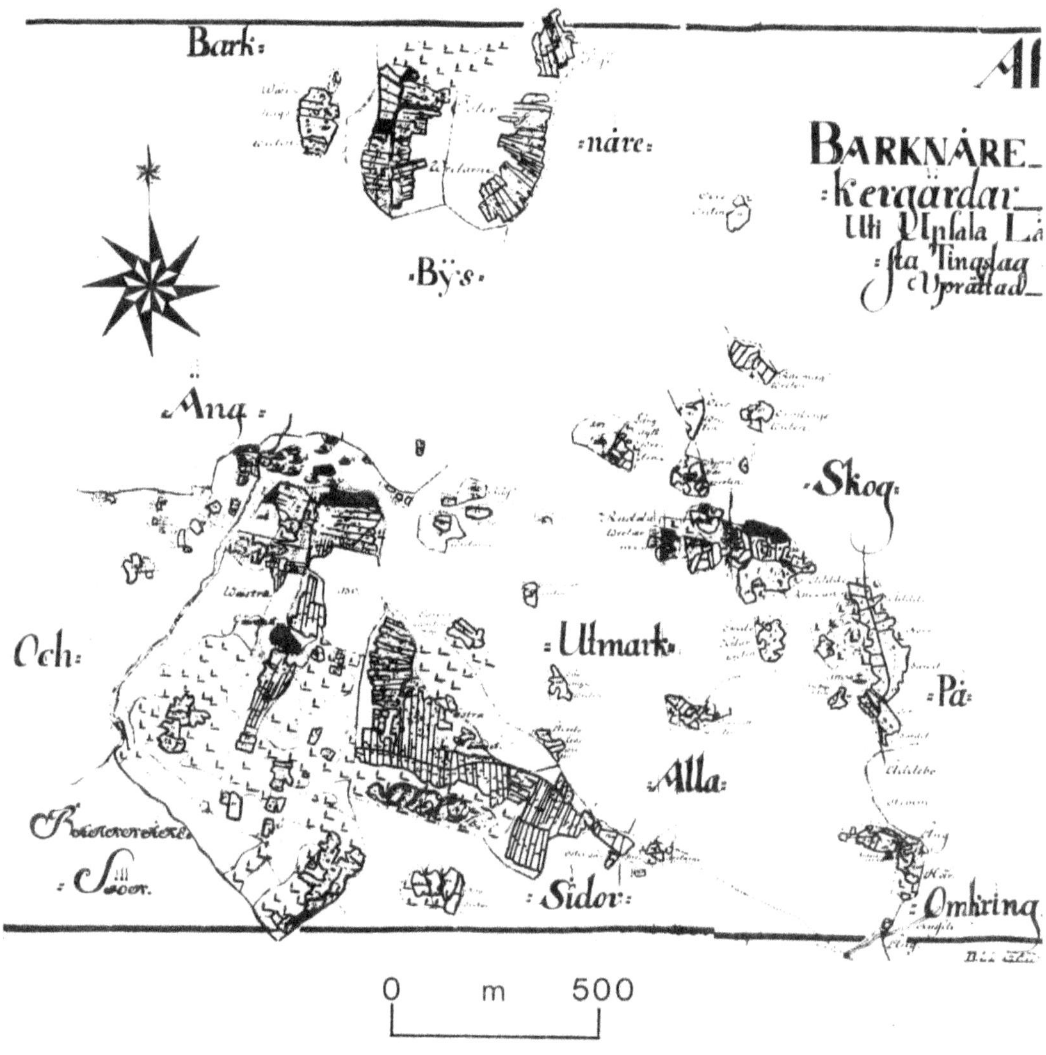

Fig. 15:2 Part of a map of 1752 showing the Barknåre primary unit. The map shows the central parts of the Barknåre hamlet. To the left of the middle of the map is the site of Barknåre hamlet. Further south are the hamlet's two fields with arable lands and meadows. To the right of the fields are some vretar, small plots in the forest area.

has developed around the units. To limit the time-demanding field-work attention must be concentrated on excavation of objects in the agrarian landscape which describe both the generality and the originality of the different settlement units. The principle for the field-work is that information from excavated 'key-objects' can be projected to the surrounding areas. This selection is possible with prior knowledge of the ideas of planning incorporated in the agrarian landscape.

In map studies and field-work special attention has been paid to preserved traces of medieval forms of organisation, mapped in the 18th century and suitable for future excavations. This concerns primarily objects such as settlements, cattle paths, field boundaries, heaps of clearance stones, stone fences and arable lands, all of which form characteristic patterns. This will also provide a starting point for studies of more spontaneous events, for instance expansion to the outlying areas of both cultivation and meadow lands. An important step in this selection is to place Barknåre in a larger context. The Swedish cadastral maps here provide the opportunity of comparison with other hamlets in the parish and in the region by synchronized map series from the 18th century.

FIELD SURVEY

In Sweden there is a national register and survey of ancient monuments including preserved physical remains from prehistoric and historic times. In spite of this survey it is necessary in a detailed study such as this to carry out a field-survey oneself. This gives local knowledge of the terrain, a useful understanding of the patterns in the agrarian landscape, and allows the discovery of new features, all providing an invaluable foundation for future excavation strategies.

Surveys done within this project, both in the parish and in Barknåre, have produced a more complicated picture of the ancient remains than the official register. The existence of newly discovered prehistoric cemetaries, fossilized fields and prehistoric settlements have created a need for reformulating the original hypothesis concerning the early settlement development. During these surveys it was especially possible to register traces of farming, either abandoned and still in use, within the fields and the meadow lands mapped in the 18th century. As a basis for survey, a modern economic map (scale 1:10 000) was used, covered by a transparency with a simplified picture of the 18th century map.

The survey showed that the features included in the map from 1752 were normally incorporated in lands used today. In exceptional cases, where these objects are abandoned, they are preserved and easily visible in the terrain. The lack of larger areas with abandoned features demonstrated the essential continuity in land use. Thus those prehistoric and early medieval features discovered in the area normally ought to have been included in, or covered by, the agrarian landscape mapped in the 18th century. The great significance of prehistoric graves as an instrument for dating the establishment of the settlement was shown above. The presence of graves also gives us an indication of the former presence of nearby early settlements, fields and fences, which may be difficult to identify in the agrarian landscape of today.

FIELD-WORK WITHIN THE BARKNÅRE HAMLET

A central problem in the understanding of the development of the primary-unit is the establishment and the growth of Barknåre hamlet together with its production lands. The hamlet is the largest settlement unit within the primary unit and probably also the oldest. Investigations of the cemeteries close to the hamlet, suggest that the hamlet was established towards the end of the Viking Age.

In two alternative interpretations of how a hamlet in the region has been developed it has been proposed that a hamlet such as Barknåre is either a result of a more or less extensive immigration or is the result of a spontaneous division of an earlier mother farm.

We have tried to test both of these interpretations. The investigations which concerned areas around the site of the settlement, were based upon certain assumptions: like those of today, the early farms in the area are presumed to have been located on high and dry areas, on sites surrounded by nearby easily drained soils for cultivation and in close contact with lands for meadows and grazing lands. According to our present-day knowledge of the land elevation in the parish, areas which today are higher than 10 m above sea level lay above the 7th century shoreline, while the shoreline of the 13th century is today 5 m above sea level. Thus areas about 10 m above sea level have been called 'potential settlement areas', where there are suitable locations for early settlement, together with required production-lands.

Within a radius of about 1 km from the site of Barknåre hamlet, six potential settlement areas were found, which today are situated partly in arable lands. These areas could have been early independent farms, where the buildings in a later phase were moved in to the site of the present day hamlet.

In each area 10-20 test-holes were excavated in a line from the highest point down the slope. Each test-hole was documented concerning the different soil layers from which soil-samples were taken. No positive traces of cultural layers from settlements were found. A possible exception is a thick soil-layer and relatively high phosphate-values in the Western

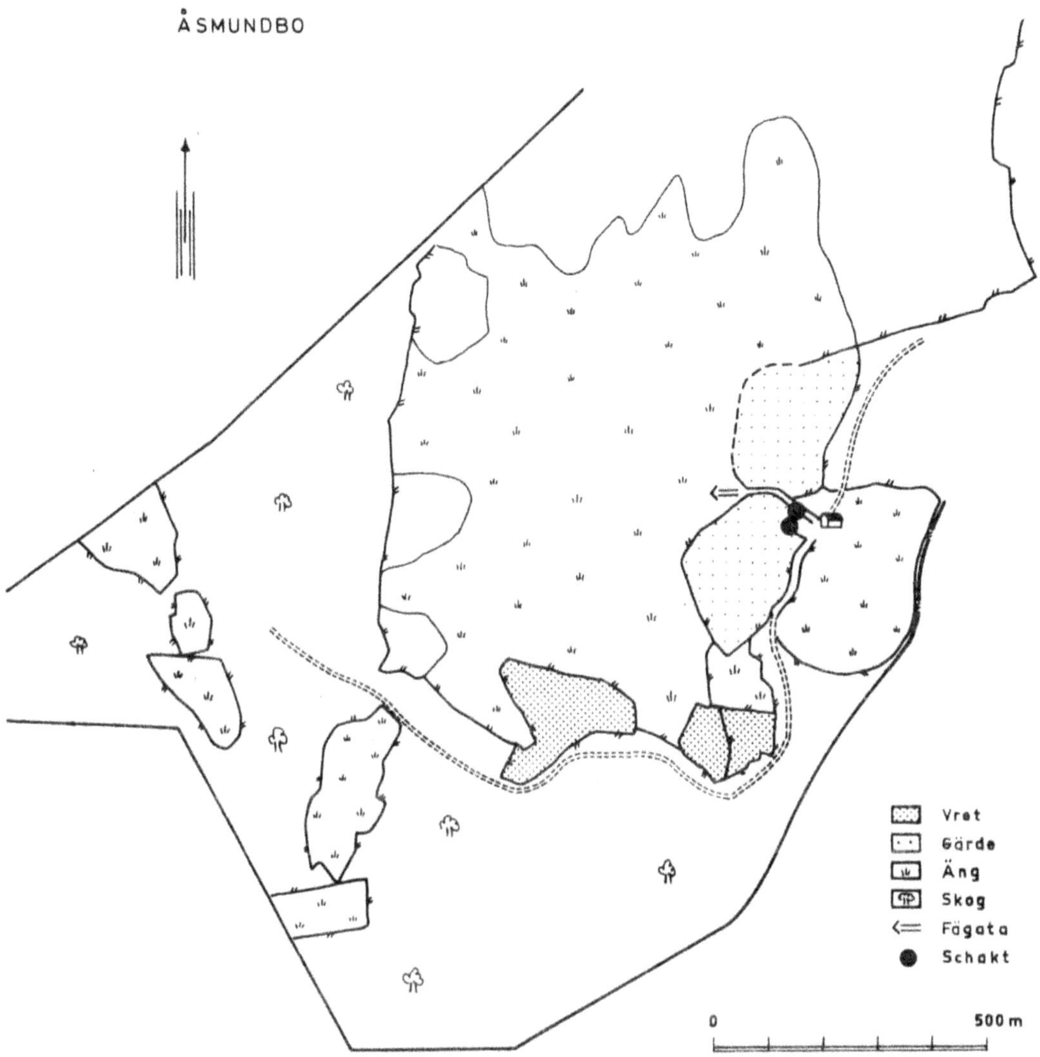

Fig. 15:3 Åsmundbo according to a map of 1752.
From at least 1570 there was one farm here. At the beginning of the eighteenth century the farm was divided into two farms of equal size. The two fields (gärde) are situated on the clay soils in the bottom of a narrow valley. From the settlement there is a cattle path which separates the fields. The location for the settlement, the cattle path and the two fields seem to have been coordinated in one planning phase. Later land elevation did not produce any more improvable land (compare Fig. 15:4).

Field of Barknåre hamlet.

Preliminary analysis of the soil samples have shown interesting differences, for instance, when concerning phosphate quantities. Briefly it can be mentioned that the concentration of phosphate in the strip-field system south of the site of Barknåre hamlet have a comparatively high phosphate concentration, while smaller and scattered plots within the two fields have lower concentrations of phosphate. Much lower still is the phosphate contained in the small and relatively recently cleared arable lands SW vretar outside the fields (see Fig.15:2). Probably these variations in phosphate for different types of arable lands should be interpreted both as chronological differences and as differences in the amount of manure used on the fields.

The 150 soil samples collected, which originate from different types of land-use, are now being studied in a more advanced way. For every soil-sample measurements of physical and chemical properties have been made, such as distribution of soil particle size, volume-weight, susceptibility (a measure of magnetic property of the soil), phosphate quantities, pH values, degree of darkness, organic quantities, charcoal contents and frequency of 'artefacts'. Variations in these measured values can depend up on differences in human activities. Through statistical research we can thus classify the soil samples and try and judge why these differences appear. This kind of research can be compared with the human geographical interpretation of the land-use.

In this way, we get independent interpretation which may later be used for a functional analysis of different types of land-use, for instance, concerning manuring of arable land.

Even if much work still remains to be done within this subject, the results from Barknåre reached to date and, above all similar work already carried out on material from Gotland have shown promising results.

With the purpose of creating a foundation for future excavations within the site of the hamlet a series of test-holes have been excavated in order to locate older cultural layers. Excavations of a site still in use are a problem. In addition to the fact that the size of the settlment area creates problems upon excavation, the inhabitants have during a long period of time influenced and changed the soil at the site. Within the tofts of Barkåre hamlet, situated on a ridge, the farmers have also filled out large areas with earth for gardens. In the test holes the soil samples taken among other things it will be possible to create a phosphate map. In order to interpret this map we have, through studies of 18th and 19th century cadastral maps, extracted information about the existence and function of early houses, along with the land-use within the site. This information has been complemented by interviews

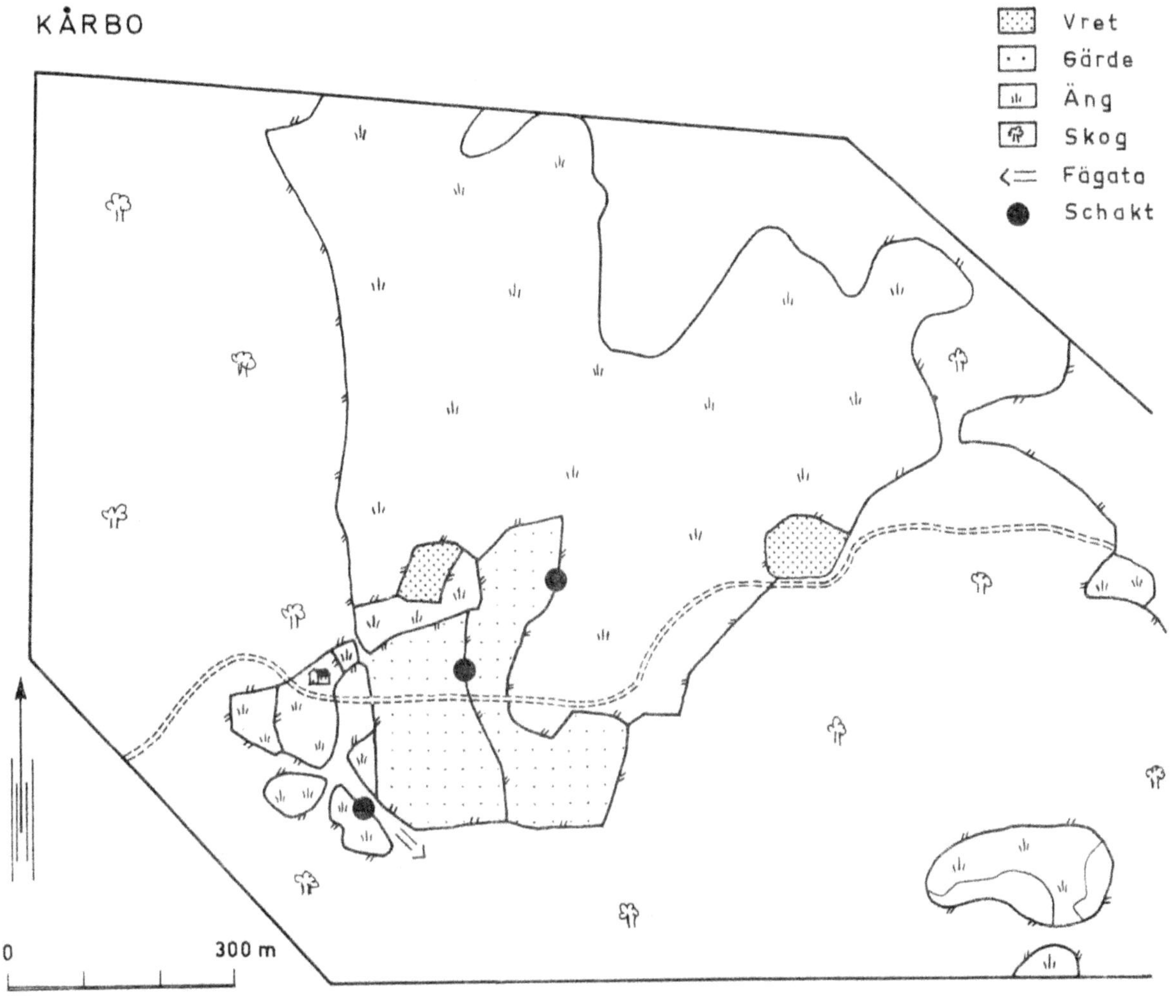

Fig. 15:4 Kårbo according to a map of 1752.
Here was one farm <u>at least</u> from 1570. During the middle of the seventeenth century the farm was divided into two farms of equal size. From the settlement, located quite high in the terrain the land slopes eastwards. Both the fields (<u>gärde</u>) lie on the slope. The meadows (<u>ang</u>) east of the eastern field lie along the bottom of a small valley which runs from north to south. The bottom of the valley lies today about 5m above sea level, that is approximately the shore-line of the thirteenth century (see the text). After the establishment of the settlement, land elevation may have provided new land which made possible the change from one to two fields (compare Fig. 15:3).

with people living in the hamlet.

FIELD-WORK WITHIN THE BODA-UNITS

The field work carried out so far has had a general character, with map studies, surveys, phosphate probing, test-holes systems and excavation of trenches. With two examples from the investigations of the boda-units I would like to show how these levels of investigation have been combined to give information about the agrarian landscape.

In Figs. 15:3 and 15:4 we can first study the organization of the agrarian landscape around the boda-units. From the settlement a cattle-path often leads through the field-area. The arable lands, usually in two-course rotation are only exceptionally in a strip-field system. The arable land areas are generally small with large quantities of clearance stones. The boda-units have been organized according to ideas recognized in Barknåre hamlet.

Preceding field work, hypotheses about the development of the unit have been formulated. For Åsmundbo we have assumed a short and well directed planning-phase after settlement. It seems that Åsmundbo right from the beginning was arranged according to the idea of two-course rotation (see Fig.15:3).

Kårbo may illustrate a boda-unit which may have had a more complicated development. Here land elevation provides possibilities of expansion to a development from a one-field to a later established two-course rotation (see Fig 15:4).

During 1981 I carried out excavations in 6 of the 12 boda-units. One to five trenches in their landscape were located as is shown in Figs. 15:3 and 15:4. In the trenches charcoal was collected for C14-dating of such important events in the agrarian landscape as clearance, field-fencing and parcelling. In these primary investigations, I have avoided excavations of actual settlement areas, where more layers will result in a greater chronological complexity.

The excavations were placed in the stone-material which, for instance, surrounds a field, forms a fence for a cattle path, makes up a heap of clearance stones or a boundary between two strip-fields (see Fig.15:5).

In these stony areas it is remarkable that the farmers normally have used wooden fences and only very seldom stone walls. The stone-material is in fact clearance stone, thrown into long rows, on which wooden fences have then been built. Under the stones is a 10-15 cm thick soil layer over the subsoil. C14- datings of charcoal from this soil layer have traditionally been used for dating stone fences. The charcoal has been believed to come from fire-clearance of the land, immediately preceeding the construction of the stone fence. The

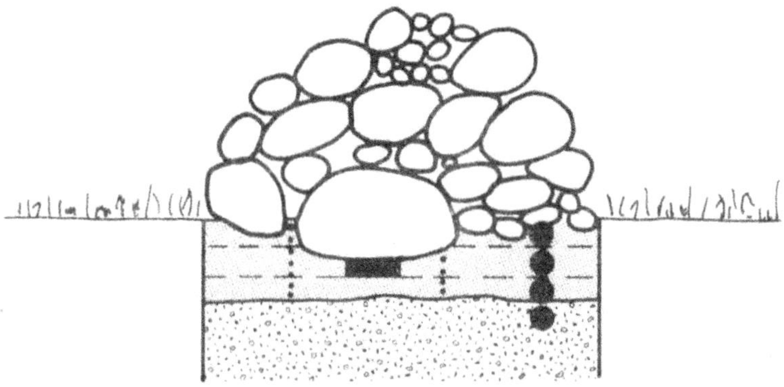

— — Sifting layers. For collecting artefacts and charcoal for C 14 datings.

Soil sample For analysis of fossilized soil under large boulders

For chemical and physical analysis

For pollen analysis

Top soil

Sub soil

Stone construction for instance a fence, field-boundary or heap of clearance stones

Fig. 15:5 Section through and beneath a stone construction.

purpose with the excavation of these trenches has primarily been to obtain datable objects, and secondly, to characterize the layers from which the C14- datings originate.

To refine this procedure, the fossilized top-soil layer under the stones has been excavated in two, or even more levels. The soil has been sifted in layers, and charcoal, bones and other artifacts have been collected. The trenches are standardized, so that the excavated areas have varied between 1 - 2 m(2) depending upon the charcoal content. The layers identified in each square are the basis for C14-datings, and it is hoped that these will show if the charcoal comes from one or several burnings. In order to characterize the levels dated by C14, more precisely, different kinds of soil analyses will be done on the layers:-

1) Analyses of the charcoal. This may show which kind of trees and which parts (branches, trunks and roots) have been burned.
2) Analyses of pollen grains in the soil.
3) Analyses of fossilized soil under large boulders.
4) Analyses of the soil samples for measuring chemical and physical properties in the soil.

These analyses will allow us to interpret the land use before the stone-fences were built. These interpretations will be combined with pollen diagrams from bog sites in which it is possible to study the more long-range changes in the use of the area, and the development of the vegetation in the parish. In addition, for the isolated boda-units in the forest area we shall also establish some local pollen diagrams. Together with the above mentioned field-work this will hopefully give us a good picture of the development of the agrarian landscape in Barknåre.

REFERENCES

BRINK, S. 1979 Bodlanden i Järvsö socken Uppsala studier i namnforskning. Ortnamn och samhälle 5 (Uppsala).

BRINK, S. 1983 Place - name evidence for rural expansion in marginal areas (present volume).

DAHLBACK, G.1974 Landhöjning och bebyggelse i nordligaste Uppland (Fornvännen).

SPORRONG, U.1983 Plains and Periphery: Settlement morphology in the Mälar basin, central Sweden (present volume).

Acknowledgments

I have to thank Phyllis Anderson, of Stockholm, for her invaluable help with translating this paper into English.

CHAPTER 16

EAST JUTLAND VILLAGES IN THE MIDDLE AGES AND IN 1683, WITH SPECIAL EMPHASIS ON THE ALLOTMENT OF LAND AMONG VILLAGERS.

E. Ulsig
(University of Aarhus)

The landscape around Aarhus along the East Jutland coast is young fertile clay moraine with a - for Denmark - rather hilly terrain (0-120m). The area under discussion consists of four medieval herreder (hundreds), the same four which still exist today, though only as a basis for the registration of landed property and topographic description. The area can be divided roughly into two zones: (1) Hasle herred and Vester Lisbjerg herred (2) Oster Lisbjerg herred (leaving out the most remote, less fertile Eastern part) and Ning herred (Fig.16:1). Zone 1 is characterized by its ancient place names, fairly big villages, and small parishes (only 8,5 km^2 per medieval parish), that represents the highest concentration of village churches in all Denmark. Zone 2 has opposite characteristics: many younger place names (thorps), mostly small settlements, large parishes, in Oster Lisbjerg herred 20,8 km^2 to each church, in Ning 14,0 km^2. The interpretation must be that zone 1 was fully settled when the parish organisation was completed in the 12th century (probably in the first half), whereas zone 2 experienced a wave of new settlement after the completion of the parish organisation, though centainly before 1250. In 1683, however, according to the Great Land Register, the whole area was very well-cultivated with nearly half the acreage being arable. The medieval pattern remained, however, clearly visible in the size of the settlements. In zone 1 the average village had 11,8 farms, in zone 2 the average was 8,0 for Oster Lisbjerg herred and as low as 5,5 for Ning. Furthermore it should be noticed that, with few exceptions, the villages of zone 1 used a regular three-field system whilst preliminary studies of zone 2 show a much greater variety in the field systems. The difference in age and pattern of settlement and in field system between the two zones was, in my opinion, due to slight, but noticeable geographical differences. The landscape of zone 2 is hilly in a rather irregular way, whilst Hasle and Vester Lisbjerg herred are characterized by long gently sloping hillsides which afford natural drainage and allow the cultivation of larger, more coherent areas.

The following investigation will be concentrated on a few selected villages. There are three subjects under discussion: (1) the development of the various categories of holdings, a central aspect of the development of the manorial system; (2) the landskifte, i.e. how were the lands (strips, selions; in Danish agre) allotted to the holdings of the village, (3) the relation between village and thorp.

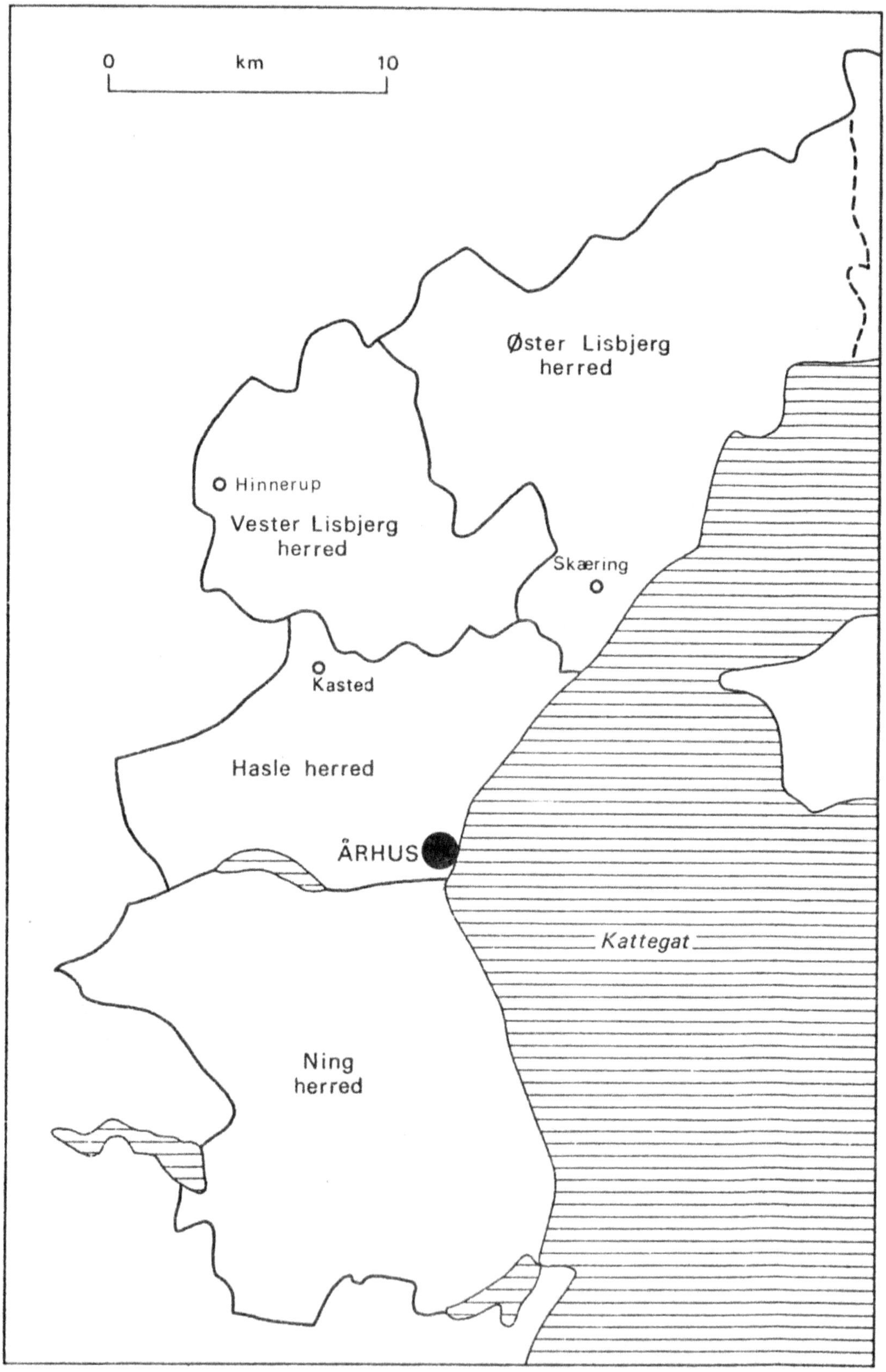

Fig. 16:1 Århus and surrounding <u>herreder</u> (hundreds).

The source material for Early Modern time is the same as for the rest of Denmark, manorial surveys and especially the Land Register (Matrikel) of 1682-83, containing field books for every village of the kingdom, giving the measurement of every land and its position in the furlong. These field books can be supplemented with village maps from the last quarter of the 18th century. If the maps contain furlong names the information from 1682-83 can be placed on the maps, the accuracy depending upon the continuity of the names of the furlongs.

The medieval written sources for the area concerned are sparse, but amongst them is one of outstanding quality (compared with other Danish medieval sources), the jordebog, i.e. manorial survey, of the lands of the Aarhus canons ca. 1315. The detailed information from this survey, and the fact that the cathedral kept its estates after the Reformation, allows a precise comparison between several villages, or parts of villages, in 1315 and in 1683.

THE CATEGORIES OF HOLDINGS.

Figure 16:2a shows the 1315 information on two villages, Hinnerup and Kasted. Both consisted of a big farm, a mansio or curia (demesne), inhabited by a villicus, and a number of adjoining small holdings inhabited by coloni The villicus payed his rent in kind, the coloni payed money, hens, and a few days annual work, their dues being more a sign of their status than of economic value. The villicus sowed 11-14 times as much as each colonus. Though the coloni could each sow 5 tønder (barrels, of 132 litres) and thus had some basis for their subsistence, they had to supplement their income by working the land of the demesne, in fact it is obvious that this was their function in the manorial system, their raison d'etre.

Figure 16:2b, for 1683, shows the complete change of the villages and the manorial system. All holdings were now of equal size, rather big, definitely all employing more labour than the peasant's own family; they all paid rent in kind and a range of other dues, and usually - but actually not in these villages - peasants at this time performed heavy labour service on the noblemen's home-farm. Unlike the coloni of 1315 they were 'middle class', having men and horses at their disposal.

Thus the manorial system had changed completely. But the extent of the village lands, of the area cultivated, remained virtually unchanged. In Hinnerup 108 tønder were sown in 1315, this means that the total arable (with 1 tønde of grain sown in 1 tønde-land, and with a two-field system) was about 216 tønder land, while in 1683 the arable was measured to 218 tønder. For the canons' share of Kasted the figures were 126

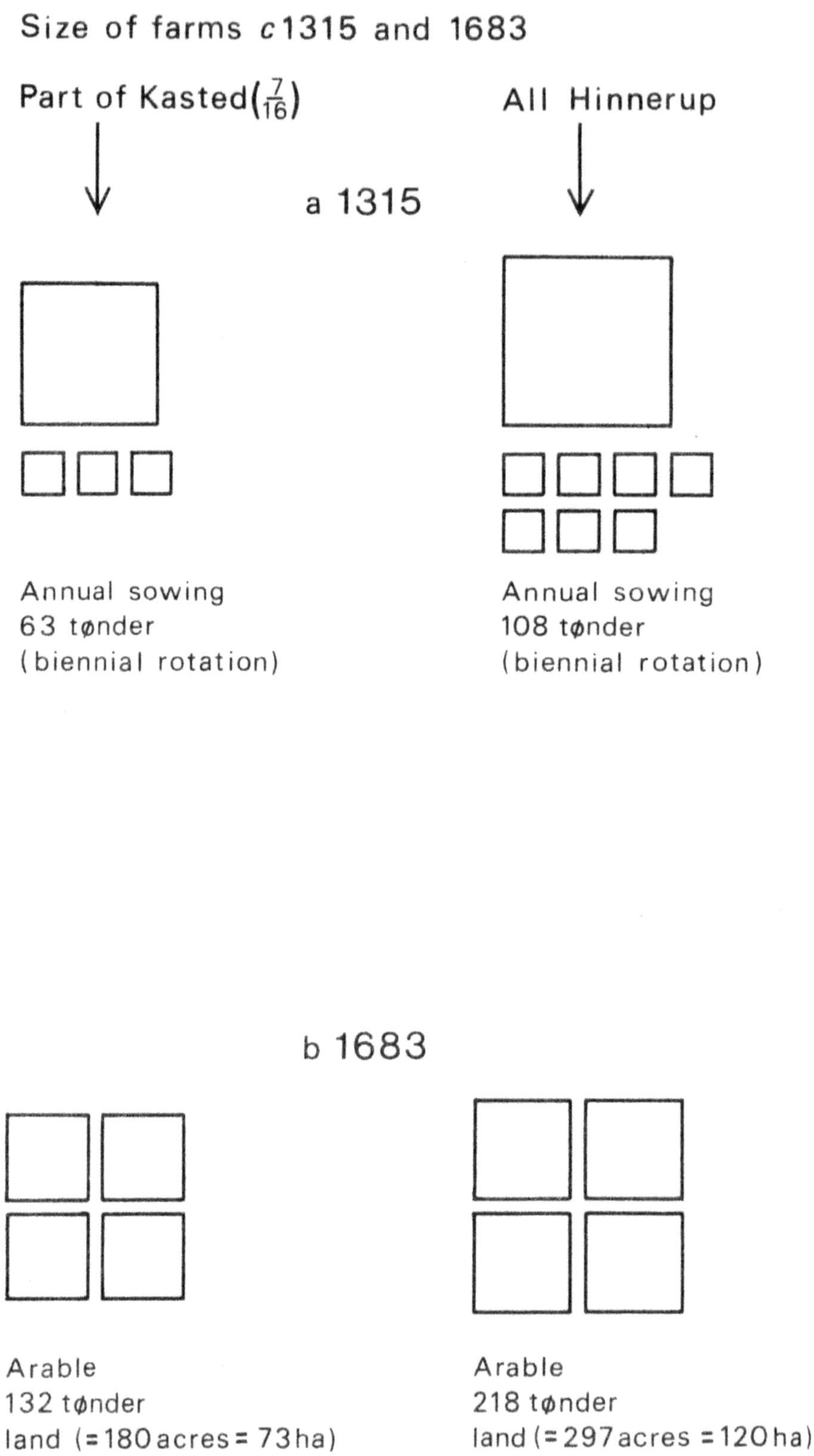

Fig. 16:2 Size of farms c. 1315 and 1683.

and 132 *tønder* respectively. Of course it may be sheer coincidence that the figures are almost the same, but it is no coincidence that the canons in 1315 possessed 7/16 (43,75%) of Kasted, and that the same property in 1683 was measured to 44,4% of the village. There was continuity in the allotment of the selions, in the *landskifte*.

To turn to the village of Skaering; the medieval owner of most of the village was the Abbey of Om. The sparse medieval source material is only known from later reports, but a survey of the monastery's estates from 1554 (shortly before Om, as one of the last monasteries, was dissolved) enumerates a big farm paying a rent of 13½ *tønde* in grain, 9 farms paying 2¼ - 7½ *tønde* and 5 small farms not paying any rent in kind; however, various dues inform the historian that the latter were indeed farms, and not houses or cottages. There were 4 other holdings in the village, but the identification of all 19 holdings in the 1683-land register causes no problem. The big farm, Skaering Munkgaard, had 176 *tønder land*, the 5 small ones only 10, 10½, 11, 16 and 16½ *tønde* while the 13 ordinary farms varied between 18 and 73 *tønder land*. The relative and absolute size of the holdings, and the fact that the small ones paid no proper rent, show us Skaering in 1683 as well as in 1554 as a medieval relic or fossil with a demesne and its *coloni*, and some ordinary farms (such farms of course also existed in the Middle Ages). The peculiarity of Skaering in 1683 stands out when one compares the small farms of the village with Hasle herred's 160 farms. Only 3 of them were small (8, 10 and 13 *tønder* land) and they were freeholders, the size no doubt an effect of partible inheritance. In fact it is quite clear, that farms which were not big enough to secure a family its living and the state its taxes were unwanted in the manorial system of the 17th century.

THE *LANDSKIFTE*

Having observed the medieval system of demesne and coloni still in existence in Skaering in 1683, it will be particularly appropriate to study the *landskifte* in 1683 in this village.

Looking at the furlongs one will notice that the lands of certain holdings usually lay together, and in the same pattern, while their proximity to the lands of other holdings was more random. Apart from the lands of the demesne farm and those of the 5 *coloni*, there were 6 such patterns or groups of selions (partition units or tenurial cycles) - with a varying number of holdings as participants, 2, 3, or 4. Only one farm took part in two groups. I shall illustrate the system with the biggest furlong in Skaering, called Langager (50 *tønder land* = 68,2 acres = 27,6 ha). If we number the lands from 1 to 67, as the land surveyors actually did in 1683, then farm X possessed no. 1-3 and 56-58, farm Y no. 4 and 59. Together the two farms thus had two big strips, each consisting of 4 lands

(Fig.16:3a). In all there were in Langager 18 big strips, each containing 4 or 3 lands (width 9-10 m, length 440 m). 6 of these strips belonged to the demesne and 12 to the 6 groups mentioned, 2 to each, while the coloni had no land in Langager (Fig.16:3b).

Viewing the system of allotment in one furlong alone is confusing, but the point is that the system was basically the same in nearly all furlongs. Together all the strips of farm X and Y formed what is called a bol, mansus (something like hides or Hufen), and the system of allotment of land, the landskifte, is termed bolskifte. The order of the big strips, the bolstrips, was accidental, being probably an effect of drawing lots, and the number of bols in Skaering was 9½. The bolstrips were probably allocated by drawing lots. Most often the bols had but one strip each (or less) in each furlong. If they had two strips (as in Langager) the furlong was usually treated as if were two furlongs. The demesne had 2 bols (133 tønder land). The ordinary farms (including a farm in the neighbouring village Svinbo, see below) shared 6½ bol, five of these were of fairly equal size, about 60-70 tønder land, with 2 or 3 farms as participants, but the last bol had 104 tønder land, shared between 4 farms, and must be reckoned (and this is seen in the pattern of the selions, too) as 1½ bol (Table 1). Finally the lands of the coloni formed - together with 10 tønder land from one of the bigger farms - their bol, though certainly a most unusual one.
They had no part in Langager, but were the sole possessors of 4 small neighbouring furlongs (Fig.16:3b). In some of the other furlongs, however, the coloni-bol participated on an equal footing with the other bols. But neither in their own furlongs nor in their bolstrips was there any observable system for the distribution of land amongst the coloni themselves.

Table 1.

The 'bols' (hides) of Skaering 1683

Bol	Size in tønder land	Number of farms
A	72	2 *)
B	68½	2
C	65	2
D	72	3
E	67½ + 36½	2 + 1½
F	58½	1½
G	133	1
H	74	5 (6)

*) One of these farms was situated in Svinbo, where it had 16 tonder land, cf. Fig. 5.

I shall attempt no specific explanation for the coloni-bol,

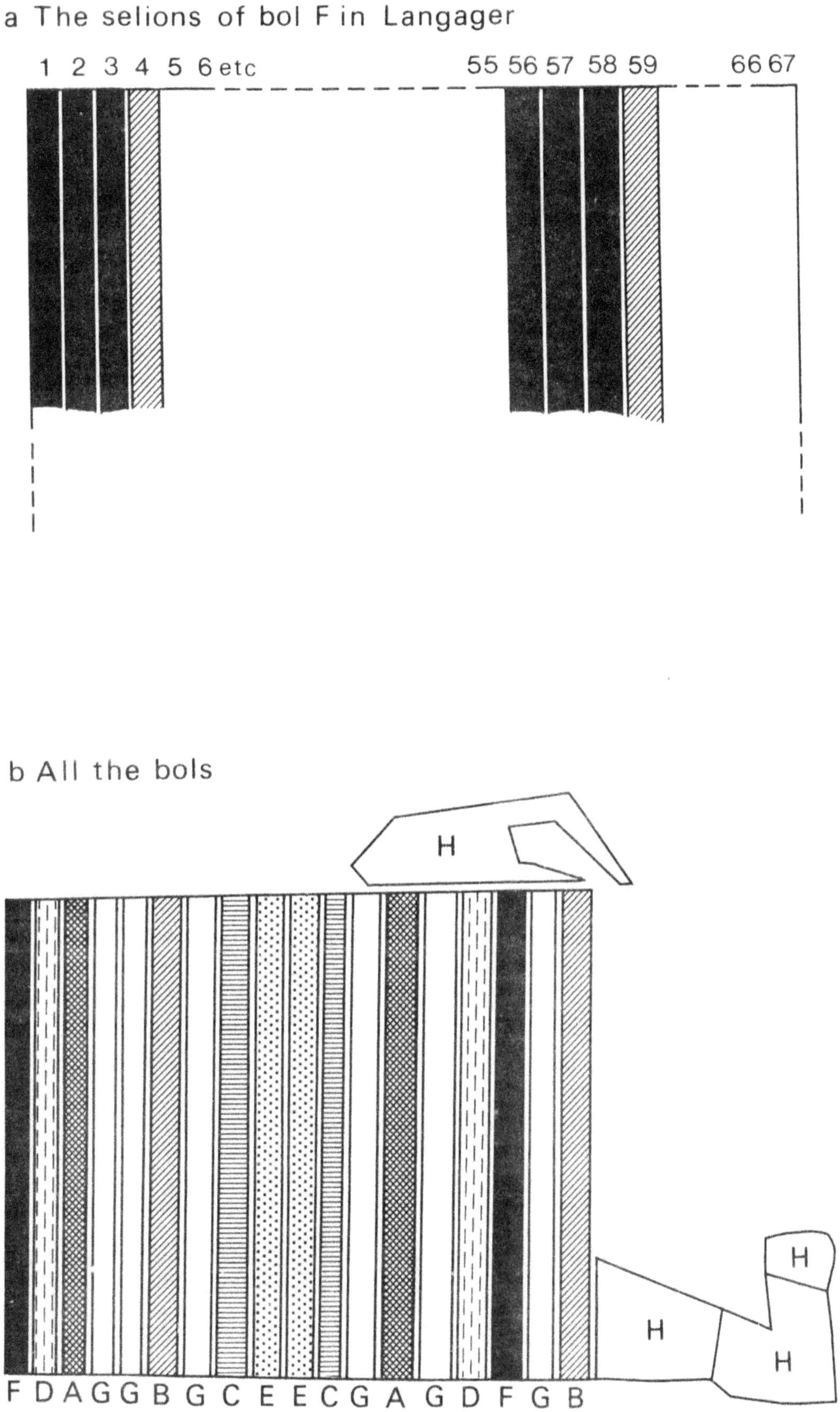

Fig. 16:3 Bolskifte. Langager with four adjoining furlongs in 1683 (Skaering).

only emphasize that it was certainly not of recent date. It might be added, however, that their often peripheral location (as at Langager) might reflect their inferior position in the village. The co-possession of 10 tønder land of the coloni-bol with one of the bigger farms was an effect of a merger, the rents and dues 1554 show that a colonus had taken over the big farm. So it can be concluded that the number of coloni at some time before 1554 was not 5 but 6.

THE THORPS

About one eighth of the village land was not included in the bolskifte, belonging either to the demesne, to the two bols which contained the two biggest peasant holdings, or to these two holdings in their own right. Most of these furlongs lay at the periphery of the village arable, and some of them were marginal lands. An area to the north-west, part of which was called Virup farmstead, and which was divided between the demesne and the two biggest farms, reveals itself (by its name and by being allotted to holdings, not to bols) as a deserted thorp. The area was small, only 11½ tønder land, but it is likely that the peasant who once lived here had land in the neighbouring Staurballe (11 tønder land).

We don't have to look far to observe thorps like Virup still in existence in 1683 (Table 2.). One of the neighbouring villages of Skaring to the North carried the name Svinbo. Here lay 3 small farms, with only 38 tønder land, and enclosed. However, they all had lands on Skaering open field, two of them possessed just a few scattered strips (5 tønder land each), but the third and biggest one shared bol B in Skaering with one of the Skaering holdings, in this way having 20 tønder land distributed throughout Skaring field and in reality being farm no. 20 in Skaering.

Table 2.

The 'thorps'

Virup	11½	?	?
Svinbo no. 1	16	20	36
- no. 2	13½	4½	18
- no. 3	8½	5	13½
Søstrup	12	38	50
Hjortshøj 5 farms			60,66,72½ 73,89
Hjortshøjlund	5	40	45
Kankbølle 4 farms			35,35,76, 108
Stenaa	2	29	31

And in Elsted parish, just west of Skaering, the single-farm Søstrup had 12 tønder land enclosed and 38 tønder in Elsted field. In Hjortshøj (just to the north of Skaering) the strips and blocks belonging to the single farmstead, Hjortshøjlund, lay intermingled with the strips and blocks of the 5 holdings of Hjortshøj, and the 4 holdings of Kankbølle and the farmstead of Stenaa were combined in the same way. In both villages land was allotted (most irregularly in Hjortshøj) seemingly without any system or pattern, obviously, however, not by drawing lots.

For the village-thorp (or single farmstead) relation I shall not hesitate to offer an explanation. The thorps represented in 1683 a relic, pointing back to a time with a looser village structure, to a time when it was still possible to make your own assart or arrange your farmstead on your outlying land provided you fenced it off from the common field and secured access to it for your cattle, as stipulated in the Jutland Law of 1241. It should be noticed that the single farmsteads were usually smaller than the holdings of the village proper, and the Virup deserted farmstead is only one of several desertions in the Aarhus-district. And it should be noticed that these thorps lay in what I have named zone 2, the more irregular moraine landscape, or at the border between the two zones.

The analysis of Skaering 1683 has shown the medieval pattern of the village, i.e. the enormous difference between the big farm of the villicus and very small farms of the coloni - as is known from other villages of the manorial survey of Aarhus cathedral in 1315. Furthermore Skaering 1683 shows the small medieval thorps, whether still in existence as Svinbo or deserted as Virup. The bolskifte as seen in Skaering in 1683 is well-known in Denmark. It is generally accepted that it points far back in time. This is confirmed in Skaering where it is obviously the central structure, added to which are the thorps and the coloni. The single farm might change with time, might be divided, or united with a farm in another bol, but the bols persisted. If we look back as far as the 12th century, we see that all landed property (of which we know) was reckoned in bol. Realising that the bol was not only a land measurement, but possessed a reality in the village field, it is tempting to accept the idea that the bols were the original holdings. Danish scholars at the beginning of this century did accept this idea, but today it is more difficult to accept such a thesis which apparently presupposes, either that settlement was organized democratically by freemen of equal status, or, at the other extreme, planned by central, i.e. military or royal, authorities.

BIBLIOGRAPHY

LARSEN, H. (1918) 'Nogle Oplysninger og Bemaerkninger om de danske Landsbyer', Aarboger for Nordisk Oldkyndighed og Historie 1918, 177-293.

AAKJER, S. (1929) 'Villages, cadastres et plans parcellaires au Danemark', Annales d'histoire economique et sociale 1929, 562-575.

MEYER, P. (1949) Danske Bylag (Kbh., 1949).

RASMUSSEN, P. (1962) 'De regelbundne landskifter, deres beskrivelse og deres bestemmelse', Bol og By 2 (Kbh. 1962).

RASMUSSEN, P. (1957) Kulturhistorisk Leksikon for Nordisk Middelalder II, 55-63 (1957).

CHAPTER 17

ESSAI D'INTERPRETATION DE LA REPARTITION DE L'HABITAT AU XVIIIe SIECLE

C. Foutrein
(Féderation Universitaire et Polytechnique de Lille)

Rendre compte de la physionomie de l'habitat dans le finage d'Arnèke en Flandre française, s'interroger sur les originalités de cet espace humanisé du Houtland, essayer de comprendre l'agencement des habitations au XVIIIe siècle pour en percevoir la progressive mise en place, tels sont les objectifs de l'étude entreprise.

OBSERVATIONS SUR LA DISPOSITION SPATIALE

Exception faite de dix maisons bâties sur une parcelle de terre labourable, un pré, de celles accompagnées uniquement de leur jardin, demeures de salariés agricoles, forge, estaminets isolés en pleine campagne, de la vingtaine d'habitations à l'étroit dans de menues parcelles accolées les unes aux autres en couronne autour de la place de l'église, les bâtiments sont établis dans une prairie bordée de haies. Les appellations utilisées dans le terrier varient, selon qu'il s'agit d'une exploitation agricole (behuysde hofplecke), d'une pâture (behuysde garsbilck) ou d'une parcelle en herbe (behuysde garsland) toutes habitées.

La lecture de la Figure 17.1 permet de circonscrire les particularités de la répartition et met en évidence un fait essentiel: la dispersion n'est pas intégrale, puisqu'un peu moins du quart des demeures sont réellement dispersées. Tout découpage comporte une part d'arbitraire, toutefois on peut distinguer plusieurs espaces habités. Une structure plus lâche individualise la portion méridionale du finage. Cependant, il ne s'agit pas de dispersion indifférenciée, puisqu'il n'existe pas de distance égale entre les habitats "quelle que soit la direction envisagée".[1] Des groupements de 2 à 5 maisons ponctuent ce secteur, le plus souvent deux habitations distantes de 150 mètres au maximum, forment un groupement. L'impression de dispersion est accentuée par l'existence de fermes isolées.

Sans qu'il y ait opposition très tranchée entre cette portion du finage et le reste du territoire situé au Nord de la Peene, affluent de l'Yser, il y a cependant plus que des nuances. Une distribution particulière individualise ce secteur. Un relative concentration prévaut dans le tiers occidental. Quatre fermes et une trentaine d'autres demeures pelotonnées au bourg forment le groupement le plus important. Deux points de cristallisation de l'habitat, inégalement distants entre eux, mais situés sur le même axe, accompagnés de groupements élémentaires, constituent une ligne de force du peuplement. Cette frange occidentale se fragmente en fermes isolées au Nord du finage ou s'intercalant entre les groupements. De forme globulaire, ces rassemblements d'une dizaine de maisons se distinguent de ceux implantés à l'Est. Ici, il ne s'agit pas d'alignements réguliers de fermes mais de deux traînées discontinues de groupes de cinq maisons au maximum.

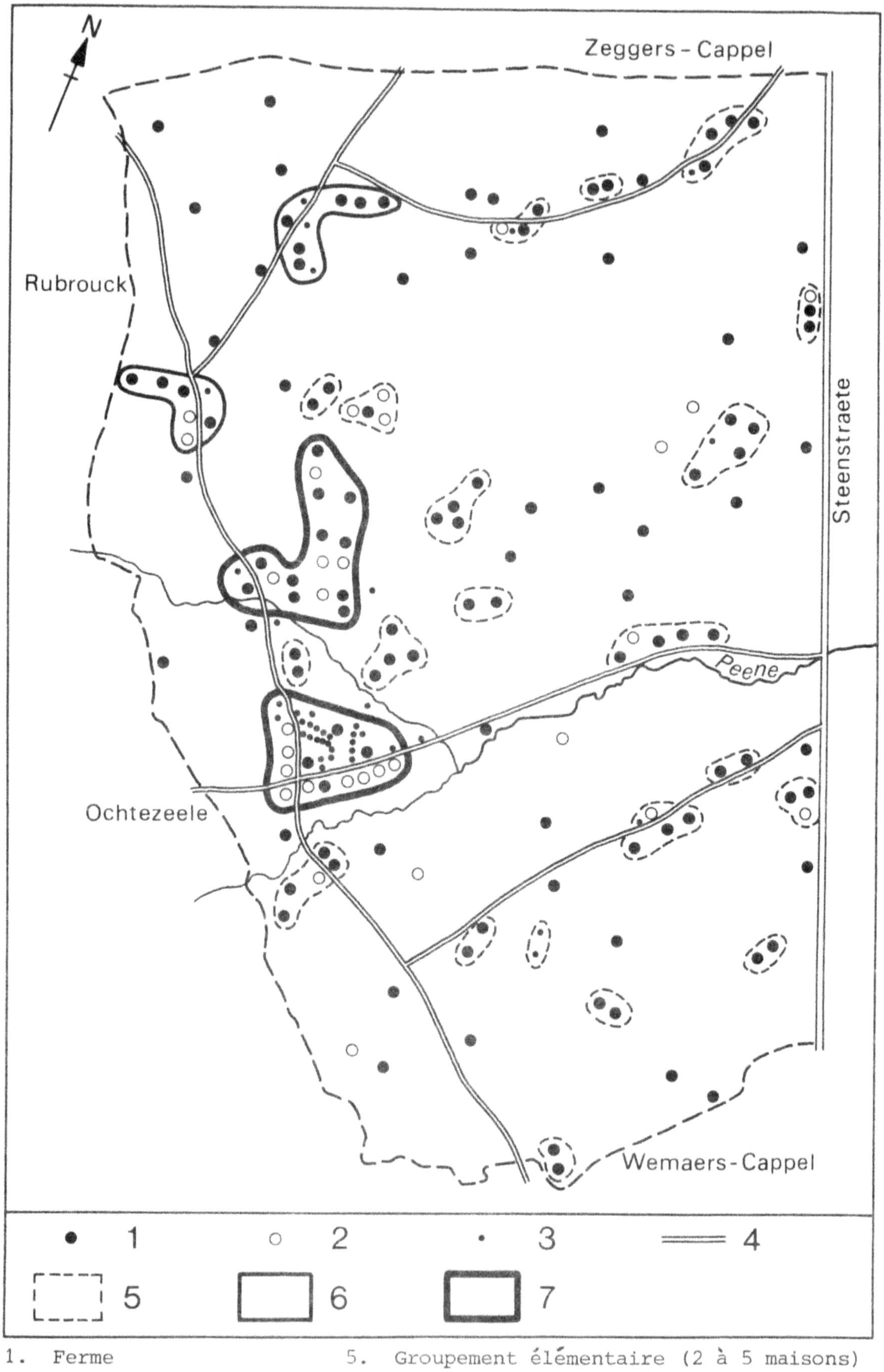

1. Ferme
2. Habitat en prairie
3. Habitat non agricole
4. Route principale
5. Groupement élémentaire (2 à 5 maisons)
6. Petit groupement (6 à 10 maisons)
7. Groupement de plus de 10 maisons

Fig. 17:1 Croquis de la repartition de l'habitat d'apres le terrier de 1774.

A Arnèke, plusieurs aires coexistent, se caractérisant par une densité plus ou moins élevée. Fermes isolées et groupements variés par leur forme et le nombre de feux y diversifient la répartition de l'habitat. L'examen de la distribution de l'habitat du point de vue de sa disposition spatiale, de ses formes, de son importance, fait entrevoir une structuration progressive et singulière du paysage humanisé.

SIGNIFICATION DES LOCALISATIONS, ESQUISSE DU PROCESSUS D'INSTALLATION DES LIEUX HABITES

Remarques méthodologiques

En l'absence de textes d'archives concernant l'habitat, il s'avère que la méthode statistique facilite l'observation de cette réalité construite par l'homme. Le quantifiable, élément descriptif, critère de différenciation porteur d'éléments explicatifs, rend possible l'analyse des relations internes à l'organisation des cellules habitées. Celles-ci ne forment pas seulement une unité spatiale mais possèdent aussi une valeur économique et sociale que le registre terrier rend observable après reconstitution des propriétés et des exploitations. La notion de taille est une valeur à privilégier, qu'il s'agisse de l'importance de l'unité de peuplement, de la superficie de la propriété où s'est fixé l'habitat, de celle des fermes et de leur finage. Les établissments ruraux seront conjointements analysés selon des critères localement significatifs au regard des hypothèses présentées lors d'études antérieures.[2] Les critères qualitatifs de parcellement et de position contribueront à permettre un jugement d'appréciation concernant la distance de l'habitat aux masses boisées, au chef-lieu, aux limites paroissiales. Enfin, les connaissances toponymiques mettront en évidence quelques faits susceptibles d'éclairer l'implantation de l'habitat.

Les exploitations dispersées

La portion de territoire située au Sud de la Peene présente une disposition relativement aérée: la dispersion y est plus importante; aux limites de la paroisse, la densité est plus faible. Ce phénomène dénote-t-il des conditions particulières d'installation propres à une époque?

Des groupements élémentaires se remarquent parmi le semis de fermes dont certaines ne sont pas très éloignées du bourg. L'une d'elles correspond en quasi totalité au domaine foncier du Chapitre Saint Pierre de Cassel et suggère l'idée d'une implantation ancienne. Serait-elle issue des terres données au VIIe siècle par le Seigneur Wambert[3] en vue de la création d'un monastère? Les documents historiques ne précisent pas la superficie léguée aux religieux, mais le domaine primitif a pu s'étendre au Sud de la Peene et favoriser ultérieurement la fixation, dans le voisinage, de plusieurs exploitations isolées. En ce cas, elles seraient héritières du domaine monastique établi pour faciliter la mise en culture. Cette opinion s'intègre dans le cadre des hypothèses qui s'appuient toutefois sur un critère d'ancienneté, à savoir, un découpage des labours en quartiers courts, dérivant de champs blocs, par la suite divisés (Fig.17.2).

1. Ferme de 33 ha éstablie en la propriété du chapitre Saint Pierre de Cassel
2. Ferme de 27 à 28 ha
3. ferme de 19 ha
4. Bois
5. Ferme de 17 à 18 ha
6. Ferme de 10 à 11 ha
7. Ferme de 7 à 8 ha
8. Ferme de 2 à 3 ha

Fig. 17:2 Fermes au sud-ouest et au chef-lieu.

Fig. 17:3 Fermes importantes des seigneuries situees au nord d'Arneke.

1. de 31 ha (Couthof)
2. de 19 à 20 ha (Bertramshof)
3. de 16 à 17 ha) (Winnezeelehof)
4. de 15 à 16 ha)
5. Bois

La densité plus faible s'explique par les relations des finages avec leurs habitats. Partout, mais plus spécialement dans les cantons méridionaux, des terres sont cultivées par des paysans qui n'ont pas leur demeure à Arnèke. Il est probable que les lambeaux de bois, vestiges d'une écharpe plus étendue, y constituaient une espèce de no man's land et isolaient les populations des paroisses voisines. Lors des défrichements médiévaux, les paysans de Wemaers-Cappel auraient contribué à l'abbatage des arbres d'Arnèke. La présence de pâtures-manoirs à la limite Arnèke-Wemaers-Cappel est en faveur de cette opinion. Le découpage des terroirs primitifs tenait sans doute compte des possibilités naturelles. Celles-ci évoluant, les limites paroissiales se seraient fixées plus rationnellement. Cette observation est en faveur d'une étude de l'habitat en fonction des terroirs et non des limites administratives.

Des analogies existent entre les exploitations isolées qui se sont fixées au Sud de la Peene et celles qui se sont implantées dans la frange septentrionale de la paroisse. Les unes commes les autres couvrent des superficies comprises entre 15 et 33 hectares, estimées importantes dans le contexte local. Elles disposent d'un finage compact correspondant assez souvent à la totalité d'une propriété dont la superficie se situe aux environs de 20 hectares. Pourtant ces particularités communes ne leur permettent pas d'être en tous points comparables. Leur parcellaire les distingue. Au Nord de la paroisse, le laniérage régulier s'explique par les conditions hydrologiques et aussi par une mise en culture plus récente, dans un cadre naturel longtemps boisé. Les fermes des seigneuries de Bertramshof, Winnezeelehof, Couthof ont un suffixe de prime abord germanique, ce qui les ferait remonter au moins à l'époque féodale (Fig.17.3). Sans prétendre y percevoir une étape précise de mise en place du peuplement s'effectuant lors des invasions franques et se concrétisant par une construction de demeures isolées dans un cadre naturel vierge, ou de végétation dégradée de landes et de bosquets, il n'est cependant pas téméraire de croire que ces fermes de seigneuries sont à inscrire dans un cadre social, sinon politique, particulier. Cependant une ancienneté plus grande est envisageable pour ces habitats puisque le suffixe <u>hof</u> est synonyme de <u>curtis</u>. Des cas comparables ont été observés en Beauce où de grandes fermes aux noms composés comme ceux des villages, c'est à dire d'un génitif et du mot <u>villa</u>, sont plus anciennes que ne le laisse supposer l'appellation.[4] En cette région de Flandre limitée par une voie romaine, la Steenstraete, une ancienneté remontant à l'aube du premier millénaire n'est pas systématiquement à exclure, d'autant que des repères métriques font penser à une centuriation possible.

L'assise géographique d'exploitations isolées de moindre importance que celle des manoirs précédemment décrits, dénote d'une implantation différente de l'habitat et font penser à une installation postérieure. Ainsi, au NO, une ferme à l'orée du bois de Bertramshof déjà en partie légué à des fins humanitaires, provient certainement du grignotage progressif du revêtement arbustif sous l'effet de l'acroissement démographique nécessitant une utilisation plus complète des terres. L'habitat agricole s'est établi dans une propriété constituée en partie de bois susceptibles d'être défrichés et convertis en terres labourables dont la superficie suffisait au bon fonctionnement et à l'équilibre d'une nouvelle ferme.

Les fermes les plues étendues des confins septentrionaux se sont établies au-delà d'une marge boisée fragmentée à l'époque de la rédaction du terrier, mais très certainement continue antérieurement. Ces fermes isolées correspondent très souvent à une propriété à classer dans les catégories supérieures de superficie. En outre, leur finage est généralement compact, les parcelles dispersées peu nombreuses. Dans quel contexte social ces propriétés se sont-elles constituées et pour quelle raison se sont-elles maintenues au XVIIIe siècle?

Dans les cas observés, un fait est remarquable: il est rare que le propriétaire soit exploitant. Cette constatation expliquerait la non-division du domaine au bénéfice de descendants pas davantage portés à l'exploitation directe que leurs prédécesseurs, les uns comme les autres n'étant pas laboureurs. En ce cas, il s'agirait donc d'une catégorie sociale particulière. A-t-elle des racines dans la paroisse? Le terrier de 1774 précise parfois le lieu de résidence du propriétaire forain, mais il est prudent de tenir compte d'omissions possibles. La consultation de documents antérieurs à celui du XVIIIe siècle, impossible de par leur disparition, aurait pu apporter quelques éclaircissements sur le phénomène de conservation d'un habitat isolé n'évoluant pas en petit groupement.

La présence en la paroisse de nombreux arrière-fiefs dont l'extension géographique n'est pas cartographiable, faute de précisions, pourrait dans une certaine mesure expliquer le conservation de petits domaines avec habitat isolé, sans fixation de demeures dans le voisinage immédiat.

LES GROUPEMENTS DE DEMEURES

Quelle est l'origine et la signification du mode de répartition qu'est la traînée dense de groupements formant l'axe de direction méridienne partant du bourg?

La disposition des habitations à l'intérieur de ces petites agglomérations, leur site particulier, l'extension de leur finage guident la réflexion. Les groupements globulaires comportent des fermes de superficie plus restreinte que celle des habitats dispersés précédemment décrits. Leur surface oscille entre deux et une dizaine d'hectares. Ces groupements sont-ils originels ou sont-ils le résultat d'une augmentation démographique entraînant une extension locale de l'habitat?

Un réflexion sur la dénomination des prairies, sur la condition sociale du paysan y résidant, l'importance des biens qu'il cultive, apporte quelques indications sur le processus de formation du groupement. Dans près de 42% des cas, ces prairies habitées (behuysde garsbilck) font partie d'une exploitation de 10 à 20 hectares. Des fermes un peu moins importantes, d'une superficie comprise entre 6 et 10 hectares, constituées au maximum de treize parcelles labourables, suffisantes à cette époque à l'entretien d'une famille, leur sont comparables. En ce cas, il s'agit le plus souvent d'une pâture habitée par un salarié agricole, sans doute employé par le fermier, ou par un descendant ayant quitté le foyer familial, mais encore employé dans la ferme paternelle. Dans un peu plus du tiers des cas (36%), l'habitat rural en prairie est à inclure dans une superficie cultivée, comprise entre 1 et 5 hectares. Sans doute l'occupant est-il à l'occasion salarié agricole, mais il diffère des autres pour avoir pu acquérir jusqu'à quatre champs qu'il cultive à son compte. Est-il en passe de devenir fermier? Nul doute que de petites exploitations ont pour origine une tenure de ce type.

Mais il ne faut cependant pas perdre de vue la division possible d'une exploitation initiale, correspondant à une propriété fragmentée, comme le prouve l'étude détaillée du registre-terrier. En plusieurs groupements, on remarque des biens inscrits sous le même patronyme, le prénom seul variant. Il s'agit de membres d'une même famille. En pareil cas, l'habitat isolé a donné existence à plusieurs fermes ou demeures agricoles, à la suite d'un agrandissement familial. Très souvent, les plus petites fermes se voient dans l'obligation de se pourvoir en terres cultivées dans tous les azimuts. Par contre, les fermes les plus étendues se distinguent des précédentes par l'avantage de pouvoir cultiver des champs dans le voisinage immédiat. Ces diverses remarques font entrevoir un développement progressif facilité par les liens personnels existant entre les membres d'une communauté familiale, et une mise en culture graduelle. Il est quasi certain qu'une ferme implantée dans un espace largement inculte, a évolué en petit hameau aux fonctions essentiellement agricoles. Une hiérarchisation se décèle dans ce groupement, tant par l'importance variable du finage de chaque unité de culture que par la dépendance perçus entre eux. Les contrastes repérés dans la répartition, fermes isolées s'opposant aux groupements, s'expliquent donc non seulement par les conditions historiques d'établissement, mais encore par les conditions sociales.

Le nombre de feux des groupements s'accroît au fur et à mesure que l'on s'approche de l'actuel chef-lieu. Dans ces groupements proches du bourg, les propriétés ne sont jamais étendues; bien souvent elles sont une portion de propriété, le reste des biens se trouvant dispersés dans la paroisse. Les exploitations moyennes et petites disposent d'un finage très morcelé; par contre, les plus grandes ont toujours le privilège d'avoir des champs moins dispersés. Certains habitats agricoles sont ceux de fermes nettement marginales, cultivant un, deux ou cinq champs au maximum. On perçoit aisément le processus d'implantation de ces petites fermes par l'observation de la forme de la pâture-manoir de l'exploitation la plus importante dont une portion a été détachée pour pourvoir en herbage la ferme la plus récente.

La configuration de la seigneurie d'Angest, écharpe de direction méridienne se gonflant vers le bourg, correspond approximativement à l'aire plus fortement peuplée qui englobe le chef-lieu et les groupements axés sur la route menant à Bourbourg. Il est donc nécessaire de circonscrire les particularités du chef-lieu afin d'élucider le problème de sa fondation, de son importance et de son développement. La faculté d'adaptation de l'homme se manifeste par le choix des sites les plus accessibles à l'habitat. Sans être toujours déterminantes, il semble qu'en cette région jadis boisée, les circonstances les plus favorables à la fixation de l'habitat aient été les vallées. Dans le Houtland, elles sont souvent le milieu d'élection du chef-lieu des communes. Pareillement, le bourg d'Arnèke se situe dans l'angle de confluence de la Peene et d'un sous-affluent de l'Yser. Ce site défensif grâce à la proximité des cours d'eau, et n'exigeant pas un assèchement important, a pu favoriser la constitution d'un noyau anciennement peuplé. En 1774, le bourg possède un fond assez dense de population. Quatre fermes se serrent autour de la place de l'église. Un ramassis de prairies habitées grossit le groupe de fermes et constitue un terroir herbager, aux parcelles encloses. Le milieu physique, le voisinage de la micro-vallée de la Peene, aux inondations possibles, favorise la végétation herbacée au détriment des labours.

L'habitat humain considéré comme élément de permanence dans le paysage rural, il est plausible que le bourg soit un des noyaux les plus anciennement habités, comme le seraient ceux qui constituent la ligne de force de peuplement précédemment décrite. L'âge exact de l'installation de ces habitats est difficile à apprécier, cependant des connaissances toponymiques apportent quelques éclaircissements sur ce problème.

Le toponyme Rantiacum qui a subi une évolution phonétique pour donner Arnèke,[5] mène à un rapprochement d'autant plus significatif que le centre actuel du chef-lieu se trouve à une distance triple de 710 mètres de la Steenstraete limitant le finage à l'Est. D'autres indices métriques relevés dans la paroisse et le long des voies rayonnant de Cassel, confirment l'hypothèse d'une occupation romaine, vraisemblablement modeste. Une étude minutieuse du parcellaire, des recherches d'ordre archéologiques, une prospection des sites d'habitats disparus, pourraient confirmer cette hypothèse d'une organisation dirigée par les occupants dans un finage où ils auraient corrigé les lignes primitives du parcellaire, ajusté la trame existante de l'habitat.

La stabilité historique fait défaut et les invasions germaniques ont été particulièrement actives en cette région. Pourtant le toponyme ne présente pas de radical ou de suffixe germanique comme la majorité des noms de lieux de la région. Une population bien implantée dans un cadre relativement aménagé a pu contribuer à maintenir les appellations anciennes malgré le passage ou l'établissement d'envahisseurs dans le voisinage. Lors des destructions liées aux invasions, des habitats ont disparu, pouvant laisser une infrastructure agraire propice à une nouvelle implantation. Sans s'aventurer dans les domaines obscurs de l'histoire, il est possible de présenter quelques conclusions qui pourront être confirmées et complétées ultérieurement.

Le nombre plus important de demeures du confluent des rivières à la limite septentrionale du finage est certainement un mouvement amorcé depuis longtemps, mais qui, stoppé à l'occasion des vicissitudes du début du premier millénaire, a repris au cours du Moyen-Age. Ce secteur du village a pu bénéficier d'un repli de la population cherchant à s'écarter de la Steenstraete, voie trop facile à la pénétration.

Les formes d'organisation rencontrées à Arnèke n'élucident pas le problème de la répartition primitive. Cependant, l'observation a permis en certains cas de percevoir nettement le développement de l'habitat par division des biens s'accompagnant d'implantation de nouvelles demeures. Mais il ne faut pas ignorer le cas de petits groupements pouvant être originels et dont les terres se trouvaient à la périphérie du site habité.

La recherche des causes de la diversité de la distribution de l'habitat est inhérente au problème de la dispersion. L'éparpillement plus accentué des demeures en certains finages de Flandre intérieure, dès la fin du XVIIIe siècle, l'évolution perçue au cours de ces deux derniers siècles, aboutissant à une dispersion pouvant souvent être qualifiée d'indifférenciée, est en faveur d'une modification de la trame de l'habitat, d'un passage progressif d'un type d'agglomérations de petite taille à la dispersion. Celle-ci serait un phénomène historique, remontant au Moyen-Age, dépendant de circonstances nombreuses, sociales et économiques.

Un essai de synthèse est prématuré mais la description des modes d'implantation de l'habitat, fermes dispersées, groupements plus ou moins importants, comparables à ceux des pays celtiques,[6] sont autant de formes dont il faudra découvrir la signification en d'autres localités du Houtland. Les facteurs explicatifs présentés au cours de cette étude, les hypothèses qui en découlent quant à la mise en place de l'habitat, devront être vérifiés en d'autres finages afin de dégager les principes d'organisation de l'habitat du Houtland et déboucher ainsi sur une typologie régionale.

NOTES BIBLIOGRAPHIQUES

1 Lexique agraire français (Comité national de géographie. Commission rurale).

2 Foutrein C. - Evolution des paysages en Flandre intérieure française (XIXe-XXe siècles), dans Géographia Polonica, 1978, no. 38, pp.97-107.

3 Flou K. de - Woordenboek der toponymie van Westelyk Vlaanderen. T.I.

4 Aufrère L. - L'habitation et le village en Beauce. CRCG, Paris 1931, T.III.

5 Gysseling M. - Toponymisch Woordenboek van België, Nederland, Luxemburg, Noord-Frankrijk en West-Duitsland.

6 Flatrès P. - Paysages ruraux des pays atlantiques. Une comparaison: structures rurales en Norvège, dans AESC, 12, 4, oct.-déc. 1957, pp.602-12.

ABSTRACT

The purpose of this paper is to describe the rural settlements in a village of Inner Flanders and to explain their formation in time and space.

In Arnèke, in 1774, the pattern of house distribution is a scattered one, with a tendency towards agglomeration mainly near the church and along the main road cutting across the village. Within the area studied, two divisions can be seen on both sides of the river Peene.

The South of the parish differs from the other part. Scattered habitations and insignificant groups of houses prevailed in this area.

The most densely populated portion is beyond the left bank of the river. Several types of groupings of dwellings can be noticed. Between the Peene and its tributary, farms and houses stand close together, not far from the church or in the pastures. The fundamental influence can be distinguished in part through natural conditions and in part due to historical events. The site was easy to defend and the settlement has been formed at a comparatively early period and has been modified perhaps during the Roman Age. The distance between the churchyard and the road (Steenstraete) established by the Romans is meaningful (710m x 3).

An accumulation of houses is also found along the high road, including farms and cottages built in enclosed meadows. But some isolated farms are a characteristic feature of the North of the parish.

A clear relationship between the size and the localisation explains the existence of single farms. The largest estates tend to produce the dispersal of homesteads. I point out that it is possible to distinguish between older and younger farms. In the SW of the village, there was a monastic estate. From this observation, we can conclude that these isolated farms were settlements of a greater age than those whose names end in "hof", situated in the North of the parish. This region of sparse population, where the available land was not fully cultivated because it was covered with scrub and forest, had been uninhabited until feudal times, and divided into manorlands. Consequently, there are interactions between man and his physical and historical environment.

Due attention must be given to the size of the estate and farm tenures which constitute the groupings. With this information as a background, a quantitative approach has been necessary to explain the problems of the groupings in relation to the human conditions.

It is supposed that most of them already existed in ancient times and the existence of the main road has been a factor in drawing the peasants near it. The groups were composed of a few prosperous farms which were probably the basic settlement. An increasing population must be taken into consideration to understand this development. The pressure of population divided the homesteads into smaller farms. The same name is found for several inhabitants. They have created a series of small agglomerated settlements, where the system of land tenure is not the same. There are landlords and tenants; farm labourers and journeymen living in the houses settled in the pastures.

The pattern varies also in response to other factors, such as the area of cultivated soil. On the one hand, the rural dwellings are in the middle of the fields, and on the other, single farms and insignificant groups of buildings could be seen round the small open-fields.

Nowadays, the pattern of house distribution is a disseminated one, except near the church. But it seems that in the past centuries, these isolated farms were not the principal characteristic of Inner Flanders.

III : Frontiers

The separate themes of frontier settlement and rural industry were the subjects of just over one third of the papers presented to the Conference: they are here assembled as the third section of this collected work. Neither theme has been in the forefront of the Conference interest in the past but as both are topical and present challenging historical problems it is to be hoped that they will be prominent in future years.

The eleven papers are disparate in time and place. There are four on Scandinavia (including Denmark), two on Germany, two on England and one each on France and Italy. A paper on Canada was the only one to take the Conference beyond its customary European context. As they all concentrate on different periods it is difficult to isolate common strands. Meetings of like academic minds usually benefit most from hearing about familiar problems tackled via new sources and different approaches. To some extent this happened but future organisers might wish to focus more on a single frontier problem or upon particular aspects of rural industry, for example the integration of agriculture and craft production, influences upon daily and seasonal work regimes, the circulation of capital or the mechanisms for marketing.

One group of papers deals essentially with the frontiers of internal colonisation at different periods. Not surprisingly, in the frontier situation, much stress is put on environmental controls; at high altitudes and in remote areas human adaptation is more difficult and to wrest a living from the land is more challenging and invariably more specialised than in the more permissive lowland. Sereno's paper on settlement retreat in the western Italian Alps puts stress on climatic change and upon its long-term effects on the growing season. She argues that the progressive abandonment of the highest settlements in the Piedmontese Alps coincided with the coldest local phases of the 'Little Ice Age' in the seventeenth and early eighteenth centuries. She also draws attention to the problems caused by soil erosion in the mountain areas and to the methods used to conserve precious patches of fertile soil. Terraces meant that remote areas could still be cultivated by valley communities long after the permanent upland settlements which gave rise to them had been abandoned. Linke, in his paper on the Harz mountains, also takes up the theme of soil erosion, here caused largely by the clearance of woodland both for pasture and tillage but also, in the fourteenth and fifteenth centuries, for mining; at this period streams were diverted and new channels and ponds constructed to drive machinery. Soil erosion resulting from these activities may account for some of the abandonment of settlements in the later Middle Ages; in this respect Linke opens up an interesting line of enquiry which has not, as yet, had the attention it deserves in other parts of upland Europe.

Cabouret, in a detailed paper on the upper limits of settlement and cultivation in Scandinavia, is concerned to show that upper limits varied with site, aspect, soils, and with the different demands of livestock and tillage. He demonstrates the variety of human responses at different altitudes in Norway and Sweden. At a much lower altitude and in a different cultural context the variety of response to the physical landscape is also a theme pursued by Unwin in his paper on the post-Roman settlement history of two shires on the margins of upland Britain. Working mainly from the evidence of Domesday Book (1086), place-names, and archaeology, he discusses the likely settlement history of this area between the fifth and eleventh centuries. He subscribes to an increasingly held view of a population decline in the early Anglo-Saxon period followed by active colonisation by Anglo-Saxons and Scandinavians in the eighth to tenth centuries. By the time of the Norman Conquest he suggests that this had resulted in very different densities of occupation even within this comparatively small area; the lowlands of south-east Nottinghamshire were populous and closely settled in contrast to the thinly peopled uplands of Derbyshire.

The evidence of place-names is as crucial to Unwin's argument as it is to Brink's interpretation of a later phase of settlement in central Sweden. In this paper he demonstrates that boda names are usually to be associated with out-settlements which were occupied at first temporarily and later permanently for the exploitation of a particular resource be it hay, pasture or iron. Brink is concerned to advance and refine the interpretation of this particular place-name element in the study of the origins of rural settlements of this kind. Sporrong, on the other hand, is concerned with more general models of colonisation and settlement as they relate to a particular part of central Sweden; his paper is an interim discussion of an on-going inter-disciplinary research project in the Mälar basin. Becker's paper, principally on a central place theme, is a further indirect reminder of the importance of place-names; when they describe terrain, economic activity and the people who promoted it (as they do in the Klondike) place-names often outlive settlements in their usefulness and may be of lasting value in the explanation of historical events and processes.

All four papers which deal with rural industries in the early modern period (by Leister, Hansen, Pinard and Hodgson) seek to demonstrate in different ways the nature of the activity and the relationships with population, agriculture and woodland. While the industry varies from place to place - coal in Durham, hosiery in Jutland, glass in parts of France, wool and flax in Hesse - all are characterised by a liability to short-term boom and slump in contrast to the relative stability of the agricultural economies in which they are founded. This was especially so in rural Hesse where the rapid decline of craft industry in the face of factory competition caused major social disruption in the nineteenth century. In Durham, however, where the social structure was entirely different, coal-wealthy landowners were also leaders in agricultural innovation. The role of individuals and of particular groups who were in a position to manipulate the system to advantage (e.g. the hosiers of Jutland) is likely to be another important theme in future research on so-called 'proto-industrialisation' and one which merits attention at future meetings of the Permanent Conference.

CHAPTER 18

PLACE-NAME EVIDENCE FOR RURAL EXPANSION IN MARGINAL AREAS; AN ONOMASTIC-GEOGRAPHIC STUDY FROM CENTRAL SWEDEN.

S. Brink
(University of Uppsala)

This paper is an attempt to illustrate how place-names can be used in investigations of the history of settlements, in this case settlements with marginal sites and, in the main, of medieval origin. Furthermore, it is a first attempt to get a collected picture of the heterogeneous boda-names for a part of Sweden, to survey the different types of settlement represented by these names, and to suggest a generalized subdivision of boda-settlements. Hence, the main aim of this paper is to clarify the picture concerning the element boda.

Rural settlement in central Sweden (Fig.18:1a) took place mainly in two different ways. First, colonisation of the kind known as "land-nam", i.e. someone from outside has taken possession of land and cleared it for cultivation - this results eventually in a dispersed distribution of settlements. Gradually the establishment of boundaries between different units becomes necessary. This boundary adjustment and organisation of the agrarian landscape makes the distribution of settlement more fixed. Furthermore only existing settlement units are then able to produce new farms. Thus we arrive at the second method of settlement expansion, which can be termed internal colonisation or boundary-fixed colonisation(1). Here new units can arise as a result of division of, or separation from, an existing farm. In my view a third method, perhaps a special case of the second one, is when a new unit arises from a habitation which was first used only periodically and then becomes detached from the parent unit and permanently settled. (Lindquist 1968:137 ff., Bylund 1956:2 ff., Bylund 1961, Bylund 1968:64 ff., Bodvall 1959:13 ff., Hovstad 1979:9 ff.)

In the north of the province of Hälsingland in Sweden the latter process has taken place whereby periodically used settlements have, by degrees, become permanently occupied. This kind of settlement is by tradition referred to in Swedish as bodland, containing the components bod 'booth, shed, a primitive habitation' and land 'arable and meadow, land (in general)'. (Fig.18:1b). To obtain a corresponding term in English one might use "boothland", but as this may give the wrong associations, I have preferred to keep the Swedish term bodland in this paper. These bodland used to form a constituent part of the rural system which existed in these parts (Fig.18:2). The most vital and important component of this system was the farm in the old village. Here we find dwellings fit for winter habitation and the whole range of farm buildings typical of an agricultural holding of the past. In

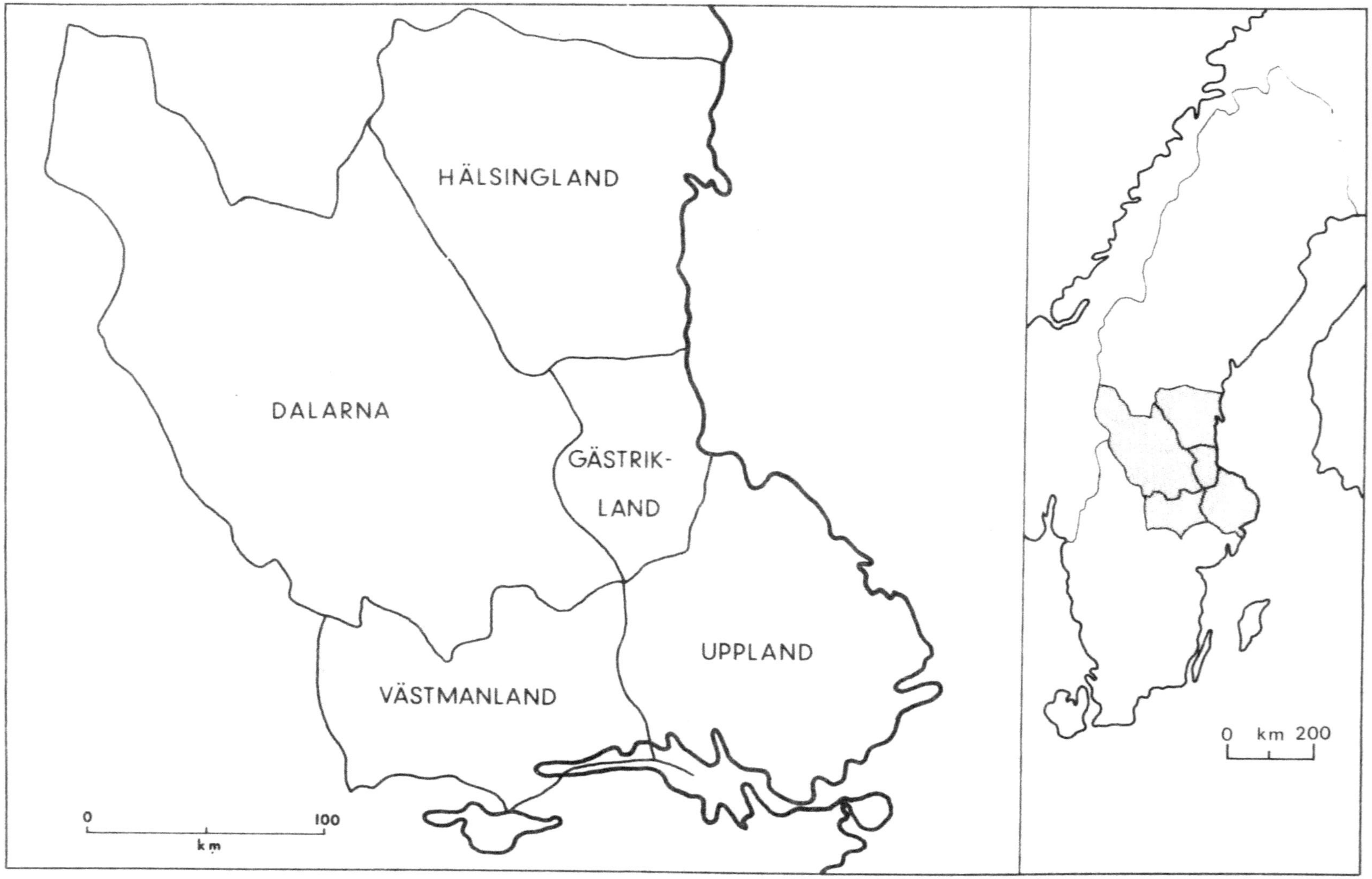

Fig. 18:1a The area of investigation: the provinces of Hälsingland, Dalarna, Gästrikland, Västmanland and Uppland in Sweden.

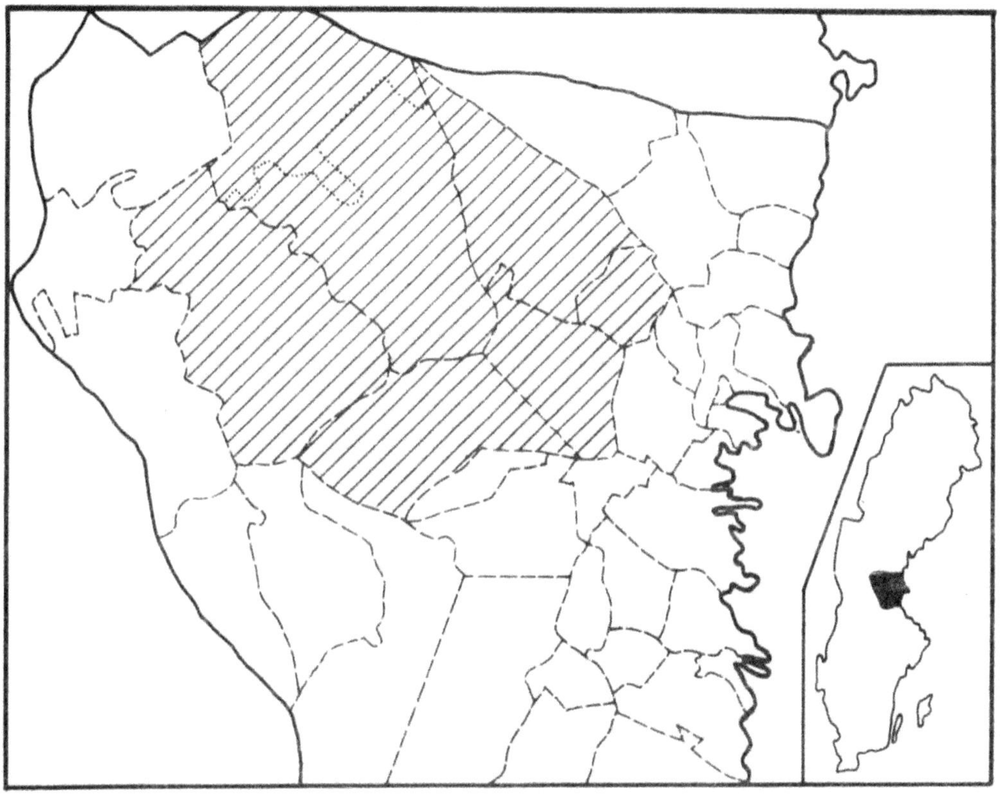

Fig. 18:1b The province of Hälsingland with the six northern parishes (from left to right) Färila, Ljusdal, Järvsö, Bjuråker, Delsbo and Norrbo which have had a known bodland system. (Bodvall 1959: 28).

addition, most of the arable land was to be found there. The second part - the intermediate stage in this agricultural system - was the bodland farm. This was often situated 5 to 10 km from the main farm, located in minor valleys and in peripheral, marginal sites around the arable land and meadows of the villages. At the bodland farm were found the more important buildings such as dwelling-house, byre and other farm buildings; it was in a way a scaled-down copy of the main farm. One important difference was, however, that the bodland farm was not fit for winter habitation - it was used periodically during the summer season only. At the bodland some of the farmsteads possessed arable land - the acreage dependent on cultivation conditions - and most of them had hay-fields. All the household and livestock of the village farm moved to the bodland as soon as spring sowing at the main farm was completed. Then the sowing of whatever arable land there was at the bodland started, and the cattle were put out to graze on the pasture in the vicinity. Just before midsummer the livestock were moved to the more remote seter, or shielding, (Swed. fäbodställe). This habitation was the third and final stage of the system.

The seter was situated in the woodlands, often a couple of miles from the main and the bodland farms. They were spread a few kilometres apart over the large area which originally had been the common land of the parish. By degrees this common land was claimed for pasture and for harvesting fodder, with the seters as bases (Fig.18:3). Only a female cowherd, either a member of the household or hired for the summer, stayed with the livestock at the seter. She had an important job: on the one hand, to take charge of the milk and to make butter, cheese etc., on the other, to herd the cattle, (Swed. dial. gäta) in the daytime. This latter task was of importance because she had to take the cattle to special pastures, (Swed. dial. gässlor) which changed every day. This was necessary to prevent the cattle gaining access to the innumerable hay-fields on wooded and boggy land. A vital task for the farmer of these parts, where cattle-raising was of considerable importance, was to ensure that there was sufficient fodder for the livestock over the long winter period. Therefore every self-sustaining meadow and bog was exploited. To increase the area of meadow land further, forest areas were cleared and burnt over, and for one or two years often yielded an abundance of hay. When the forest began to re-encroach on the clearing, it became excellent pasture (Jonsson 1970:113 ff., Montelius 1977:34, Kardell 1980:12, Westin 1930:49 ff.) At the end of the summer the cattle were moved back to the bodland where the household stayed during the autumn until the hay harvested during the summer was used up. In the late autumn or sometimes early winter, the household and livestock moved back to the main farm in the village.

The bodland settlement type is first described in some topographical books from the 18th century. In 1749 Pehr

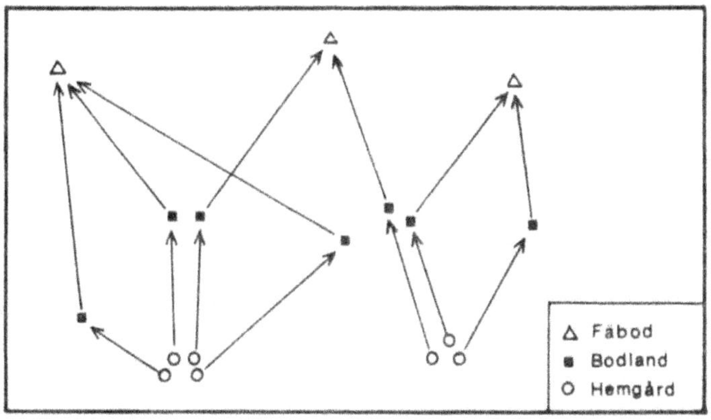

Fig. 18:2 Sketch showing basic principle. The lines are intended to show the interconnection between the different elements of a farmstead. The rural system of these parts (Northern Hälsingland) had the following constituent parts: the home farm (<u>hemgård</u>) the <u>bodland</u> farm, the seter (<u>fäbod</u>). Community in a home village did not necessarily mean co-operation on <u>bodland</u> or seter. (Bodvall 1959: 15).

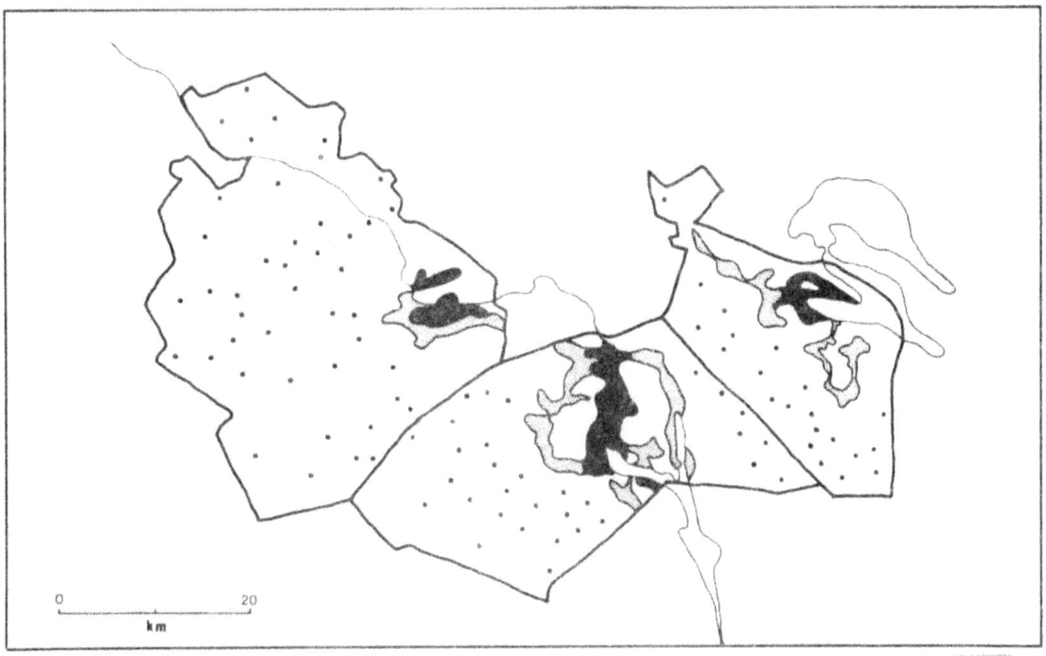

Fig. 18:3 The distribution of arable land and the location of the seters in the three parishes of (from left to right) Färila, Järvsö, and Delsbo in Hälsingland. The black areas refer to the arable land at the home villages, the shaded areas to the arable land at the <u>bodland</u>. The seters (black dots) are located all over the woodlands of the parishes, originally the common land, so that the whole vast area was exploited - on the one hand for pasture, on the other for collecting hay-fodder - with the seters as bases.

Schissler writes:(2) "(the peasant farmers)...have 3 separate dwellings to maintain on their farmsteads, namely first the main farm, which is the foremost, ... second the *bodland* farm, which is somewhat inferior and can be situated between ½ and 1 mile from the main farm, which the farmer must maintain because of the land he owns there; because many can have more land at the *bodland* than at the main or home farm... The third dwelling is called the *seter* farm and is often situated as much as 3 miles from the main farm in the woods" (Schissler 1749:22f.). From the parish of Delsbo the Reverend Lenaeus tells us that: "...almost every farmer in the parish must have and maintain three cribs (or stalls) for every cow he owns, the first at the home farm, the second at the *bodland* and the third out on the *seter*..." (Lenaeus 1764:154).

Sources that mention *bodland* before the 18th century are very rare. In a couple of letters from the mid 16th century some villages are referred to as *bodland*. From the Middle Ages there is no definite evidence other than a regulation of 1425 which says that the clergymen of Hälsingland could possess a number of farms, which they could occupy and between which they could move their livestock (Hedblom 1960:58). From dialect and ethnological records of this century we know that the farms had been permanently settled and turned into independent units (Bodvall 1959, Hedblom 1947:130 ff., Hedblom 1958:61 ff., Brink 1979:7 ff.).

If we examine where the *bodland* land was located we see that to a great extent it was a continuation of previously cultivated areas (Figs.18:3 & 18:4). But in general terms it was found in areas where natural conditions allowed cultivation, in the case of Hälsingland mainly in the valleys. The cultivated area expanded along rivers and along the shores of lakes. It is easy to see that *physiographic conditions* have been of decisive importance for the location of settlements. Of further interest is that it has been possible to establish *bodland* in relatively recent times - potentially cultivable land has existed and this marginal land has gradually been exploited.

It is of special interest that many villages which in earlier days were *bodland* have names ending in *-bo(da)*. This is to be seen against the fact that the appellative for *bodland* in these parts was Swed. *bodarna*, i.e. the definite form, plural of *bod*, Eng. *booth* (a loan from the Danish) < Gmc *bōþō- 'a primitive hut etc.'. Today we often encounter a form which has been eroded to *-bo*, which has gradually assumed a suffix-like function with an unclear meaning.

It turns out that most of the names of *bodland* villages can be divided into two large groups: on the one hand those which are originally nature-names e.g. *Sandvik* 'the sandy bay', *Alsjö* 'the lake where alder grows', *Sandnäs* 'the sandy peninsula', *Löräng* 'the meadow by the foamy stream', *Lunga*

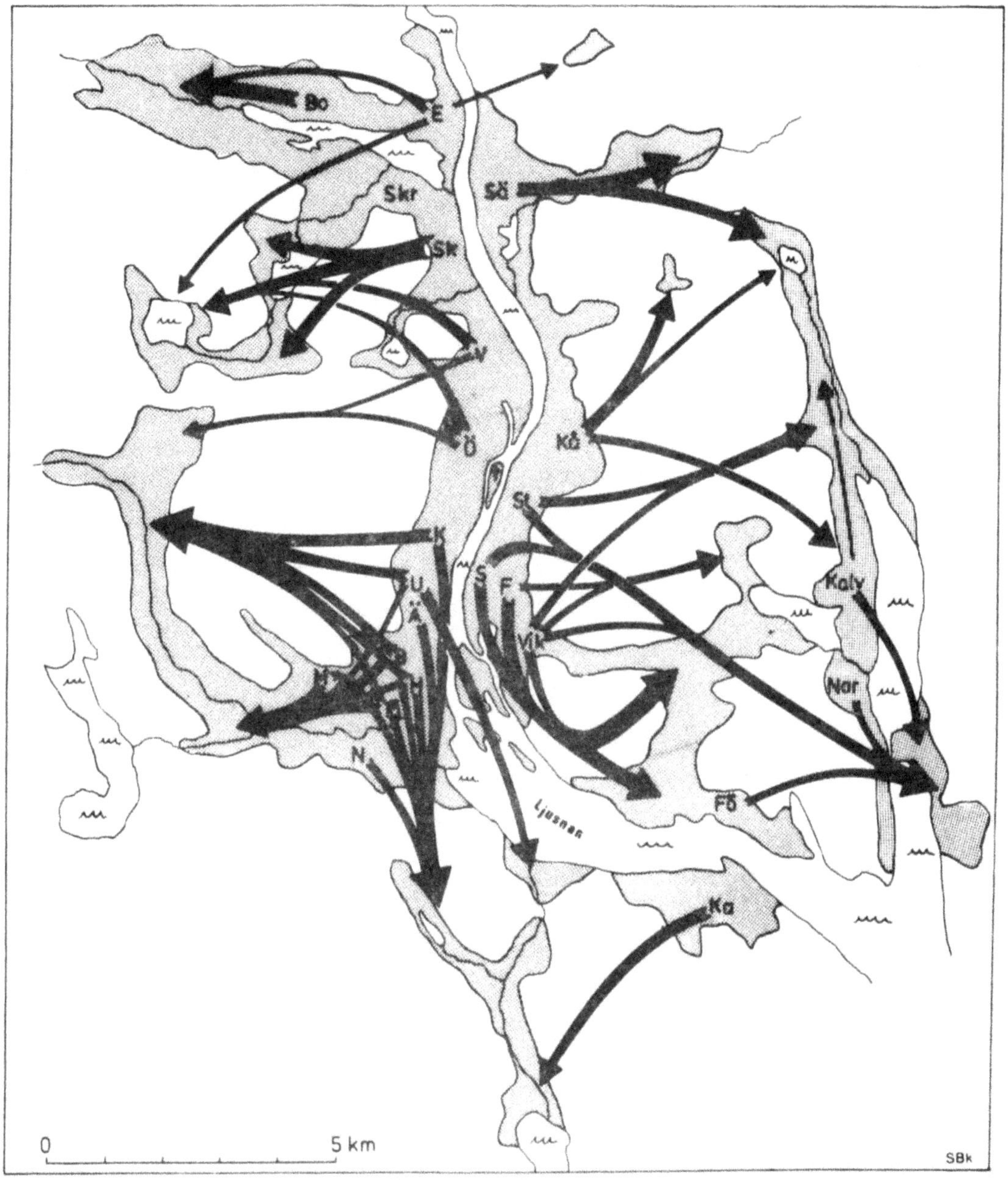

Fig. 18:4 The central part of the parish of Järvsö in Hälsingland. The shaded areas show the extension of the arable land at the home villages and at the bodland. (For the distribution of arable land between the home villages and the bodland, see Fig. 18:3). The arrows indicate where the home villages (capital letters) had their bodland. One can see that the bodland is a continuation of the cultivated land of the home villages. (Brink 1979: 15).

from the appellative Swed. dial. lung(er) 'sandy land etc.', Stråsjö 'the reedy lake' etc. On the other hand there are village names with the element -bo(da). This group has varying first elements. Common are personal names eg. Häknorrbo (< OSwed. Hākonarbodha) 'Håkan's bodland, Simanbo (<OSwed. *Sighmunda(r)bodha) 'Sigmunds bodland', Gunnarsbo 'Gunnars bodland', or trade-names as in Skommarbo (<Skomakareboda) 'the shoemakers bodland', Smedsbo 'the blacksmith's bodland': in these cases the name-giver is presumably the one who established the bodland. Some names give us an insight into how these bodland were cleared for cultivation e.g. Svedjebo 'the burn-beaten bodland', Brandbo (same meaning). Others are what one might call relation-names e.g. Gryttjesbo 'the bodland by the lake Gryttjen' (<grytug 'stony'), Yckelsbo 'the bodland by the lake Yckelsbosjön' (<OSwed.* Yklir 'the lake that easily overflows'), Sjöbo 'the bodland by the lake', Bäckebo 'the bodland by the stream' etc.

There are relatively few villages with names ending in -bo (da) to be found in older sources. From Halsingland I have only found 18 references to names ending in -bo(da) in medieval sources (i.e. before 1500). The boda-villages which are found later as independent units in 16th-century land registers are more numerous, about 60. For instance, there are no boda-villages from the parish of Järvsö mentioned in medieval sources, but there are four in the 16th-century land registers. By comparison there are now 12 bodland villages with names ending in -bo(da) in this parish. And these appear, judging by their location and size, to be of medieval - or at least 16th century - origin. This can be taken to mean that a great number of boda-villages - which were bodland - were at this stage not yet independent units, but were included in the main farms in the old villages for tax-assessment purposes. Hence the settlement picture of the late Middle Ages which is reflected in letters, documents and fiscal records is distorted if we do not take these circumstances into account.

The relatively few bodland names which formally do not belong to either of the two above-mentioned groups, often tell us that we are dealing with what were originally clearings and outfields, e.g. Rossla 'the clearing' (<OSwed. rubsl, a derivative of the OSwed. verb ryþia 'to clear land'), Sveden 'the burn-beaten land' (<OSwed. sviþ), Lock from the appellative lock neut. with an older meaning 'a smaller enclosed outfield' (<OSwed. lŭk, lŏk <Gmc *lŭka-, derived for the Gmc verb *lukan 'to close, to enclose'; this appellative corresponds to Eng. lock, OEng. loc 'an enclosed place') (Brink 1979:30 ff.).

From this it is clear that in the bodland parishes in northern Hälsingland there is an obvious correspondence between bodland villages and village-names ending in -bo(da).

Now we find villages with names ending in -bo(da) not only in the northern part of the province but in almost all of Hälsingland (Fig. 18:5). In the parishes to the east by the coast - where we find the overwhelming majority of Hälsingland's ancient monuments, indicating extensive Iron Age settlement - we only find a few villages ending in -bo(da), and some parishes lack them altogether. However these boda-villages have similar sites to those in the north, that is, they appear as border settlements around the older villages. Without exception the boda-villages lack prehistoric monuments. These villages are therefore likely to be of medieval origin and they bear witness to an expansive phase of medieval colonisation (3).

A closer examination reveals that the use of bodland was more widespread in Hälsingland further back in history than in recent times. A letter of 1546 tells us about a "bodland called Longboda" in the parish of Forsa, where the use of bodland is otherwise unknown. A hypothesis that all the boda-villages in Hälsingland are of the same origin as the northern boda-villages, and hence that a bodland-system in some form existed in earlier times in the whole of the province, appears very likely.

So much for the province of Hälsingland. In Gästrikland, its neighbour to the south, we find nearly three dozen boda-villages in the 16th-century land registers. Today we have about 80. These villages have, in the same way as those in Hälsingland, peripheral, marginal sites in relation to the central, old villages. Of vital interest for understanding the settlement history of the boda-villages in Gästrikland is a letter of 1553 in which the Swedish king grants permission to a farmer in the parish of Hedesunda to clear and maintain "a bodland called Gunnaboda" (Hedblom 1958:61). The scholar Folke Hedblom, who has studied the boda-names and also the bodland phenomenon most thoroughly states that the boda-villages of Gästrikland - in the same way as those in Hälsingland - have without any doubt undergone a development from a bodland stage.

From the neighbouring province of Dalarna in the west, we can in the same way draw attention to an old document from 1528 which mentions "a bodland called Lilessarbode" in the parish of Grytnäs (Dipl. Dal. suppl. p. 34 or.). Hence we find an existing use of bodland here, at any rate as late as the 16th century, and we can see a correspondence between bodland and -bo(da) here also. The picture for Dalarna and especially for the parishes north of the lake Siljan is unfortunately not so clear. Here we find a rather complicated rural system (see e.g. Montelius 1977, Montelius 1975, Frödin 1925, Erixon 1918:43 ff).

In Västmanland the circumstances are similar: we find in this province, in the same way, evidence in older sources that

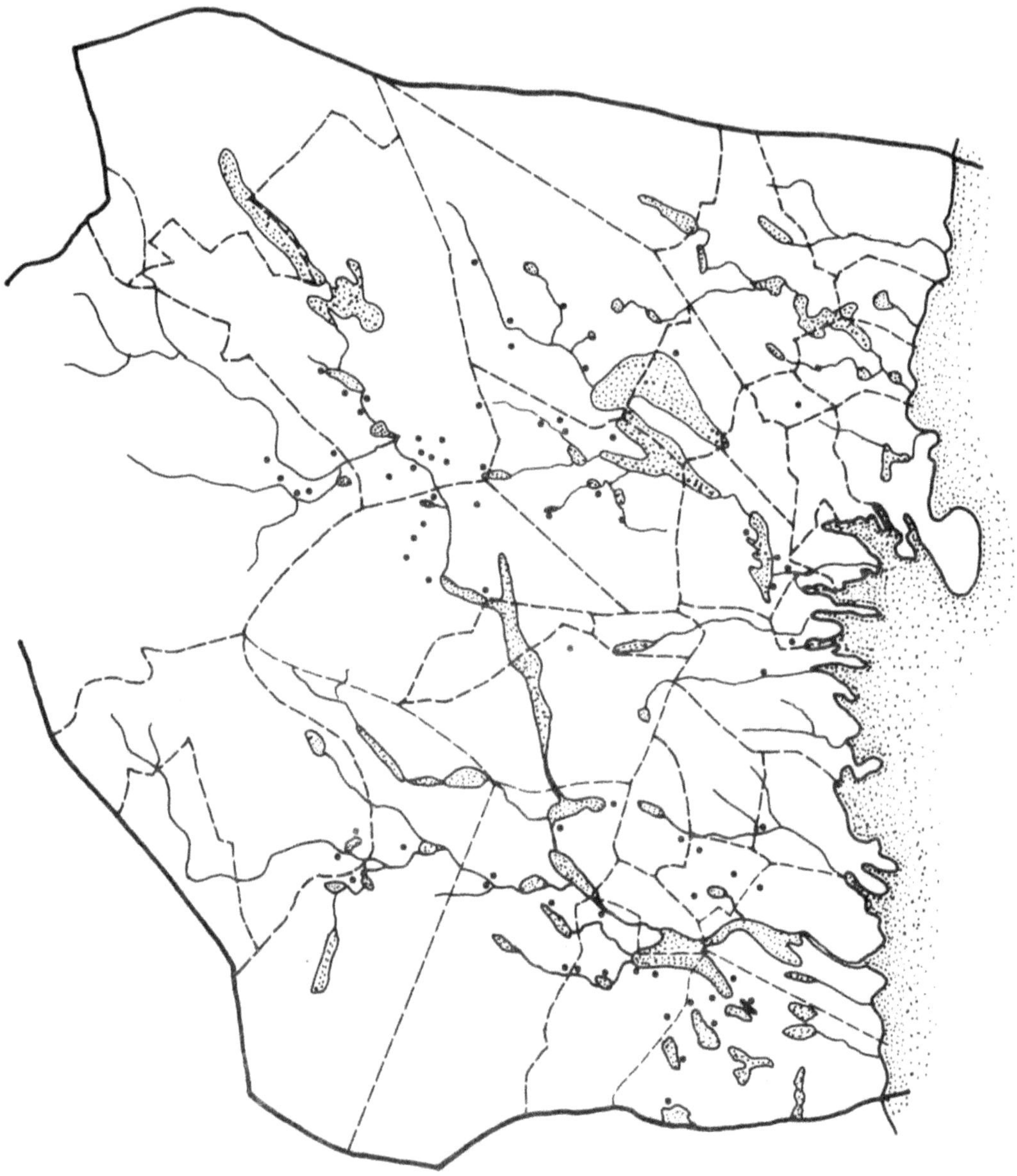

Fig. 18:5 The medieval boda-settlements in Hälsingland.

bodland were used at one time and that to a great extent they have names ending in -bo(da) (Hedblom 1947:138).

And finally to the province of Uppland. Here we find many instances of the place-name Bolandet (< Bodalandet). Further examination reveals that these villages or farms can appear under an alternative name, containing the element -bo(da) (Hedblom 1947:139). In the same way as in Hälsingland the boda-villages in Uppland appear to have been border settlements around the older villages. The boda-villages of Uppland are particularly to be found in the northwest part of the province, which, in comparison with Central Uppland, has a marginal location within the province. Hence the boda-villages bear witness to an extensive and expansive medieval colonisation.

There is thus no reason to doubt that we have been dealing with one and the same settlement phenomenon so far. The boda-villages of Hälsingland - which have a demonstrable connection with the so-called bodland - have basically the same historical background as the boda-villages in the other provinces mentioned. Furthermore these are in some cases, at least from the 16th century, known to be bodland.

Unfortunately the boda-names are a heterogeneous group of place-names. Although we have seen that there is a connection between boda-villages and bodland, not all settlements with names containing -bo(da) need originally be of the bodland kind. In the province to the north of Hälsingland, Medelpad, we find a hamlet Boda in the parish of Liden. This hamlet is situated on the River Indalsälven and is mentioned as far back as 1273 as "Fiskabua", i.e. the fishing hut or booth. The origin of this hamlet must be connected with the extensive salmon-fishing at this place, which goes back a long way in history. Thus, here we find what could be called periodically exploited "fishing booths" (Hvarfner 1964:23). Villages and hamlets with a name ending in -bo(da), and of similar origin are to be found all along the eastern coast of Swed-n.

A quotation from a letter from Dalarna dated 1452 - "the afore mentioned old werpae bodhe foundry" - indicates perhaps that the hamlet Varbo, as it is known nowadays, started with a foundry for producing some kind of metal (SD 6 p. 413 or; Stähl 1960:70). One scholar, the archaeologist Åke Hyenstrand, makes the tentative suggestion that in Dalarna there could have been what he calls "iron booths", i.e. periodically inhabited settlements where iron was produced. These could by and by have turned into purely agricultural hamlets and gradually become permanently settled (Hyenstrand 1974:218). This is an interesting idea and it seems to find support in the above quotation about the hamlet of Varbo. However, it should be stressed that geographically, these "iron booths" should mostly be found, if we can count upon their existence in the past, in that area which is known as Bergslagen.

The problem is further complicated by the fact that in the greater part of the Swedish seter area, seters have been given the appellative bodar (Hedblom 1945:6 and 1977:40). In other words we find a great number of seter-names with the element -boda(rna). These seter habitations can, however, be distinguished fairly easily as a settlement type from the boda-villages which can be assumed to have a bodland origin (Hedblom 1958:65). At the seters, as was stressed above, cattle-farming was the primary occupation and thus access to large grazing areas was decisive in the location of the seter. As a result the seters were located in the vast woodlands, the common land. As for the bodland, access to hay-fields and arable land - rather than access to pasture - primarily determined its location.

We can thus note that boda-names constitute a heterogeneous group of place-names. This has also been pointed out in the onomastic literature. Often the analysis has not been taken any further, however. In my opinion one can probably go a step further. Knowing the different settlement origins that the boda-villages can have, it is possible in a surprising number of cases to suggest what the origin of the settlement must have been by means of what is often a simple analysis of the sites of the boda-settlements in the area.

If the site is on the outskirts of the cultivated land in a settled area, but connected with the arable and meadow, we can conclude with a high degree of certainty that in this case we are probably dealing with a settlement with a "bodland" origin. If we find a settlement in the woodlands, perhaps some miles from the old villages, we should in most cases regard this as having a seter origin etc.

A common feature of this vast heterogeneous boda-name group is the fact that in all the settlements concerned some natural resource such as pasture, hay, iron etc. was exploited, initialy probably periodically. Gradually they were then permanently settled and finally became independent units. They were without doubt an asset during settlement expansion and were able to absorb a growing population.

Finally, to sum up, I would propose the following hypothesis: Most of the boda-settlements, which are not of seter origin, have in Hälsingland - and probably in the whole of Central Sweden - passed through a kind of bodland stage. These boda-villages are thus a consequence of so-called internal colonisation and were to a great extent originally exploited periodically.

NOTES

1. This phenomenon has been given different terms in Swedish geography. S-O. Lindquist uses the term "internal colonisation" (Swed. inre kolonisation) (Lindquist 1968:144 ff.) This term has also been used in a more general sense (Jonsson 1970:114, Bylund 1956). I. Jonsson uses the term "boundary-fixed colonisation" (Swed. gränsbunden kolonisation) for this particular phenomenon (Jonsson op. cit).
2. All quotations in this paper are translated, as literally as possible, from Swedish into English.
3. It is not possible, in this paper, to discuss the different place-name elements that denote medieval settlement expansion and intensified cultivation. I simply just refer to the following works: A general discussion of these phenomena e.g., Pamp 1974:48 ff., Ståhl 1976:83 ff., Lindroth 1946:45 ff., Hellberg 1967:468 ff., Hellberg 1980:36 ff., Ståhle 1954:17 ff., Helmfrid 1962:54 ff., Sandnes 1971:36 ff., Larsson 1964:81 ff. Works discussing different place-name elements e.g., -bo(da): Hellberg 1967:183 ff., Ståhl 1960:148 f., Envall 1960:29 ff., Hedblom 1958:57 ff., Ståhl 1957:30 ff., Brink 1979:19 ff. -torp: Hellberg 1954:106 ff., Hellberg 1980:68 ff., Ståhle 1954:17 f., Eriksson 1967: 28 ff. -måla: Ödeen 1927. -ryd: Ejder 1979, Holmsen 1961: 217 ff. -täkt: Ståhl 1960:150 ff. -hester: Ekenvall 1942. -arvet: Ståhl 1941:33 ff. -böle: Thors 1952:11 ff., Hellberg 1980:55 ff. (-säter: Hedblom 1945).

REFERENCES

BODVALL, Gunnar 1959: Bodland i norra Hälsingland (Geographica 36). Uppsala.

BRINK, Stefan 1979: Bodlanden i Järvsö socken (In: Uppsalastudier i namnforskning. Ortnamn och samhälle 5). Uppsala.

BYLUND, Erik 1956: Koloniseringen av Pite lappmark t.o.m. år 1867 (Geographica 30). Uppsala.
- 1961: Theoretical considerations regarding the distribution of settlement in Inner North Sweden (In: Geografiska Annaler 1960:4).
- 1968: Generationsvägor och bebyggelsespridning (In: YMER 1968).

Dipl. Dal: Diplomatarium Dalekarlicum. 1-3, suppl. Stockholm & Falun 1842-53.

EJDER, Bertil 1979: Ryd och rud (Skrifter utg. genom Dialekt- och ortnamnsarkivet i Lund 2). Lund.

EKENVALL, Verner 1942: De svenska ortnamnen på hester (Studier till en svensk ortnamnsatlas 1). Uppsala.

ENVALL, Petrus 1960: Benning och Boda (Ortnamn och kulturhistoria 2). Stockholm.

ERIKSSON, Manne 1967 : Tallen och det farliga torpet (In: Svenska landsmål och svenskt folkliv 90, 1967).

ERIXON, Sigurd 1918: Bebyggelseundersökningar. Öfversikt. Periodiska bebyggelsetyper. Fäbodväsen. (In: Fataburen 1918).

FRÖDIN, John 1925: Siljansområdets fäbodbygd (Skrifter utg. av Vetenskaps-Societeten i Lund 5). Lund.

HEDBLOM, Folke 1945: De svenska ortnamnen på säter. En namngeografisk undersökning (Studier till en svensk ortnamnsatlas 2). Uppsala.
- 1947: Boland (In: Namn och bygd 35, 1947).

- 1958: Gästriklands äldre bebyggelsenamn (Från Gästrikland 1957-58).

1960: Fäbod (In: Kulturhistoriskt lexikon för nordisk medeltid från vikingatid till reformationstid 5). Malmö.
- 1977: Ortnamnsproblem inom det svenska fäbodväsendet (In: Nordiskt fäbodväsen. Forhandlingar vid fäbodseminarium i Älvdalen, Dalarna, 1-3 sept. 1976). Stockholm.

HELMFRID, Staffan 1962: Östergötland "Västanstång" (In: Geografiska Annaler 1962:1-2).

HELLBERG, Lars 1954: Studier i de nordiska torp-namnens kronologi (In: Namn och bygd 42, 1954).
- 1967: Kumlabygdens ortnamn och äldre bebyggelse (Kumlabygden 3). Kumla.
- 1980: Ortnamnen och den svenska bosättningen på Åland (Ortnamn och samhälle 2. Ortnamnen och den forntida svea-staten 1). Uppsala.

HOLMSEN, Andreas 1961: Norges historie, fra de eldste tider til 1660. 3rd. ed. Oslo - Bergen.

HOVSTAD, Håkon 1979: Gårdsgrenser som hjelpemiddel ved bosetningshistoriske undersøkelser? (In: På leiting etter den eldste garden. Ed. R. Fladby and J. Sandnes). Oslo.

HVARFNER, Harald 1964: Fiskaren, laget och redskapet. Stockholm.

HYENSTRAND, Åke 1974: Järn och bebygelse. Studier i Dalarnas äldre kolonisationshistoria. Falun.

JONSSON, Ingvar 1970: Finsk kolonisation och bebyggelseexpansion i Norrland. (In: Migrationen mellan Sverige och Finland. Nordisk utredningsserie 14/70). Stockholm.

KARDELL, Lars, Dehlén, Rune, and Andersson, Birger 1980: Svedjebruk förr och nu (Sveriges Lantbruksuniversitet. Avdelningen för landskapsvård. Rapport 20.) Uppsala.

LARSSON, Lars-Olof 1964: Det Medeltida Värend (Bibliotheca Historica Lundensis 12). Lund.

LENAEUS, Knut Nilsson 1764: Delsboa Illustrata eller Delsbo Socken i norra Helsingland. Stockholm.

LINDQUIST, Sven-Olof 1968: Det förhistoriska kulturlandskapet i östra Östergötland (Studies in North-European Archeology 2). Stockholm.

LINDROTH, Hjalmar 1946: Bohusländska ortnamn och bohuslänsk bebygelsehistoria (Göteborgs och Bohusläns fornminnesförenings tidskrift 1945. Göteborg.

MONTELIUS, Sigvard 1975: Leksands fäbodar (Leksands sockenbeskrivning 7). Leksand.
- 1977: Fäbodväsendet i övre Dalarna. Stockholm.

PAMP, Bengt 1974: Ortnamnen i Sverige. 4th ed. (Lundastudier i Nordisk språkvetenskap. Serie B, nr 2). Lund.

SANDNES, Jørn 1971: Ødetid og gjenreisning (Skrifter utg. av norsk agrarhistorisk forskergruppe 1). Oslo - Bergen - Tromso.

SCHISSLER, Pehr 1749: Hälsinga Hushåldning. Stockholm.

SD: Svenskt Diplomatarium (Diplomatarium Suecanum). 1-. Stockholm.

STÅHL, Harry 1941: Bebyggelsenamnen på -arv(et) i Falutrakten (In: Namn och bygd 29, 1941).
- 1957: -bodha (In: Kulturhistoriskt lexikon för nordisk medeltid från vikingatid till reformationstid 2). Malmö.
- 1960: Ortnamnen i Kopparbergslagen (Kungl. Vitterhets Historie och Antikvitets

Akademien, Filologisk-filosofiska serien nr 7). Stockholm.
- 1976: Ortnamn och ortnamnsforskning. 2nd ed. Stockholm.

STÅHLE, Carl Ivar 1954: Ortnamn och bebyggelse i Västra Vingåker (Sörmländska handlingar 18). Nyköping.

THORS, Carl-Eric 1952: Om namn på -böle, med särskild hänsyn till de finländska namnen (In: Namn och bygd 40, 1952).

WESTIN, Josef 1930: Kulturgeografiska studier inom Nätra-, Näske- och Utbyåarnas flodområde samt angränsande kusttrakter (Meddelanden från Lunds univ. Geografiska inst. Avhandlingar 2). Lund.

ÖDEEN, Nils 1927: Studier i Smålands bebyggelsehistoria. Ett bidrag till svensk ortnamnsforskning (Namn och bygd, Bilaga C). Lund.

CHAPTER 19

PLAINS AND PERIPHERY: SETTLEMENT MORPHOLOGY IN THE MÄLAR BASIN, CENTRAL SWEDEN.

U. Sporrong
(University of Stockholm)

THE RESEARCH PROJECT

For a number of years I have been studying landscape development in the Lake Mälar basin in Central Sweden. The research project's proper name is 'The Mälar landscape between history and prehistory - a poorly documented period in the agrarian landscape's development'. The aim has been to start from today's landscape and study how specific features have evolved. Several familiar, characteristic forms in the landscape have their roots in the distant past and despite the dearth of sources it has been found that the early medieval period is an interesting phase in the history of the Swedish landscape. But the lack of evidence does pose problems. The traditional archeological source materials cease to occur around this time and written sources have not yet come into existence. It seems, nevertheless, that the period involved a number of events which had a marked impact on the morphology of the rural landscape. There were, for instance, changes in social structure, and changes in forms of settlement and cultivation, the conversion of the population to Christianity and the ongoing struggle between feudal and central powers. The retrospective approach, which calls for identification of the chain of landscape changes, right up to the present day, has proved advantageous. The benefits are of two kinds: the purely scientific results of a specific nature, and the possibility which this 'dynamic' procedure affords of explaining the background to most of the landscape's current features. Continual changes in the landscape have been generated by social upheavals, technological innovations, legislation and much besides. This in turn means that all the threads which go to make up the present landscape fabric cannot be disentangled without insight into the complete pattern of their development - a complex, troublesome form of research that tends to require a multidisciplinary assault. One is liable to lose sight of the overview. Each temporal cross-section has to be studied within the frame of its own particular conditions and circumstances. This is also why people today, with their very different way of life have such difficulty in perceiving the significance of spatial relationships within the landscape. In the past, particularly in the Middle Ages, a society, and its landscape, was more of a closed unit. People were probably more aware of the relationships which prevailed in their local surroundings and which affected their daily life, partly because the economic system in which they existed had such a high degree of self-sufficiency. Without under-estimating the complexity of medieval agrarian society, it seems reasonable to suppose that it is far more difficult for modern man to obtain an overview of the factors behind the landscape's changing morphology. This means that research also has a pedagogic role (HELMFRID 1980, LINDQUIST 1980). In order to perform these tasks of adding to and communicating knowledge one needs the support of a general viewpoint. At present, landscape research lacks general theory. For this reason we shall follow one of the more closely reasoned theoretical works (HUDSON 1969).

First, however, I should mention that the theoretical argument is to be exemplified in another context, with results from an extensive research program of a practical, specific nature which has been undertaken with the aid of field data, old maps and fiscal records. This material is to be published as soon as possible in the form of a monograph. The present paper simply provides some examples of the region's settlement characteristics. These illustrate the link with the physical geographical conditions rather than the importance of administrative influences on the landscape. The physical landscape constitutes the primary resource which man has exploited (SPORRONG 1976). Various types of land use have been developed, with varying success, depending on the particular properties of this primary resource. The man-physical landscape interaction is accordingly an important starting point for the study of settlement patterns. But of course, it also has to be combined with a theoretical viewpoint that explains development. What were the fundamental purposes behind the 'strategies' that people adopted in different contexts?

The aim of this presentation is therefore twofold:

a) to discuss one model concerned with the development of changes in settlement morphology in the Mälar basin from the year 1000 to the present, and
b) to use this model in a search for general features of development as a foundation for a 'regionalisation' of our findings, that is to say, to describe the Mälar region's settlement 'structure'.

THE LOCALISATIONAL PATTERN

Geographers have worked for a long time to explain the laws behind particular, current spatial distributions. The basic concepts can be said to have been concerned with form/process or geometry/movement interactions. This approach has certainly led at times to rather 'sterile' accounts of distributions but in recent years there has been a movement towards a more integrated approach - combining the viewpoints of geography and the other humanistic or social sciences. The age of classification can be said to be over but in many instances the geometric patterns do serve as good starting points for analyses. Examples of the criticism mentioned above are to be found, for instance, among modern geographers with a Marxist orientation (ANDERSON 1980).

In the present set-up for practical research, however, it seems to me that geographers are justified in appearing under the general label of spatial analysts. As far as research into settlement history is concerned, this view might be summed up as follows: the settlement pattern is the spatial expression of a continuous change in the population and its composition as a consequence of altered habits and needs concerning dwellings and productive land (Hudson 1969).

As a frame for a presentation of landscape development in the Mälar basin we shall therefore employ and develop the above-mentioned geographical perpective. This will be done with the aid of established geographic localisation and distribution theories, what might be called an integrated space/time model.

Colonisation

This term implies that a population invades a new area and in the process exceeds the boundary of its earlier settlement. As an explanatory concept the term has its roots in biological research, in which colonisation involves the utilisation of a particular given niche in the landscape, the boundaries of which are determined by a series of physical or physiological factors (heat, cold, light, moisture, etc.) Naturally, in cultural terms, there are other limiting factors e.g. the transport factor in relation to the nature of production. The niche's boundaries are not absolute, but rest on a balance struck between a number of basic factors (STENEROT-SILLEN & STJERNQVIST 1950). The main categories are: population levels, resources and their distribution, technology, social organisation, values, interests and conflicts and external relationships.

A change in any one of these (e.g. in the technology employed) alters the balance, and consequently the form and extent of the niche in question; an example, of such changes is seen in the construction of a road in pathless territory, which greatly enhances the transportation potential and cuts the time factor. Of course there are other, less drastic changes that are equally far-reaching. The introduction of the potato crop in northern Sweden seems to have permitted an appreciable shift in man's 'marginal position' there. Similarly when it was learned how to increase the flood meadows' yield of Carex hay by up to tenfold, it must have decisively underpinned stall feeding (BYLUND 1960). The biological aspects underscore the interplay of factors. Nevertheless, there is often a 'minimum' factor which tends to limit development. This may be clearly visible in the landscape, e.g. the tree line imposed by altitude.

The original colonisation creates an initial population in a given spatial position (niche). This population becomes a starting point for studying the subsequent development of the landscape. The border zone around the population is the most sensitive, changeable area. One can perhaps speak in terms of centre and periphery, the centre invariably being the most densely populated part.

Diffusion

Following the original colonisation we get a new generation a short distance away from the primary unit. 'Vacant' areas nearby are filled first. The settlement agglomerations grow and demographic pressure increases.

The process of dispersion is sometimes referred to as diffusion: - to start with pressure increases in a given core area, stimulating subsequent colonisation, which involves a change in the size of the populated area. Eventually, the original population pressure diminishes - a natural consequence of the colonisation process. During the dispersion phase the settlement units tend to gather into clusters; the system strives towards equilibrium but imbalance prevails (expansion or contraction) as long as the clustering tendency lasts. Colonisation and dispersion may occur simultaneously.

Bylund (1960) is one of those who have studied the course of colonisation in the inland of northern Sweden. He has constructed a series of models with increasingly complex assumptions and then compared them with what actually happened according to the written sources. An important observation is that the process is characterised by gradual, limited dispersion from the primary units, without marked 'waves' or 'jumps'.

Stability and competition

The weak units succumb while the strong survive and develop towards a regular settlement pattern. The stabilisation phase implies that for some reason the conditions for a continued dispersion of settlement units have changed. Usually the colony has come up against some physical barrier, e.g. a limit to the land that can be cultivated for internal needs (due to a lack of cultivatable land, climatic obstacles, etc.) Dispersion may be arrested by other types of barrier, e.g. price changes for grain products. The deviation from continuous growth also heralds a phase characterised by some degree of stability - at least at first. The remaining units compete for available 'space' and concentrate initially on retaining what they have secured, though later they also try to expand at the expense of weaker competitors. Growing shares of the land suitable for cultivation are concentrated in the hands of a diminishing number of owners, farms or settlement units, assuming that the situation remains viable. A lack of alternative occupations may lead to a fragmentation of property (GEERTZ 1971).

This means that, ideally, the surviving units (hamlets) tend to resemble one another in size, conformation and social organisation and that the distance to neighbours (i.e. the distance between settlement units, usually villages or hamlets) tends to be roughly the same throughout the greater part of the cultivated landscape. One might conceive an optimal relation between distance and settlement size. Such 'mature' landscapes are to be found in the Swedish plains in, say, the 17th and 18th centuries. Several instances of this level of development can be cited from the Mälar basin, and in central Sweden, generally, the geometrically regulated hamlet can be regarded as a manifestation of development's mature, stabilised phase.

The system tends to function in such a way that when the 'goal' of development has been achieved, i.e. when stability has been established, the entities present do all they can to preserve their positions. In other words, this stability is underpinned by socially determined decisions concerning diverse regulation to do with land ownership and land use. As long as equilibrium prevails, the community in question can be said to be static - there is a relative absence of change and development. However, when the equilibrium is disturbed one can often discern this in altered settlement patterns. The regularity is transformed into clustering or agglomerated distributions (LINDQUIST 1968, 137: SPORRONG 1969).

The development cycle outlined above can of course be repeated, move off course or even be reversed. The causes of such aberrations are legion but usually they lie in external infuences or in planned, often centrally directed, intervention, for example, the medieval North German colonising ventures and today's Swedish national planning.

These ideas, focussing upon pattern of localisation were inspired originally by the principles of biology. They seem more attractive as an explanation for landscape development than, for instance, ideas of the von Thunen type, which propose a gradient function, though this does not always decrease or increase linearly. Furthermore, the traditional, geometric approaches within geography tend to be static, and do very little to explain the dynamic and differentiated development of the landscape - a development that in central Sweden at least, is strongly linked to the landscape's natural resources.

THE MÄLAR LANDSCAPE: A CASE STUDY

The region around Lake Mälar is largely a typical basin, enclosed by higher land except to the east. The more elevated land is pierced, particularly to the north, by a series of river

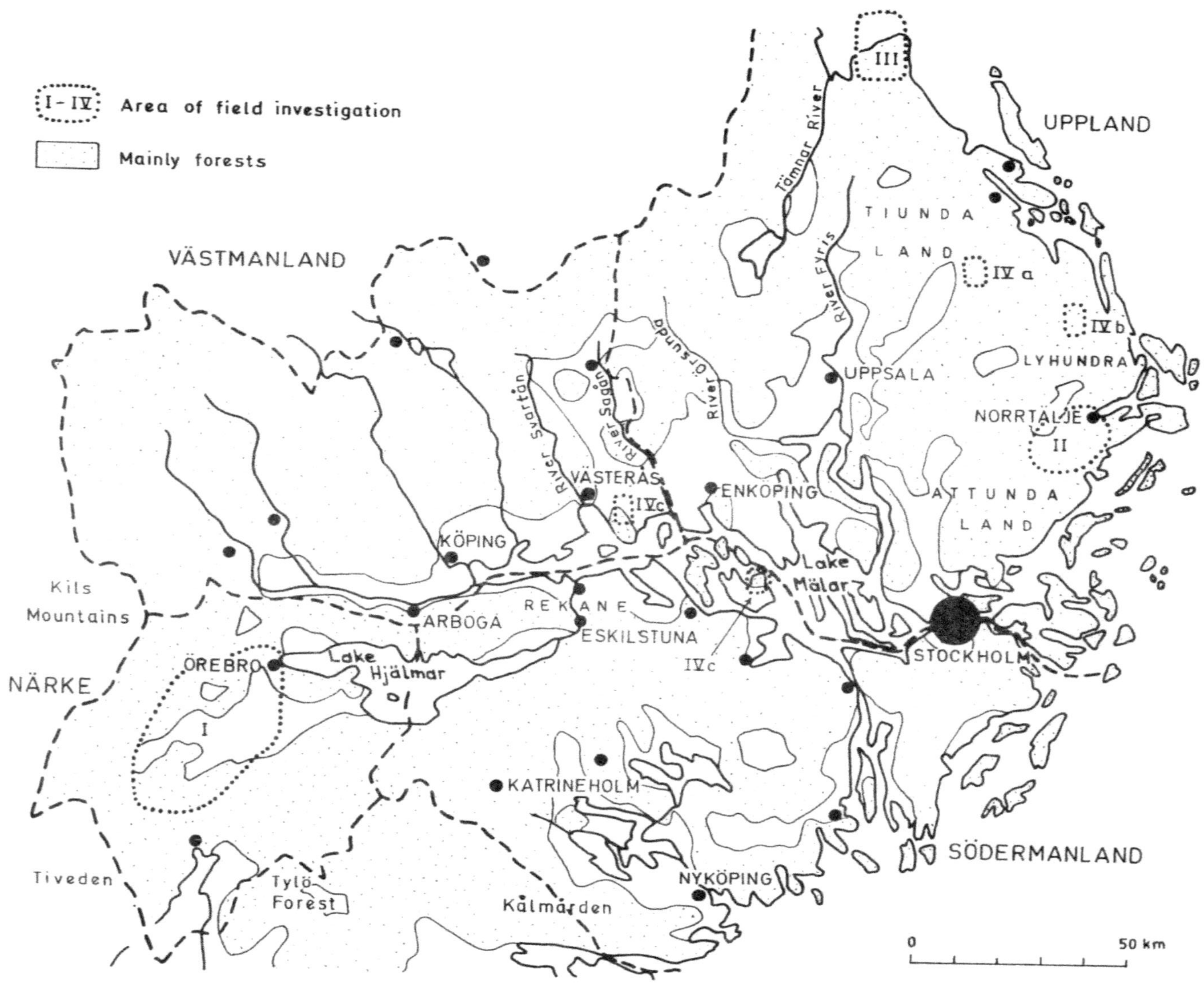

Fig. 19:1 Lake Mälar and surrounding region.

valleys of varying breadth (Fig. 19.1). On closer inspection the ground is characterised by what are usually known as fissure valleys - a type where clays in low-lying parts contrast with moraine hills of varying extent. This combination, together with favourable climatic circumstances, seems to have suited early farming; dwellings were built on the heights, and the clay soils were cultivated. The region saw expansion throughout the period studied. Relatively large plain settlements arose at the intersections of the fissure valleys, while the rest of the region mainly consists of a mosaic-like landscape with small hills, the typical fissure-valley form. Here and there tree-covered moraines and rocky ground form lacunae in the cultivated landscape or were colonised very late. The prehistoric finds suggest that settlements were developed in the first place in the region's central parts, Södermanland included (HYENSTRAND 1974), but the river valleys also have a long tradition of settlements. The forest lands in central and northern Västmanland and in north-west and northern Uppland seem to have attracted their first settlements in medieval times, i.e. they were used for permanent dwellings and cultivation only at a relatively late stage. Between these two 'poles' there is a varied transitional settlement. It, too, has a long tradition.

These relations can be presented in other ways. Archeologists claim that the size of settlement units is a function of their age. In the Mälar region the largest units (villages) lie in the central parts but these areas are not always those with the most plentiful prehistoric remains. This suggests that post-prehistoric expansion and development partly occurred in 'new' areas outside the settlements with a wealth of prehistoric finds.

So much for the main features of colonisation and expansion in the Mälar basin. It can be noted that in the period 900-1850, which is largely the span covered by the research project, the productive agrarian landscape underwent a series of changes that are common to the entire region. In the Middle Ages, and possibly earlier, rules were established for the inheritance, procurement and ownership of farm land. The land was assessed and could thus be taxed; it could also be leased out. During the Middle Ages new techniques for cultivation were introduced with such vitality that the entire Mälar region had a uniform fallow system (mainly two-course rotation) with fragmented holdings in the form of strips (predominantly long, narrow parcels). Later, in the 19th century, a laga skifte, or enclosure was undertaken thoughout the region and finally breaking up the field patterns which had been built up on principles which were considered right and proper in medieval times. The study period can accordingly be said to embrace many of the important phases in the course of development discussed above in more theoretical terms. The phases of colonisation and expansion were already, by A.D. 900, complete in the central settlements but were under way in the periphery. Expansion in the Middle Ages led to the establishment and development of completely new regulatory systems. The consolidation phase took over and in general the central farm areas did not change until their stability decreased in the 19th century with the dissemination of many new ideas in politics, economics and production technology. This altered the basic circumstances and the agrarian society was transformed.

There are also examples of landscape features which do not apply throughout the region. One concerns the geometrically regulated hamlets. While these can be seen as a function of the regulatory system from the Middle Ages, they can also be classified as a phenomenon which was not implemented in full, because in most places it was not needed. The storskifte (strip consolidation) of 1749 can likewise be counted among such abortive events (HELMFRID 1960). The primary reason was no doubt the local agrarian community's lack of interest in the measure, which was often seen as unneccessary (only about 50% participated). The laga skifte of 1827 was a very different matter just because the farmers were prepared; indeed, it was they who demanded it.

Development in the Mälar region - three examples

Returning now to the physical geography of the Mälar region and simplifying somewhat, one can say that there are three main types of landscape. One type comprises the plains and major river valleys. These areas were already being cultivated to a relatively high degree in the Middle Ages and arable farming is the predominant form of production (GÖRANSON 1977). There have been strong incentives to adopt the organised field systems and land parcelling arrangements. As a result, the agrarian landscape underwent several changes towards more concentrated and regulated forms, ultimately acquiring stability.

A second type is the genuine fissure-valley landscape (in Sörmland and eastern Uppland). Here the resources for arable farming are much more restricted - the terrain is far more varied. As a result, the settlements display a marked adaptation to the terrain and the units are accordingly smaller throughout. There was less need to implement the medieval regulatory ideas and there are few instances of units being amalgamated into regular hamlets, a phenomenon that is common in the plains. The landscape's physical conditions accordingly bring development to a halt during the dispersion phase, i.e. during the internal colonisation of the landscape.

The third type, finally, consists of the forest settlements. The resources are such that the forms of settlement here are much less developed - they are close to the pattern of pure colonisation and markedly adapted to the terrain. The regulatory impact of arable farming is absent because these settlements concentrate more on livestock or forms of mixed production involving forestry and fishing.

(i) Plains and river valleys

In these areas a physically well-organised rural landscape was already present in early medieval times.

a) Plain settlements. These areas are - and probably have always been - the most attractive in the Mälar region. Prehistoric finds indicate that they were colonised at an early stage. There is a good resource base and access to it is fairly uniform over the area. The dispersion phase can therefore proceed almost exactly according to the biological principles discussed earlier. The primary units spawned offsprings a suitable distance in the surrounding landscape. This is an instance of favourable conditions for an expansive utilization of cultivatable land.

In time, however, conflicting interests arose within the primary unit-offspring system as well as externally, between colonisation ventures. There is then a transition to the phase which we called 'stability and competition' in the theoretical account; in other words, there is a regulation of further development, e.g. the introduction of boundaries. It is clear that this phase involved a fairly marked process of contraction in the agrarian landscape. The settlements in these areas are often concentrated to the centre of the production area and cultivation becomes more intensive (Fig. 19.2). The geometric regulation of the settlements is a manifestation of this. One then arrives at the mature, stabilised phase of the development cycle, which involves competition between the growth of the surviving units. But this is also a natural part of the contractive phase. The agrarian landscape acquires established forms and fixed rules that are so integrated with the social organisation that a 'free' development is rendered impracticable. The landscape takes on a relatively stable form and it is not until a new, liberal attitude to the rules for inheritance and procurement, as well as new forms of management, emerge in the 18th and 19th centuries that it becomes more fluid again. Typical examples of settlement areas of this kind are to be found along the north shore of Lake Mälar (from Arboga in the west to the Enkoping district in the east), around Uppala, in the Rekarne district and to some extent in Lyhundra härad in eastern Uppland.

b) River valleys. Development here is similar in principle to that in the plains. As the cultivatable land occurs in 'strips' in this type of terrain, land use is adapted to this pattern. The hamlets are located in a particular zone between the coarser and the finer types of soil and the linear pattern of the watercourse leaves its mark on the landscape as a whole. Good examples occur along the River Tämnar and in the valleys of Västmanland but perhaps the best instance is the

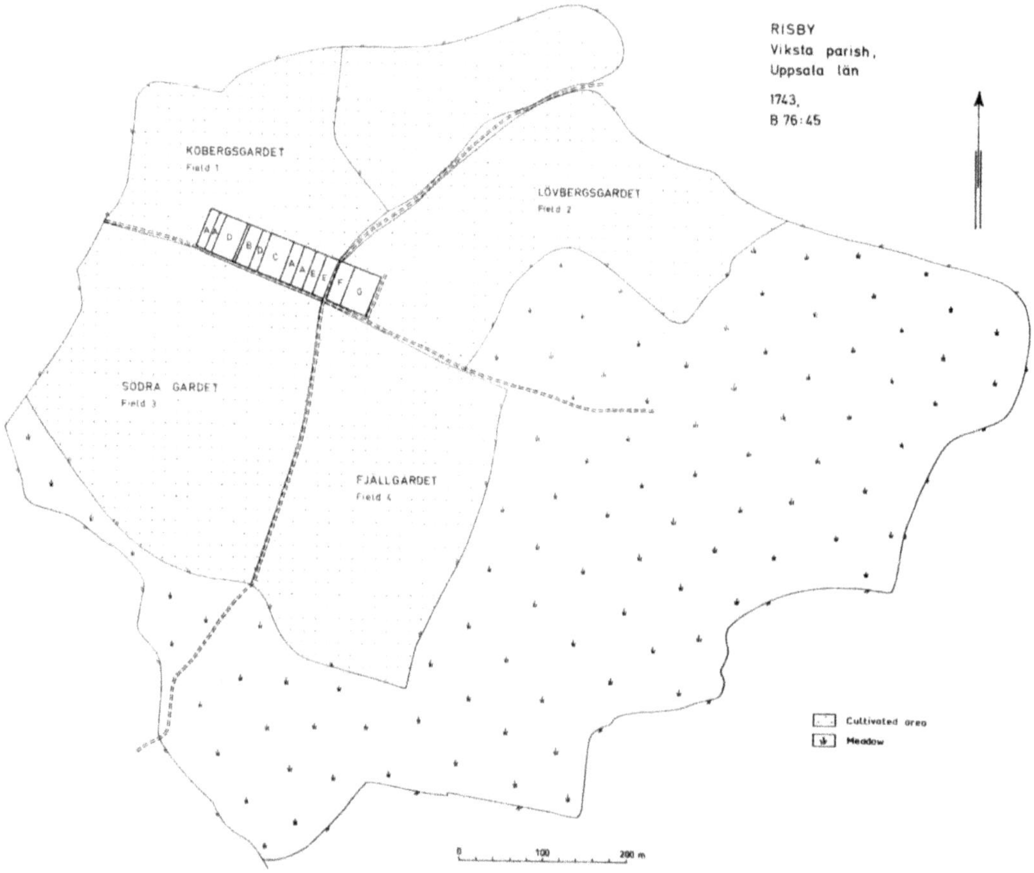

Fig. 19:2 Risby, 1743: settlement at the centre of its cultivated area.

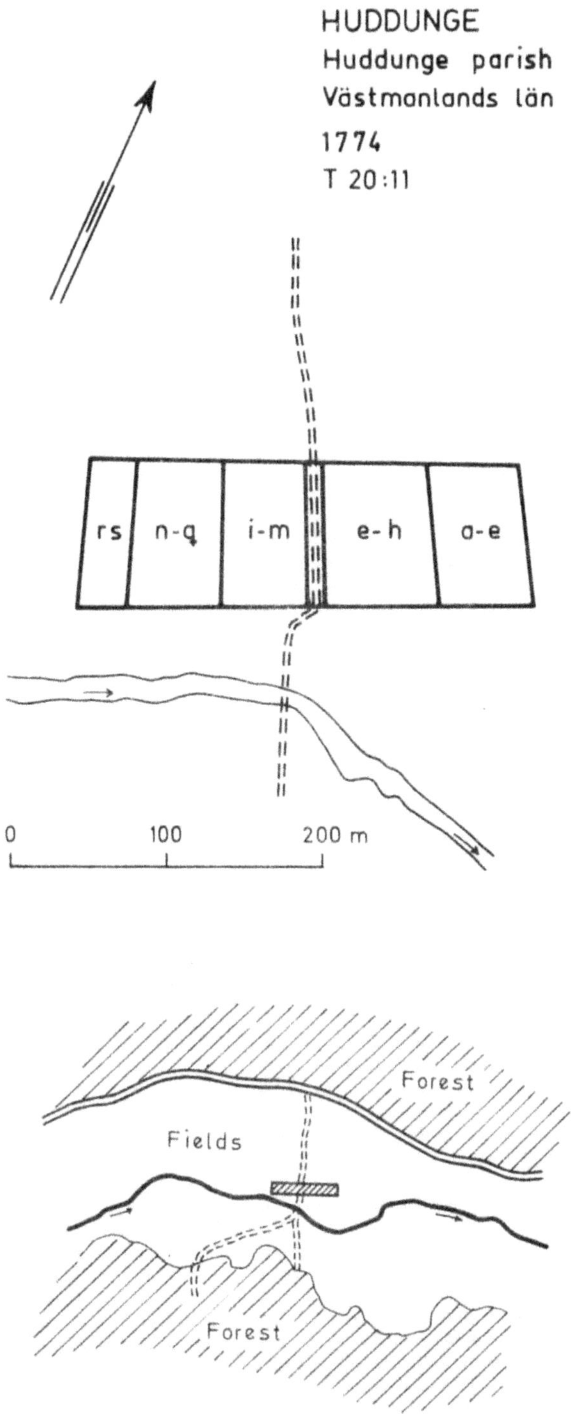

Fig. 19:3 Huddunge, 1774: hamlet with regulated layout.

upper valley of the River Örsunda (Västetlövsta parish).

One of the hamlets, with a characteristic regulated site, is shown in Figure 19. 3. Metrologic analysis of the plan suggests that site regulation, i.e. the 'final' phase in the development of the early settlement, occurred in the late Middle Ages.

(ii) Fissure valleys

Access to cultivatable land is considerably more finite here and the production system is more varied, i.e. arable farming is not so predominant. Even so, the colonisation phase parallels that in the plains, while the subsequent expansive phase takes different forms and growth is slower. The offspring of the primary units, instead of 'budding off' the latter, are dispersed over wider areas, depending on the occurrence of cultivatable land. The external conflicts are resolved in the same way as the plains in principle but here the landscape does not permit a proper 'construction' in the physical sense - the supply of resources is too dispersed. This results in more numerous but smaller settlement units in the typical fissure landscape. One might say that development was arrested in the course of the dispersion phase (Fig.19.4). There are, of course, no large, dominant agricultural enterprises - instead of a series of adjacent farm settlements one gets a scatter of micro-settlements that afford little opportunity for further 'rationalisation' in the form of a concentration into large villages. Many of these small settlements do admittedly belong to extensive manor-farm systems (in Södermandland) but this has not affected the actual pattern of settlements, a circumstance that is typical of conditions in Sweden.

The growth rate can accordingly be said to have accelerated relatively early in these districts. The farming communities in the fissure-valley landscape survive in a small-scale form that is one of the hallmarks of the Mälar region's agrarian landscape. The most typical areas in this respect lie in Södermanland and Uppland.

(iii) Forest settlements

In the more marginal parts of the study region there are undeveloped forms of settlement that still bear traces of the colonisation phase. It is only during the last two centuries that arable farming has come to be dominant feature in the landscape - not least on account of major lake drainage schemes in the middle of the 19th century. Before that livestock management and the related pasturelands predominated. Here one can say that the settlement units developed 'spontaneously', in response to the terrain, and never reached the intricate forms of the hamlets down on the plain - at all

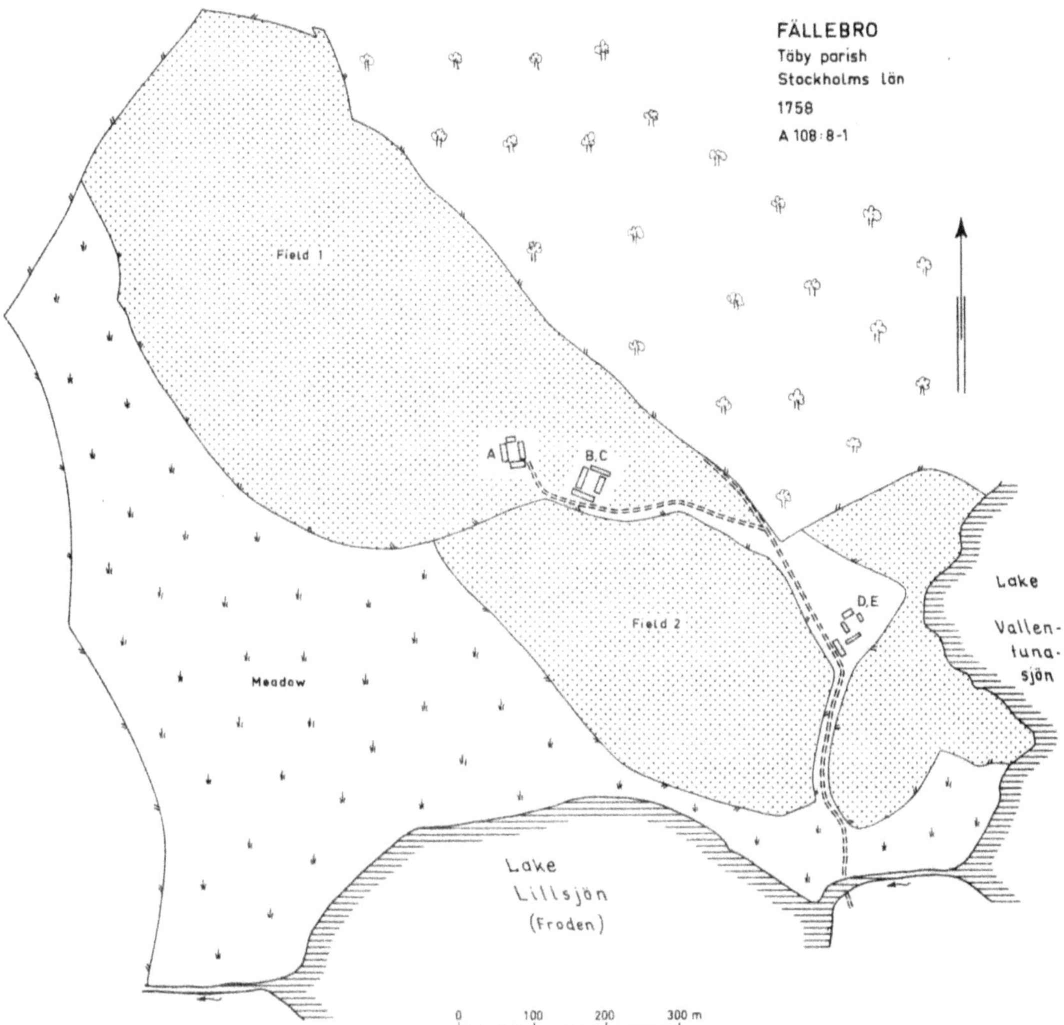

Fig. 19:4 Fällebro, 1758: dispersed settlement.

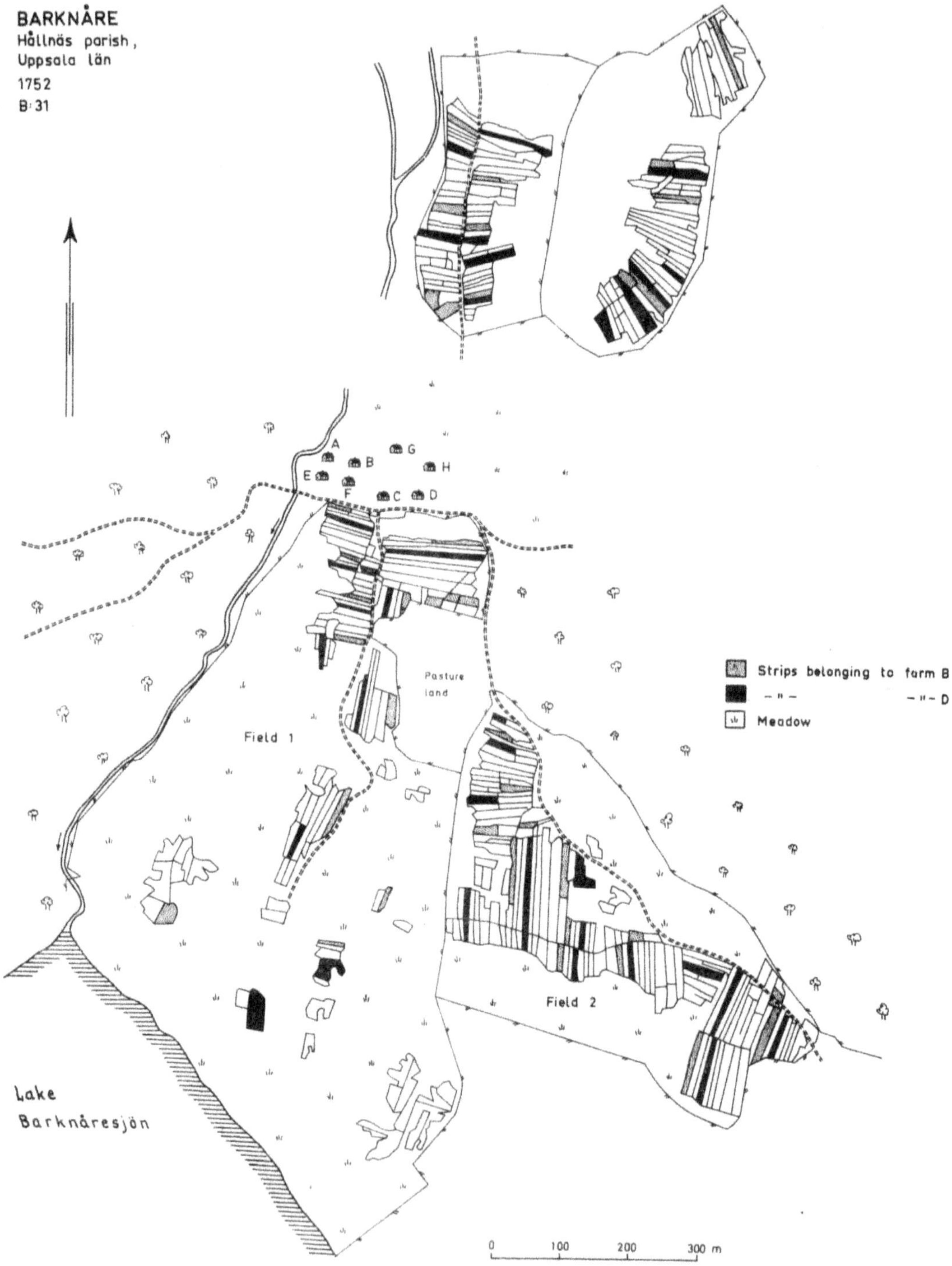

Fig. 19:5 Barknåre, 1752

events not as regards the actual buildings and the arrangement of building sites. But the forms of cultivation were often the same as in the plains. The systems for land parcelling and fallow farming were the same, though physically simpler here. These marginal areas contain many relic features, preserved as 'live' or abandoned forms to the benefit of agrarian research. Figure 5 shows a hamlet in northern Uppland where marshy land was brought into cultivation fairly recently, together with small patches of drier terrain. The settlement lies quite unsystematically along a stretch of road and has developed through successive subdivisions of the original unit. The arable area in the right foreground (one of the hamlet's two fields) was brought into cultivation - judging from the field material - in the early Middle Ages. However, development was arrested early on - there are no opportunities for further rationalisation and improvement. Today, the fields have not even been adapted to modern mechanical operations, still retaining traces of lines of stones resulting from field clearance. Unlike the examples above, this shows an agrarian landscape in regression.

SOME CONCLUSIONS

My examples show that in central Sweden there was fair correspondence between the physical resources of the landscape and the characteristics of the settlement patterns and forms which evolved. This is because in this particular environment physical circumstances rigorously condition the availability and quality of the arable and meadow lands. Furthermore, the observable patterns and morphologies can be related to Hudson's model of settlement development, involving well-defined phases in which a dominant process operates, colonisation, diffusion and stability with inter-settlement competition, and in each case it can be seen that within each terrain type physical circumstances provided a powerful constraint upon the expected sequence if socio-economic events, framed with reference to the model.

REFERENCES

ANDERSON, J. 1980 Towards a Materialist Conception of Geography. Geoforum vol. 11.

BYLUND, E. 1960 Theoretical considerations regarding the distribution of settlement in inner north Sweden. Geografiska Annaler.

GEERTZ, C. 1971 Kolonialism och jordbruk-sutveckling. (Lund)

GÖRANSON, U. 1977 Kulturlandskap och samhällsutveckling. En undersökning av rumsliga förändringar av kulturlandskapet i Torstuna socken i Uppland från 0 till 1650. (Stockholm).

HELMFRID, S. 1960 The Storskifte, Enskifte and Laga skifte. General Features in Sweden. Geografiska Annaler.

HELMFRID, S. 1980 Kulturlandskapet och forskningen. Vitterhetsakademiens konferenser 4. (Stockholm).

HUDSON, J. 1969 A location theory for rural settlement. Annals of the Association of American Geolrapheus. 1969 s.265.

HYENSTRAND, A. 1974 Centralbygd - Randbygd. Strukturella, ekonomiska och administra huvudlinjer i mellansvensk yngre järnålder. (Malung)

LINDQUIST, S -O. 1968 Det förhistoriska kulturlandskapet i östra Östergotland. (Stockholm)

LINDQUIST, S -O. 1980 Tankar kring ekonomiska kartan. In "Inventori in honorem - en vänbok till F. Hallberg. (Stockholm)

SPORRONG, U. 1969 Kulturlandskapsutveckling och administrativ indelning. Ymer.

SPORRONG, U. 1976 Mälarlandskapen mellan historia och förhistoria. En källfattig period i kulturlandskapets utveckling. Dess betydelse för dagens landskapsbild.

Meddelanden från Fars nr 30.
(Kulturgeografiska institutionen
Stockholm).

STENEROTH-SILLEN,
M. -STJERQUIST, PER. 1980 Rattens roll i landskapsomvandlingen. Delegationen for långsiktsmotiverad forskning. Rapport nr 30 S.

CHAPTER 20

MEDIEVAL DESERTED FIELDS AND DESERTED VILLAGES IN THE LOWER HARZ MOUNTAINS - AN INTERIM REPORT

M. Linke
(Martin Luther University, Halle - Wittenberg).

As part of a general investigation into the genesis of the cultural landscape of Halle, there was in the Sixties a considerable amount of historico-geographical research in the Lower Harz Mountains. Krause in his dissertation of 1967 investigated the water network and its reorganization to meet the needs of medieval mining. Neuss (1971), - from a historic point of view and - Käubler (1968) and Mey (1964) - from geographical ones paid special attention to the abandoned land in the Lower Harz mountains. Medieval mining and industry were investigated by Oelkes (1966) and Flechtner (1964) and the author followed up the question of soil erosion in the historic period and, connected with it, the effects of human disturbance on soils and relief. After a gap of some years these historico-geographical investigations have been revived in the geography department at Halle and it is now possible to give a first interim report.

Physical background

The Harz mountains were lifted and folded in the Tertiary period as a result of disturbances, which led to the dominant form of a lopsided fault block. Near to the border with the Federal Republic of Germany the granite of the Brocken (1142m) rises steeply against the northern foreland of the Harz mountains and against the neighbouring peneplains. At a line running almost north-south, the upper Harz mountains change to the Lower ones, which are the subject of this paper. The height of the Lower Harz mountains declines from about 600m at the bottom of the Brocken massif to about 250m in the eastern part. The slightly undulating plateau of the Lower Harz mountains is characterized by greater relief only where there are monadnocks or where the plateau is dissected by rivers, for example, the Bode draining to the north, the Wipper to the east, and the Selke and Thyra, which flow southwards. Only here does the landscape have a mountain character. Almost everywhere soils are thin and poor in nutrients. Only on small areas of limestone near Rübeland, and where there is a thin loess cover on the crystalline rocks are soils better for agriculture. On the elevated plains loam and sandy-loam soils predominate. Organic soils and sediments of loam soils are found near springs in the uppermost parts of the valleys. Climatically, too, the Harz mountains can be clearly distinguished from their forelands. From over 1000mm at the bottom of the Brocken massif near Schierke rainfall diminishes to 500mm at the most eastern edge of the Harz mountains. Further east, in the rain shadow of the Harz, rainfall is below

500mm. With an average annual temperature of 5.2°C in the west (Schierke - 613m) and 6.5°C in the east (Harzgerode - 399m) the Lower Harz mountains are significantly cooler than their forelands (Eisleben - 120m, 8.8°C).

The vegetation of the Lower Harz mountains has been greatly influenced by human activity. On the eastern plateaux fields and meadows predominate, and only on the steeper slopes of the valleys do we find continuous woodland. Woodland increases in extent westwards. Besides clearing for arable land it was mining above all which removed the natural vegetation. The original broad-leaved forests of the Lower Harz mountains especially in lower regions consisted of oaks and hornbeams. At a height of between 400 and 450m they changed to red beeches, which, up to a height of about 1000m were dominant at the beginning of our millenium. But by the 12th and 13th centuries spruce has began to extend. After a short re-colonization by beech, which can be explained by the temporary retreat of mining, spruce began to spread again in the Lower Harz mountains.

The economy of the Lower Harz mountains in the Middle Ages.

There is evidence from a few sites on the edge of the mountains that in the Paleolithic the Lower Harz was visited sporadically by hunters and gatherers. Mesolithic artefacts are more frequently found and from further into the mountains, but more extensive settlement of the mountains took place only in the Neolithic. Probably since the end of the Stone Age local mineral ores have been exploited. There is evidence for settlements in the early Bronze Age but after this finds decrease and by the early medieval period there is no evidence of settlement in this region.

Maybe the four villages whose place-names end in the suffix -ingen. Schnebelingen in the land of Königerode (Neuss 1971, p. 333), Elfingen and Pferdingen near Tilkerode (NEUSS 1971, p 83. p 260) and Redlingen near Abberode (NEUSS 1971, p. 284) were founded by northern Swabians in the 6th century as Neuss thinks, but this cannot be proved. Even if it is true for these places one cannot say that the lower Harz mountains have been extensively settled by northern Swabians, for there are no other names which end with -ingen.

We know from different sources, that the settlement going from east to west did not start before Carolingian time. One of the most important sources is the early ninth-century Hersfeld tithe levy which lists several settlements in the southern Lower Harz mountains. This did not include Crown land, which was situated mainly in the western Lower Harz mountains and where there were no settlements even in the later Middle Ages. There existed only the royal hunting lodges Bodfeld, Hasselfeld, Selkenfeld or Siptenfeld. From the supposition that the settlement of the king's land started only at a late period it follows that the places west of the Ilfeld-Gernrode-line are younger than those to the east of it. (Exceptions are the names ending with -feld and -ingerode: the latter may go back to foundations by the North Albingians in the 11th century.) It also follows, that the -hain and -hagen place-names which indicate clearance of woodland to the west of this line must be younger than the eastern place-names with endings in -rode. It must be noticed also that east of this line occur single -feld and -hain or -hagen place-names from the 10th and 11th centuries (Neuss 1971, pp. 263, 370). The -berg and -stein place-names of the western Lower Harz mountains might be younger, for, as we know, this royal land was given up step by step only in the 12th century.

By the end of the 9th century a few -rode names already appear to indicate the spread of a rural population into the Harz forests. Clearing of woodland increased more and more and probably reached its greatest extent about 1200. 117 or 41 per cent of a total of 283 place-names of this period are names ending with -rode. Only 11 and 19 are respectively -schwende and -hagen names. The following table gives a comprehensive survey (MEY 1964, p.67):

Table 1

Place-name endings	settlements in the Lower Harz mountains about 1200	number of deserted settle-ments	settlements in the Lower Harz about 1500	desertion quotient
names of uncertain origin	9	2	7	22%
-leben	4	3	1	42%
-ungen-ingen,	6	5	1	83%
-hausen,-heim	4	3	1	75%
-feld	17	12	5	71%
-dorf	36	27	9	75%
-burg	11	9	2	82%
-berg,-stein	16	5	11	31%
-bach	12	9	3	75%
-rode	117	79	38	68%
-schwende	11	7	4	64%
-hagen	19	17	2	89%
-eck	14	12	2	86%
slav.place-names	7	6	1	86%
	283	196	87	

It must be said of the medieval period that hamlets and single homesteads were typical forms. Käubler has identified some small one- or two-line linear villages (1968, p. 52 ff.)

In all probability colonizing farmers discovered mineral veins at the surface, and as a result there developed some mining activity. Certainly mining began to develop in the eastern Lower Harz mountains at an earlier period but it never reached the intensity of mining in the western Lower Harz, so that the most eastern settlements did not originate from mining. The attempt to divide the settlements founded by the 14th century according to the employment structure and to their function into older (agriculture and animal husbandry) and younger (mining and metallurgy) can be accepted only with reservation, for the most eastern settlements have been concerned with mining too. In the middle Lower Harz mountains it is likely that there were mining settlements before 1000 A.D., for in 933 there is mention of a mint in Harzgerode. By 1100 in Gernrode in the northern Harz mountains mining must have taken place, but widespread mining did not exist before 1200. For the upper Selke valley prospection permits for minerals were granted about 1300. Archival references do not permit closer localization of this old mining. About 1300 there were ironworks at the edge of the Harz mountains between the streams of the Ecker and Holtemmel. In 1355 the ironwork in Tanne is first mentioned. But by this date we come to a period when, on the one hand, there was more desertion of settlements and, on the other, a great mining boom, which by necessity must have led to the foundation of new settlements. In the county of Wippra, near Königerode and Horla, near Harzgerode, Günthersberge and Stolberg, mining was carried out, and near Stolberg and Strassberg were iron works (BRUNING 1926, p. 97, ff.)

In the middle of the 15th century, there were a lot of iron works in the western Lower Harz mountains, which naturally had settlements in their direct neighbourhood such as Rübeland, Altenbraak, Trautenstein and Neuehütte. Mining was carried out near Büchenberg and Hüttenrode. The Stahlberg mountain near Neuwerk was already supplying the ironworks in 1344. With the development of the mines and ironworks the need for wood increased and sawmills were established. The oldest of them is mentioned in 1471. We especially point out this fact, because undoubtedly the desertion process, on the one hand, and the mining followed by the building-up of settlements, on the other, should be brought into closer relation, for mines, hammermills, ironworks and mills required many workers. The cultural landscape of that time was directly influenced by the many hydraulic structures within the Lower Harz mountains. The reconstruction of river courses, the building of artificial ditches (as, for instance, the Worms ditch which existed before 1465 and which connected Wormke and Zillierbrook across a watershed) and adits, the building-up of ponds to drive mines, mills and so on made changes in the landscape which are still visible today. Artificial ditches still carry water; parts of the ponds are used for bathing or drinking water. However, some of the hydraulic structures of the Middle Ages have disappeared or lost their original function.

Resuming we can say that many people in the 14th and 15th centuries earned their living in the Harz mountains both from agriculture and from mining. The woodlands disappeared in favour of cultivated fields. These were the settlements of farmers and miners with fields and meadows. The first important mining phase of the Lower Harz mountains ended at the end of the 16th century. The Rhineland then became more important and the mining of the Harz mountains could not stand up to this competition.

The deserted fields and villages in the Lower Harz mountains

The period of desertion in the Lower Harz mountains, as in other parts of Germany, led to a great decrease in the number of settlements (see Table 1). About 1500 there still existed 87 settlements: the greatest number of desertions had been in the group of -rode settlements, only 38 of which still existed at that time. What were the reasons for this? In many areas of large villages in the Lower Harz mountains today there exist fields of deserted villages which are still cultivated or which have been either wholly or partially overgrown by forests. In Königerode there are the abandoned lands of 9 villages including their fields which in spite of different times of foundation in the high Middle Ages existed at the same time and have been deserted at different times. The great nucleated village of Königerode may have been developed after the desertion of other places, for the original field of Königerode is much too small for the village of today. The fields of Dankerode, too, include those of three deserted villages. From these settlements only rarely do any structures survive. But often in documents it is said for decades after the desertion that the fields were cultivated by farmers of neighbouring villages. So in 1533/34 four farmers of Königerode cultivated $5\frac{3}{4}$ hides of the deserted Kensdorf (NEUSS 1971 , p. 168), and about 1530, 6 farmers of Braunschwende still cultivated 6 hides of the deserted village of Ellerod (NEUSS 1971, p. 5). The fields of the deserted Atzelschwende were lent to the farmers of Dankerode (NEUSS 1971, p. 13). These and other examples suggest that these farmers would be the offspring of farmers who had moved from deserted places to another village but proof for this hypothesis cannot be given.

As elsewhere many settlements in the Harz mountains may have been deserted because of the disadvantages and difficulties of their sites. If one analyses the sites of the fields in deserted villages as has been done in detail for the communes of Königerode and Dankerode, and near Wippra, it becomes clear that some settlements could not protect their better fields from soil erosion especially during the fallow period, so that the three-field system of rotation proved itself to be disadvantageous with regard to soil preservation. Von Zückert (1763, p. 161 ff.) wrote in the 18th century in his book on the natural history of the Lower Harz mountains of how the water carried off the rich earth of the slopes, and one must certainly conclude that such processes also took place in earlier centuries. This was especially so in the deserted places of Weimerode and Vitzenhagen near Königerode which are situated in a very sloping area, for around the headstream valleys, and near the deserted places there are thick slope-bottom sediments which cannot be explained except as the result of relative recent soil erosion. Here especially can be found loess-rich soils.

Studies of recent soil erosion in the Lower Harz mountains have shown that after 50 years the thin loess cover which was still there to be mapped during the soil survey has since been completely carried off from many sloping fields (LINKE 1968, p 35). One can therefore say that the thin loess cover existing to-day was more widespread and thicker in the Middle Ages and that there has since been continuous soil erosion.

It is not possible to determine exactly the dimensions of the medieval erosion or of the accumulation on fields which are still cultivated but in the case of Schaubesfelde in the eastern Harz mountains, where deserted fields and villages of the middle of the 15th century lie under forest, the well-formed slope-bottom sediment can certainly be considered to be the result of historical soil erosion. The deserted field is next to the eastern slope of a small stream and must have suffered regularly from soil erosion. A lot of other deserted fields in sloping areas are characterized by isohypsoparallel step-ridges under forests. Analagous to recent step-ridges they are also interpreted as a result of soil rearrangement.

Unfortunately, over the last decades and as a result of agricultural co-operation in farming a lot of small step-ridges within the agricultural fields of deserted villages have been removed so that the full extent of these micro-relief forms cannot now be shown.

A special case are the fields of the deserted village of Hohenrode situated at the southern Harz edge, which have not been cultivated at all since 1535. In the immediate neighbourhood of this small hamlet, excavated in the Thirties (GRIMM 1939), was found the single Gewann of this place (Gewann = communal field of a village with equal divisions which is registered or known by one specific name) - clearly visible as ridge and furrow. The ridges and furrows in the upper and the middle slopes are seriously damaged by erosion. On the lower slope they have disappeared almost completely under the well-formed slope-bottom sediment. In spite of great efforts no further examples of ridges and furrows have been found in the Lower Harz mountains.

An open question till now is why the farmers of Hohenrode deserted a Gewann on this slope, because directly to the west of this place is a nearly flat area free of ridges and furrows.

Here are not neighbouring deserted fields which might have been a hindrance for the extension of cultivation for the people of Hohenrode. It cannot be clarified which parts of this area belonged to the fields of Hohenrode (altogether 18 hides) and why other parts of the fields of Hohenrode had not been in ridges and furrows. (NEUSS 1971, p. 138). Beside such field relics and the signs of historical soil erosion there are further proofs of human interference in several investigated fields. Some deserted villages are now surrounded by banks sometimes with ditches on their outside, for example Schaubesfelde (Plate 20:1 & 2) and Lichtenhagen. Possibly, these earth banks, whose material was taken from the ditch, were topped by a timber or wattle fence, but nothing is evident now. The houses of Schaubesfelde probably stood in a line parallel to the stream. The circumvallation included the church and the churchyard. A further way of discovering sites of deserted settlements is from hollow forms (? ponds) which are sometimes found in the longitudinal courses of stream beds dry today, especially if in their immediate neighbourhood there also occur field names which indicate deserted land. One example is Lichtenhagen, which was deserted at the beginning of the 15th century. Here, too, neighbouring villages as, for instance, Friesdorf, Gorenzen and Braunschwenda (NEUSS 1971, p. 196) cultivated parts of the fields before they were overgrown by forest. Parts of the population of this place, known already in the 9th century, migrated to the above-mentioned neighbouring villages.

The position of the deserted village can be localized as well by potsherds found, for instance, at the "church meadow" and by the remains of the church itself. The houses stood at both sides of a stream which is now usually dry. An excavation in a depression considered to be the village pond clearly showed a sediment layer of about 30cm consisting of silt and clay. This and the form of the depression where the ground is deeper than the ground of the old stream course suggests that we have found here the position of the earlier water.

Another possibility is that settlements and fields were given up because of the poor quality of the soil. This could be valid, for instance, for deserted areas with slight relief, where such typical signs of the historical soil erosion as step-ridges and slope-bottom sediments do not exist. The deserted sites at Volkmanrode near Stangenrode in the catchment basin of the Eine river, and at Fleckenrode, Bettlershain, Borneck and Brumbach near Wippra in the catchment basin of the Wipper river, to name only a few, probably are in this category.

In the valleys of these little Harz rivers are sedimented loams (WILL, 1957), which, for instance, in the Wipper valley reach a thickness of 0.75m (Mücke 1966, p.157). Either the original thin loess layer of these abandoned lands was eroded

Plate 20:1 Schaubesfelde: ruins of the church of the deserted village.

Plate 20:2 Schaubesfelde deserted village site: perimeter bank and ditch.

after having been cleared of woodland (then the settlements and fields were later given up) or the settlements were really faulty ones situated on a priori bad soils. Even if the answer remains undecided, the loess-layers found in the eastern Lower Harz mountains, and the loams in the neighbouring valleys argue for desertion caused by soil deterioration. In spite of the fact that the final abandonment of the areas of strong relief was finally the result of extreme soil erosion followed by soil deterioration, the above differentiation is considered to be necessary. A study of the sites of the deserted settlements shows that many small settlements whose fields were on steep slopes or on bad soils, were given up. No physical/ environmental reason, however, can explain those desertions which are situated in shallow valleys in the plateau and which were surrounded by good agricultural areas. Here we must consider the water requirements of such settlements. Contrary to the Erzgebirge, where the Waldhufen villages are located in the valleys, the settlements of the Lower Harz mountains are mostly found on the elevated plains around the sources of many small streams. After the mapping of potsherd finds etc. we are able to conclude that there were only single houses, little hamlets and small one- or perhaps two-line villages. That is why their water requirements were met sufficiently by the springs so that originally from the hydrographical point of view their location was not faulty. In the Harz mountains, however, mining led to a great consumption of wood. The extensive logging led to streams running dry or to a decreasing discharge so that the villages had to be given up. In some cases the settlements were reconstructed further downstream on areas with greater discharge. This is to be seen, for example, at the deserted places of Althorla and new Horla. The old Horla, 410-420m above sea level was situated about 1.75km to the west of the church of the later-founded Horla of today, and the population went to the new Horla. It was not a poor site as Neuss assumes (1971, p.148) which made the inhabitants re-settle further down the valley, but the decreasing water discharge in the spring area and in the upper section of the Horla valley.

To sum up we can say that in the Lower Harz mountains villages and fields have been given up for reasons of relief, soil and hydrographical conditions. Part of the rural population certainly went to neighbouring settlements. As aforesaid, however, in the Lower Harz mountains mining existed at an early period and it too absorbed part of the rural population. Therefore for our purposes it may be useful to ask the question whether mining settlements or agricultural settlements came first. It seems to be very plausible that farmers, who came from villages being given up, settled in mining settlements. But we know pure mining settlements, too, for instance, to the south of the mining locality of Strassberg. There are, for instance, Birnbaum (where Birnbaum-pond still exists), Hammersbach, mentioned in 992 and deserted about 1450, and some other settlements.

Migration for the above-mentioned three reasons into neighbouring settlements, which grew up rapidly after the desertion period, and into existing or newly-built mining settlements cannot provide a sufficient explanation for the fact that there arose such a lot of deserted villages and fields. There exist enough deserted villages in the Lower Harz mountains, which at the time of their foundation did not have any unfavourable conditions to make clear that beside natural conditions social causes, too, played a considerable role. Therefore we must seek other causes for the desertion of such important villages as, for instance, the early Franconian Lichthagen which was a parish village, to which belonged nine other smaller villages (NEUSS, 1971, p.198). One of them, Friesdorf, still exists at present. All the eight deserted places can be defined as expansion settlements, which have been partially given up at an early period. Expansion settlement does not imply any disadvantage of position in these cases, because the area in a wide circle round about the parish village provided good natural conditions.

It is not rare in the Lower Harz mountains to find large places which were given up. An example in the west of the Lower Harz mountains is Schwidersschwende. This was recognized in the last century, when Meyer writes of it: "many parish villages ceased to exist and the waste land of Schwidersschwende according to old reports was a well-built-up and strongly inhabited small market town" (MEYER, 1871, p.259). We cannot absolutely agree with the idea of Reischel that the size of a place did not have any influence on the possibility of it becoming deserted (REISCHEL, 1926, p.272).

To find other social reasons for the desertion of so many places in the Lower Harz mountains beside the resettlement in mining places, we must take into account wars and catastrophes and the theory of agricultural crisis often repeated in the recent literature.

In the northern Harz foreland at Quedlinburg near the edge of the Harz it has been conclusively shown that the farmers went for reasons of safety to the town and that they cultivated their fields from there. Why should one not take this into consideration for the mountains of the Lower Harz, too?

It is not known if, and to what extent, epidemics and illnesses decimated the population of the Lower Harz mountains, nor, how the problems of prices and wages discussed by Abel effected the desertion of villages in this region (ABEL, 1943, p.78 f). A widespread phenomenon in Germany in those years may have found its expression here. The regularity of good yields was endangered by the altitude of this region.

Finally we will discuss briefly the idea of Mortensen that

the population of small settlements congregated in large villages and that this led to the desertion of smaller settlements. In the Lower Harz mountains this can be shown to be true in the example of Königerode. This agglomeration presupposed "a changed will for settlement caused perhaps by a variation of the economic will at the given time, i.e. the late Middle Ages". The three-field system which was "more efficient in a great association, than in a little settlement" we might consider to be such a changing of the economic will (MORTENSEN, 1961, p.221). We would like to modify the concept of Mortensen as far as the "economic will", which serves him as an argument; it includes so many reflections about planned desettlement and resettlement that it probably cannot be regarded as a general symptom. For the Lower Harz mountains we accept the agglomeration process, but we think that the agglomeration itself is caused largely by natural factors (for instance, soil erosion) which vary from case to case.

All together these reflections show that in the Lower Harz mountains many problems in the history of settlement still await final solution and that certain ideas are still strongly hypothetical. Nevertheless this landscape, too, can be incorporated into the general discussions both of the medieval settlement and of the desertion of our cultural landscape.

BIBLIOGRAPHY

ABEL, W. 1955 — Die Wüstungen des ausgehenden Mittelalters. 1st edit., Jena 1943, 2nd edit., Stuttgart. 1955.

BRÜNING, K. 1926 — Der Bergbau im Harze und im Masfeldischen. Braunschweig, Hamburg.

FLECHTNER, J. 1964 — Die anthropogenen Veränderungen vonBoden und Relief in den Germarkungen Königerode und Dankerode. Martin Luther University, Institute of Geography, unpubl. diploma.

KÄUBLER, R. 1968 — Geographische Nachweise des Charakters mittelalterlicher Dörfer im Unterharz. in: Wiss. Z. d. Univ. Halle, XVII 68 M.

KRAUSE, K.H. 1967 — Anthropogene Veränderungen der hydrographischen Verhältnisse des östlichen Harzes, dargestellt an Beispielen. Martin Luther University, Math. -Nat.-Faculty, dissertation.

LINKE, M. 1968 — Bodenerosion und Bodenerosionsgefährdung im östlichen Harz. in: Wiss. Z. d. Univ. Halle, XVII 68 M.

MEY, W. 1964 — Gang der historischen Besiedlung des Unterharzes bis zum Ende des Mittelalters. Martin Luther University, Institute of Geography, unpubl. diploma.

MEYER, K. 1871 — Dir Wüstungen der Grafschaft Stolberg-Stolberg, Stolberg-Rossla und der Grafschaft Hohenstein. in: Zeitschr. d, Harzvereins, 4th annual.

MORTENSEN, H. 1961 — Ober einige Probleme deutscher historischsiedlungs-geographischer Forschung. in: Geografiska Annaler, vol. XLIII.

MÜCKE, E. 1966 Formen und Genese der Hochfläche des östlichen Harzes. Martin Luther University, Math-Nat.-Faculty, unpubl. probationary treatise.

NEUSS, E. 1971 Wüstungskunde der Mansfelder Kreise, 2 numbers, Weimar.

REISCHEL, G. 1926 Die Wüstungen der Provinz Sachsen und des Freistaates Anhalt. in: Jahrb. d. Histor. Komm. f.d. Provinz Sachsen und f. Anhalt, vol. II.

WILL, K.H. 1957 Untersuchungen über die Entstehung des Auelehmes im Gebiet der Harz-Wipper. Martin Luther University, Math.-Nat-Faculty, unpubl. dissertation 1957.

ZÜCKERT, J.F. Von 1763 Die Naturgeschichte einiger Provinzen des Unterharzes nebst einem Anhang von den Mansfeldischen Kupferschiefern. Berlin.

CHAPTER 21

ECOLOGY AND MARGINAL LANDS; A FORMER PHASE OF RETREAT OF SETTLEMENT AND AGRICULTURAL LIMITS IN THE WESTERN ITALIAN ALPS.

P. Sereno
University of Turin

Rural depopulation, the lowering of agricultural limits, the desertion of villages, the marginalization of high valleys seem to be phenomena typical of our century, according to a sociological interpretation which imputes these facts to urban growth and to industrial polarization. If this sociological thesis gives us a reliable interpretation of the present phase of regression of Alpine settlement, it doesn't on the other hand take into account the historical process of marginalization of the mountains, which started in Piedmont long before industrialization.

I believe one can detect three major phases of depopulation in Piedmontese valleys (Fig.21:1): the present phase, which is the best known; the one which took place between the end of the XIXth century and the beginning of the XXth, often wrongly ascribed to an urban and industrial attraction not yet demonstrable, for it is very likely that a decisive role in the matter was played by the heavy restrictions of local autonomy concerning the use of common pasture, woods and waste land; and finally a phase of regression of settlement and agriculture, which we can date to between the end of the XVIIth century and the beginning of the XVIIIth and which, it seems, was largely due to biogeographical causes.

The three phases have different dynamics and origins, but maybe they are not unrelated: the present fragility of the mountain kind of life compared to the hegemonic economical system of the plains might have been prepared and accelerated by a slow process of marginalization, the roots of which go far back in history. This matter needs further analysis. In this paper I shall attempt only a first examination of the oldest phase of regression. The picture obtained will necessarily be a general one and will offer only a synthetical view, which must be tested by further research.

The few statistical records we hold on migration in Piedmont during the ancien régime (1) suggest a low propensity to emigration of the Alpine people. If we conduct some deeper researches on parish registers of a few Alpine communities, we see that, between the XVIIth and the XVIIIth centuries, there was a tendency to leave not so much the villages, but more the hamlets scattered above a certain altitude, a fact which, up to now, has not had a convincing explanation. Professor Blanchard has supported the thesis of a relative stability in

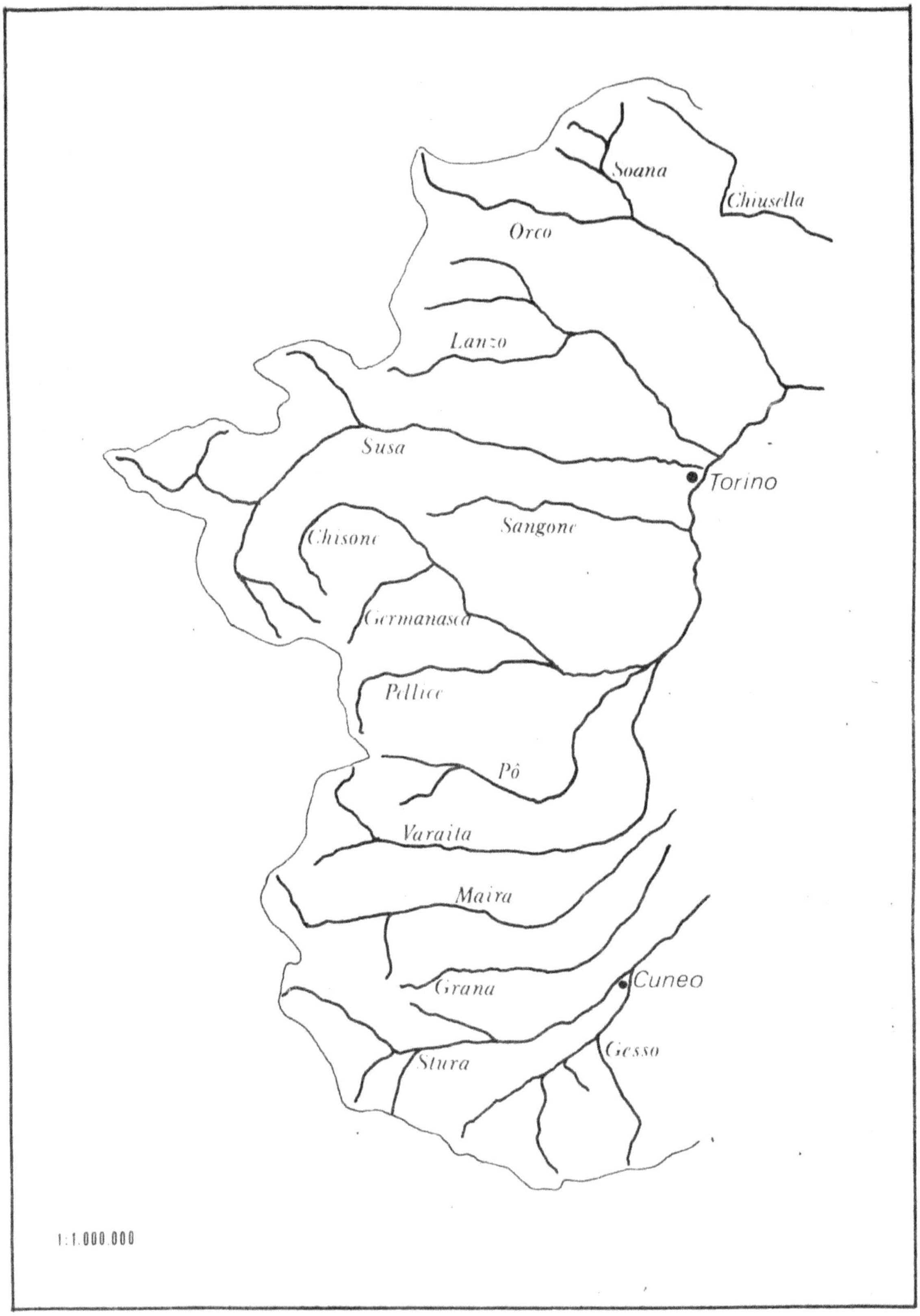

Fig. 21:1 The Piedmontese valleys of the Western Alps.

the climatic condition of the Piedmontese Alps; contradicting Monterin, he has vigorously asserted that "nous sommes convaincu que les limits d'altitude n'ont pas bougé à l'époque historique" and, with reference to the flood of Cesana in 1409, he has come to the conclusion that "la montagne piémontaise ne peut passer pour un pays de catastrophes"(2). On the basis of historical evidence I cannot support Blanchard's opinion. The 17th and early 18th centuries belong to the period known to climatologists as the Little Ice Age characterized by a progressive worsening of climatic conditions. Both Monterin and Le Roy Ladurie have brought forward some interesting examples of the effects of glacial expansion on the settlement and agriculture of Aosta Valleys (3). Piedmontese valleys are farther than the Aosta ones from the great Alpine glaciers, so that the effects of the climatic changes on agriculture seem to be less sensational, though maybe no less important. I have analyzed the climatic change in the plains of low Piedmont mainly using phenological series and weather records from the Academy of Sciences of Turin (4). As for the Alps, research is more difficult. We are not able yet to apply in our area the interesting method used by Dr Parry in Scotland (5), as weather records started too late to be used for comparison (6); phenological observations, different from the ones based on vintage proclamations (for example the ascent to the alpine summer pastures and the chestnut gathering) are too sporadic to allow serial analyses.

We do, however, hold a wide series of written sources; although they do not facilitate a quantitative reconstruction of climatic change, they do give us knowledge of its effects on the territory (7). In addition we hope to be able, as soon as possible, to explore dendroclimatological analysis.

As for the plains, we have estimated that during the Little Ice Age the highest decrease in temperature must have been of about 1.5°C and we have identified rainfall as the limiting factor on agriculture. The rainfall appears distributed in every season except winter, registering two annual maxima in spring and autumn. Its volume has been calculated at about 50% more than at present, with an average risk of floods every three years and peaks of one flood every year during the decade 1721-1730. To this we must add the extraordinary frequency of storms.

On the mountains the rainfall regime appears complicated by the very frequent snowfalls and by the long persistence of the snow on the ground. We can assert that the present temporary snow-line of five months a year (8) corresponds during the Little Ice Age to a period of seven months of snow; as a direct consequence there was a remarkable reduction in the growing period, a factor that effects the relationship between man and mountain much more than the risk, although strong, of floods. Of course, even fluvial and stream floods represent an important ecological constraint, but in most cases we can see the use of agricultural techniques to counter their effects. The floods generally occur in spring (May), more rarely in autumn and are caused both by rainfall and by snow melt. It must be said here that the years in which floods occur are generally the most favorable for the thermic regime of the mountains because during those years the snow melt takes place early. A high frequency of floods is responsible for occasional rock slides and earth flows and for soil erosion, the effects of which are well described in the sources: rejuvenation of soil profiles for 'regressive evolution', soil chains, and the making of rankers and lithosols. In theory, the humification is rapid, as the sheetflow washes the soil so quickly that a ripening of humus is not possible. The weathering from a cold and wet climate has been recorded also by the presence of musks of the ground, often described in the sources, particularly in the northern valleys. The records of natural disasters and the inquiry for the General Cadastral Survey allow us to calculate an average degree of agricultural land, a third of which was exploited between the XVIIth and XVIIIth centuries.

The erosion was countered, however, as far as possible by changes in the rural landscape: the records describe, for certain communities, agricultural hydraulic fixing works for slopes, seldom previously used, and the building up of stone walls on the terraces on which every year the eroded soil was put back. Elsewhere, as in Val Maira, we notice during the XVIIth century a real enclos with functions of climatic protection, previously unknown in the valley. As the same time in some alpine communities, as we can gather from Communal Acts, people started to pay regularly for a person to sound the bells at times of 'bad weather', and to guard open ditches which conveyed the water coming down from the snowmelt.

The most limiting ecological constraint was, without doubt, the shortening of the growing period, which made the farmers interviewed during the visits for natural disasters unanimously assert that (and this is also reported in several Communal Acts) the risk of fallanza, i.e. of harvest failure, may occur every other year, caused by early snow falls in September, or late snow melting in the summer. This is the fundamental cause of the retreat of the limits of farming and settlement.

As for the limit of cereal cultivation, (which in this area was essentially rye) from the interpretation of our sources I think we can say that it was rarely above the 1600m contour line; beyond this and except in certain microclimatic conditions, no farming was possible.

We must here remark, though, that farming at this altitude underwent rather high risks of harvest failure, as we can gather from the communities of the Cesana Valleys (Susa Valley) which in 1740 frequently experienced (9) early snowfalls which prevented the regular harvest of oats, barley and tramois rye (i.e. spring-rye, even today called in the valley sustrameisa). In these cases the growing period is obviously less than the five months which, at that altitude, represents the minimum for successful ripening. At an altitude of 1500-1600m cereal farming could be continued during the 'ancien régime' only by the jachère climatique, anticipating the autumn-sowing at the end of July and delaying harvesting until the end of August or the beginning of September of the following year. The two-course rotation of crops was therefore necessitated by an ecological constraint. Above 1600m early snowfalls made the risk of harvest failure extremely high, while profits were very low compared to the quantity of seed required, which in most cases was of a proportion of one to two. We can therefore consider the 1600m contour line an absolute bioclimatic limit, thereby emphasising that the ecological limit for agriculture during the 'ancien régime' was much lower than during this century, when cultivation reaches altitudes of 1800m and exceptionally of 2000m. Some field names, concerning land-use, indicate that there had been a higher expansion upwards during the period preceding the first half of the XVIIth century, when our sources begin to describe the process of regression of agricultural limits.

At an altitude of 1300m cereals must be in the soil for about eleven months (10), being seeded in September and harvested at the end of July. It is only at an altitude of 800-1000m that the period of occupation of the soil decreases to nine months (October-June) allowing a three-course rotation based on the alternation of rye and barley or, more often, winter rye and spring rye followed by a year of fallow. In certain valleys, particularly in Val Venasca, hemp was included in the rotation of crops and was used to make clothes and sails for sale in Genoa and Venice (11). Potatoes were diffused in the alpine region only in the second half of the XIX century; only at Venaus (Susa Valley) did the cultivation of tartifle, i.e. the small mountain potatoes, take place.

The regression of the climatic limit of cereal farming had different influences on the mountain people. We still know too little about the nature of the demographic increase in the Piedmontese Alps during the XVIth century (12), but there is no doubt that this increase was the cause of the mountain colonization beyond the permanent settlement limit at that time. In the Waldensian Valleys (Chisone, Germanasca and Pellice valleys), the XVIth century was a period of religious persecutions and of a high demographic decrease; only after the 'grande rentrée' at the end of the XVIIth was there a policy to increase the population and its effects can be seen during the XVIIIth and the XIXth centuries when the increase produced a geographical expansion that can be well studied thanks to available documentary evidence. I have recently reconstructed these processes for Rorà, a village in the high Pellice Valley (13).

The colonization of the mountain in that period occupied lower altimetrical levels than the ones attained in other valleys during the XVIth century; this was obviously due to climatic reasons. It is therefore interesting to consider the dynamics of colonization and in particular the problem of the relationship between land and family. The splitting up of the rural property, due to the custom of dividing the single parcel of land and the house into equal parts among the male heirs, soon made the infield saturated with holdings. Families then began a process of exclusion of family members, as is recorded by the notarial acts. These strategies did not involve expulsion from the community (which occurred only at the end of the XIXth century), but were based on the occupation of the outfield. At Rorà, for instance, in the second half of the XVIIIth century, the fourest i.e. the scattered settlements on common land, only seasonally used for the summer pasture, became permanent settlements. Their structure is the mas one, a unit of a small house-stable-hay loft, kitchen garden, arable land and grass. The permanent occupation of the outfield and the formation of a scattered settlement at a higher level than the village is not characteristic of the Waldensian Valleys. At the end of the XVIIth century several muande, i.e. mountain spring pastures, were permanently inhabited. In some of the southern valleys, above the present limits of permanent settlement, remains dating back to the XVIth century can still be found, the architectural characteristics of which suggest that they were the buildings of permanent settlements (Plate 21:1); the same houses were recorded in the XVIIIth century land-registers as ruins.

If we accept the hypothesis that the history of Pellice Valley population can be considered a good example,though retarded, of the Piedmontese alpine valleys colonization pattern, we must admit that the new settlements are more susceptible to the effects of climatic change, than the basic settlements from which they derive. The colonization process forms horizontal land-use patterns, infringing the traditional and functional vertical organization of the parishes, by means of which the topoclimatic asymmetry supports the agro-silvan-pastoral economy. Wherever upward expansion has overcome the limit altitude of 1600m, or even where it has occupied, though at lower altitudes, the ubac (the shady side), the scattered settlement underwent a crisis beginning in the XVIIth century. While in the Waldensian Valleys a gradual and moderate ascending colonization of the mountains took place, together with a process of settlement scattering and of occupation of the common lands, in the other valleys the opposite tendency becomes evident: the secondary scattering decreases, reaching in some cases the definitive and total desertion of hamlets, particularly of those most isolated and farthest from the villages. The research I made on parish and land registers of some communities is certainly not exhaustive, but I do believe it can be considered valid. The regressive process in the settlement limits from valley to valley seems to be a rather

Plate 21:1 Archaeological evidence from the Italian Western Alps: a scattered settlement deserted in the eighteenth century.

slow and gradual one. The worsening of climatic conditions had no immediate effect on settlement; we witness, on the contrary, a gradual change, not particularly remarkable at the beginning, but increasing from about 1640 on, as a reaction to cumulated ecological effects. An interesting census of the deserted houses at Pamparato in 1718 points out the total desertion of two hamlets and the partial desertion of several others between an altitude of 1100m and 1300m of the ubac, while many other houses scattered high up on the mountains cannot even be visited any more as the mountains have become inaccessible (14).

If we consider the dates of desertion, we notice that they can be divided mainly into two periods, namely the XVIIth century and the first thirteen years of the XVIIIth century; these periods show a high degree of correlation with the ones which dendroclimatological analysis by Dr Pellegrini for Valtellina indicates as the coldest periods of the Little Ice Age (15). The retreat of the secondary settlement limits causes the emigration of whole peasant families. The migration field is generally quite narrow: the main characteristic of the retreat seems to be a re-concentration of the population in the sub-mountain level, a process which allows us to consider that form of internal migration as a lowering of the altmetrical level.

Ancient field-names show that in the past agriculture was practised in a higher zone which was deserted in the XVIII century. Nevertheless in that period a zone of marginal land was shaped at or above the new climatic limit of cultivation: the fields were temporarily exploited, about one or two years in every ten, and prepared on common land by burning (campi roncati). This practice was resorted to by the peasant landholders faced with a decrease in the permanently cultivated area where climatic change had imposed a yearly fallow (jachère climatique). Thus although the high settlements were deserted, some of the higher land on some valley slopes was periodically cultivated by villagers of the permanently peopled zone. This infield-outfield system was structurally weak and reached crisis when the ancient local rights governing the use of common land were not recognized by the central rule. This contributed to the first great migration abroad in the second half of the XIX century and to the formation of internal socio-economic frontiers in the region.

NOTES

1) G. PRATO, La vita economica in Piemonte a mezzo il secolo XVIII, Torino 1908, 41-57; M. DOSSETTI, 'Aspetti demografici del Piemonte occidentale nei secoli XVII-XVIII' BSBS, LXXV, 1977, 127-238.

2) R. BLANCHARD, Les Alpes Occidentales, Grenoble-Paris 1952, vol. VI, t. 1, p. 273 and t. 2, p. 408.

3) U. MONTERIN, Il clima sulle Alpi ha mutato in epoca storica?, Bologna 1937 and E. LE ROY LADURIE, Histoire du climat depuis l'an mil, Paris 1967, 102-215.

4)P. SERENO, 'Annus fructificat non tellus. Considerazioni preliminari sulla piccola età glaciale nelle campagne del Basso Piemonte', Proceedings of the Conference on Agriculture and rural Life in the Province of Cuneo (Fossano 1981), in press.

5) M.L. PARRY, 'Secular Climatic Change and Marginal Agriculture', Trans. I.B.G., 64, (1975), 1-13.

6) On the early weather stations in the Italian western Alps see F. DENZA, Il Commodoro M.F. Maury e la corrispondenza meteorologica delle Alpi e degli Appennini Italiani, Torino 1875

7) RECORD OFFICE OF TURIN, Sez III, art. 470,472,473; Sez. Riun., I a., Inondazione, Corrusione e Gragnuola, m. 1-4; I a., Provincie; II a., c. 21, reg. 104-122 and 151-157; II a., c. 79.

8) On the present snow-line see C.F. CAPELLO, Limiti nivali e manto nevoso in Piemonte nell'inverno 1960-61, Torino 1963.

9) RECORD OFFICE OF TURIN, Sez. Riun., I a., Inondazione, Corrusione e Gragnuola, m. 3, fasc. 17.

10) G. EANDI, Statistica della Provincia di Saluzzo, Saluzzo 1833-1835, 162-177.

11) F.A. DELLA CHIESA, Relatione dello stato presente del Piemonte, Torino 1635, p. 39.

12) R. COMBA, 'Vicende demografiche in Piemonte nell'ultimo medioevo', BSBS, LXXV, 1977, 39-125.

13) P. SERENO, La terra e il fardello. Dinamica della proprietà fondiaria e sistemi parentali in una comunità valdese delle Alpi Occidentali, in press.

14) RECORD OFFICE OF TURIN, Sez. III, art. 471, m. 4.

15) M. PELLEGRINI, *Materiali per una storia del clima nelle Alpi lombarde durante gli ultimi cinque secoli*, Bellinzona 1973, 217-246. The periods are: 1661-1668, 1667-1681, 1693-1696, 1699-1713 (Valtellina) and 1663-1668, 1678-1681, 1693-1696, 1701-1713 (Pamparato).

CHAPTER 22

NOTES SUR LES LIMITES SUPÉRIEURES DE L'HABITAT DANS LES MONTAGNES DE LA PÉNINSULE SCANDINAVE

M. Cabouret
(Université de Metz)

Le trait le plus marquant de la répartition altitudinale de l'habitat permanent d'abord, temporaire ensuite dans les montagnes de la Norvège du Sud (au sens large au sud de la dépression de Trondheim) est que les établissements humains montent relativement haut étant donné la latitude entre 58° lat. N. pour l'extrémité méridionale de la Norvège et 64°. Certes, la population qui vit de façon permanente au-dessus de 600 m représente de très faibles effectifs mais les altitudes atteintes par les sites de certaines fermes sont assez remarquables vu les rigueurs du milieu naturel. Si 40% de la surface du pays (323,878 km^2) sont situés au-dessus de 600 m d'altitude, il ne vivait vers 1800 qu'environ 13,000 personnes au-dessus de cette limite.[1] Et si au-dessus de 900 m, l'habitat permanent est tout à fait exceptionnel, il convient de rappeler qu'environ le cinquième de la superificie de la Norvège est situé au-dessus de cette altitude (19,4% de la surface totale entre 600 et 900 m, 12,1% entre 900 et 1,200 m, 8,2% entre 1,200 m et 2,469 m au Galdhöppigen le point culminant de l'Europe du Nord.

Il y a une relation assez évidente entre la limite supérieure de la forêt qui monte le plus haut en Norvège orientale (Østland) où elle peut atteindre exceptionnellement plus de 1,000 m et celle de l'habitat permanent. Comme celui-ci, la limite supérieure de la forêt descend vers l'ouest en Norvège du sud (au sens large) et aussi vers le nord car l'extension du pays en longueur est assez considérable (1,750 km).

I - LA LIMITE SUPERIEURE DE L'HABITAT PERMANENT

A) SUR LE VERSANT OCCIDENTAL ET MÉRIDIONAL DE LA CHAINE DES SCANDES EN NORVÈGE DU SUD (AU SENS LARGE)

L'habitat ne s'élève ici qu'exceptionnellement au-dessus de 600 m et ceci seulement dans la partie amont des fjords du Vestland. En fait, la limite supérieure se tient fréquemment autour de 400 m. Dans le Sogn, l'habitat dépasse à peine 470 m en dehors des communes de Borgund et d'Aurland. A Borgund, il dépasse 600 m, ainsi 636 m à la ferme de Kleivi dans Smeddalen au nord-est et 637 m à celle de Bergstölen au sud-est dans Mørkedalen. L'auberge de montagne de Maristova est à 811 m dans Smeddalen et celle de Breidstöl à 1,023 m dans Morkedalen.[2] Dans la vallée d'Utla(dal) à Årdal, le groupe de fermes le plus haut situé à l'heure actuelle est celui de Vetti, à 317 m (c.t.). Dans la commune voisine au nord, celle de Luster, dans Bergsdalen, la ferme d'Opptun est à 397 m et dans Fortunsdalen, la ferme la plus haut située est Övreli à 450 m (c.t.).[2 bis] Dans la commune d'Aurland, dans le "canton" ("bygd") d'Aurdalen, les fermes les plus élevées à la fin du XIXe siècle étaient Nesbø et Övstebö (820 m env.) (c.t.). Elles ont été abandonnées depuis (c.t.).[3]

Toujours dans le Vestland, dans le Hardanger interne, l'habitat a une tendance à s'élever. Ainsi à l'amont de Hjölmodalen dans la commune de Eidfjord, le groupe de fermes de Viveli atteint 877 m, plus haut situé que certains chalets situés à l'aval dans la même vallée.[4] Dans une position analogue, c'est-à-dire dans une vallée mûre de la surface "paléique" du plateau de Hardangervidda, qui se termine à l'aval par une grande rupture de pente dans le profil longitudinal avec une vallée en auge encaissée en contre bas et emboîtée dans le plateau et une superbe gorge de raccordement, avec la chute d'eau célèbre de Vøringen (147 m de dénivellation), il y a des fermes haut perchées. Cette vallée s'appelle Bjoreidalen et les exploitations agricoles sont situées au "voisinage" de Sysendalen à 780 m d'altitude, les fermes de Maurset étant les plus élevées (c.t. Hardangerjökulen 1/100,000e).[5] Mais dans le reste du Hardanger, il n'y a pas d'autres endroits où l'habitat permanent monte si haut que dans cette vallée.

Au sud du Vestland, la limite supérieure de l'habitat est plus basse qu'en Hardanger; dans le Rogaland interne, à Suldal, les plus haut situées des fermes montent jusqu'à 380 m seulement à Fjellberg dans Ulladal. Il y avait dans la même vallée au début du XXe siècle, deux exploitations à 500 m d'altitude environ qui ont été abandonnées depuis (c.t. Sand 1/100,000e)[7] car elles étaient trop écartées des autres secteurs habités ici.

A Sirdal, dans le Vest-Agder, l'habitat permanent atteint 610 m à Ådneram et à Suleskar dans la vallée principale, celle de la Sira, tandis que la ferme de Kvinen dans une vallée annexe, celle de la Kvina se situe à 713 m d'altitude.[8] Toujours dans le Sørland, dans la vallée du Setesdal, dans la vallée principale, parcourue par l'Otra, quelques fermes plus ou moins isolées montent jusqu'à 770 m à Hovden dans la partie supérieure de la vallée et plus loin encore à l'amont la ferme de Bjåen est à 900 m environ. Dans une vallée affluente coulant de l'ouest à l'est, la ferme de Vatudal atteint 715 m et sur le haute surface des plateaux ("heie") Breive est notablement plus haut.[9]

B) LA LIMITE SUPÉRIEURE DE L'HABITAT PERMANENT SUR LE VERSANT ORIENTAL DE LA CHAINE DES SCANDES EN NORVÈGE

Dans le Telemark, l'habitat a tendance à s'élever. Dans Fyresdal, il y avait plusiers fermes à la fin du XIXe siècle qui s'élevaient sur les hautes surfaces de l'ouest de la commune à 700 m., Øysa par ex., Slystöl, Kvift.[10] Les fermes les plus hautes du voisinage d'Eidsborg avec une vieille église en bois debout qui indique l'ancienneté de l'installation humaine, situées dans la commune de Lårdal, atteignent presque 700 m. Un peu plus à l'est, toujours à Lårdal, les fermes de Smörklepp sont à 670 m. Il en va de même pour le groupe de fermes de Skafså un peu au sud dans la commune de Mo (c.t. Vinje (Tokke) 1/100,000e). Si à Vinje, sur le flanc sud-est du grand plateau de Hardangervidda, l'habitat relativement continu ne dépasse pas 530 m dans la vallée de Grungedal qui donne accès à la haute surface, des fermes isolées montent plus haut (Vågslid, Boltr, Flothyl). A la fin du XIXe siècle fut élevée sur la route du Telemark à Röldal dans le Vestland une ferme-auberge à Haukeliseter située à 980 m d'altitude.[11] Dans la commune de Rauland, l'habitat relativement d'un seul tenant (les fermes étant dispersées plus ou moins mais les finages étant attenants) monte plus haut qu'à Vinje. La nappe d'eau du lac atteint 687 m et dans le "voisinage" de Rauland les fermes dépassent 760 m comme dans celui de Haddeland mais à Rauland un des records d'altitude de l'habitat permanent pour l'ensemble de la Norvège est détenu par des fermes qui sont situées sur les bords du lac de Mösvatn (nappe d'eau à 918 m) à près de 1,000 m d'altitude (c.t. Rauland 1/100,000e).[12] La ferme d'Arjehovd à la pointe Nord-ouest du lac Mösvatn est à 930 m environ. A l'ouest du grand lac Mös(vatn), dans la commune de Seljord, l'habitat permanent monte aussi assez haut: 845 m à Kvammen dans la partie supérieure de la vallée d'Åmotsdal (c.t. Rjukan 1/100,000e)[13] et 828 m à Kvambekk dans une vallée affluente (c.t. Kviteseid 1/100,000e). Dans la magnifique auge glaciaire dite Vestfjorddalen où s'est installée la ville industrielle de Rjukan dans le nord du Telemark, certaines fermes sont aussi haut situées comme Berge à 866 m ou celle de Fosse (c.t. Rjukan 1/100,000e).[14] Enfin toujours à Tinn, commune où se situe Rjukan, la vallée affluente du grand lac Tinnsjö parcourue par le cours d'eau Mår a dans sa partie amont Breisetdalen une ferme Steinsböle qui est située à plus de 750 m d'altitude (c.t. Mår 1/100,000e).[15] Dans Tessungdalen, autre vallée affluente du Tinnsjö, les fermes de Skålbö montent aussi au-dessus de 750 m (même c.t. Mår) sur le versant oriental des hautes surfaces de la Hardangervidda.

L'habitat a tendance à s'élever plus haut dans la série des grandes vallées du versant oriental de la chaîne des Scandes: Numedal, Hallingdal, Valdres, Gudbrandsdal et Østerdal. Dans le Numedal, une des fermes les plus haut situées sinon la plus élevée est celle de Staupet à 844 m dans Jöndalen (c.t. Mår 1/100,000e) une vallée affluente de Uvdal, la vallée principale du canton du même nom situé à la partie supérieure du Numedal.

A Dagali, communauté rurale située sur les rives du fleuve principal du Numedal, le Numedalslågen, la ferme de Hallen monte à 883 m (c.t. Dagali 1/100,000e) et celle de Björkli à 931 m.[17]

C'est dans le Hallingdal supérieur que l'habitat permanent atteint sur la façade orientale des Scandes son niveau le plus élevé avec des exemples assez nombreux. Jusqu'en 1909 la ferme la plus haute de toute la Norvège utilisée en habitat permanent était Nygård à 1,025 m au bord du beau lac de Nygårds (vann) à l'amont de Haugastol, sur la voie ferrée Oslo-Bergen. Dans Ustedalen, la ferme la plus haut située dans les années 1950 est Slåtto à 1,024 m en amont de Geilo sur la voie ferrée d'Oslo à Bergen (c.t. Hallingsskarvet 1/100,000e). Plus au nord, toujours dans la commune de Hol dans la vallée de l'Åm affluente de celle d'Ustedalen, sur les larges surfaces situées à l'ouest du lac Strandafjord sont placées plusieurs fermes, ainsi celle de Bakko à 971 m (c.t. Hallingskarvet).[18] A Ål, commune limitrophe de Hol, dans le Hallingdal supérieur, dans la vallée de la Votna la ferme de Randen tout-à-fait à l'amont atteint 857 m (c.t. Hemsedal 1/100,000e).[19] Dans la vallée d'Hemsedal proprement dite, d'assez nombreuses fermes dépassent 800 m, ainsi dans le "voisinage" de Liann dans Mörkedalen et dans celui de Gröndalen, dans deux vallées qui confluent pour former celle d'Hemsedal. Sur la haute surface qui s'étend depuis le versant nord-est du Hemsedal jusqu'à la grande vallée de Valdres, il y a un groupe de fermes relativement important qui atteint 900 m à Lykkja, un des "voisinages" les plus élevés de toute la Norvège (c.t. Gol 1/100,000e).[20] La ferme auberge de Bjøberg à l'amont de Mørkedalen est à 1,013 m. Toujours aux confins du Hallingdal, plus précisément de la vallée de Hemsedal et de Valdres une autre des grandes vallées de la Norvège orientale sur la haute surface de la vidda, il y a des fermes à plus de 800 m dans la commune de Gol comme Dalen (858 m), Haug (851 m), Möni (870 m), Klantent (860 m) et au-delà de la limite administrative Flageslett à 958 m. Il y en avait d'autres à plus de 900 m jusqu'à un passé récent (c.t. Gol 1/100,000e).[21]

Dans la partie supérieure de Valdres, à Aurdal dans la vallée de la Regna, axe principal de cet ensemble régional, le peuplement atteint 725 m au-dessus du groupement de fermes d'Aurdal en position d'adret dans la commune de Sør Aurdal et une altitude équivalente plus à l'amont dans la vallée au-dessus de Fagernes, sur le versant qui domine la rive septentrionale du lac Strondafjord dans Nord Aurdal. Dans Øystre Slidre, l'habitat permanent agricole atteint 780 à 800 m dans les "voisinages" de Beito et de Robøl.[22] Beito est situé au nord du lac Øyangen lui-même situé au Nord du massif de Slettefjell et Robøl est au sud-est. Dans le "canton" supérieur de Valdres, Øye, à l'extrémité occidentale du lac Vangsmjösa, la ferme la plus haute Uvdal (Opdal) n'est qu'à 609 m ce qui n'est pas particulièrement haut.[23] La partie supérieure de la vallée de l'Etna qui se termine dans le lac Randsfjord constitue une petite cellule régionale traditionnellement rattachée au Valdres. Sur la haute surface qui encadre les versants, certains groupes de fermes sont installés à près de 800 m, tels le "voisinage" de Byfelli où la ferme la plus élevée Ludnigård est à 765 m et Granheimbygda, plus à l'amont où Lauvhagen est à 811 m (c.t. Aurdal 1/100,000e).[24]

Dans le Gudbrandsdal, l'habitat en moyenne n'est pas très élevé et se tient en règle générale au-dessous de 600 m mais en pas mal d'endroits il atteint cependant des altitudes notablement plus fortes. Le long du lac d'Espedal dont le plan d'eau est à 722 m, il y a des exploitations agricoles donc entre 725 et 750 m au-dessus du niveau de la mer.[25] Dans la vallée du cours d'eau Vinstra, affluent de la vallée principale du Gudbrandsdal, dans le "canton" (bygd) de Skåbu la plupart des fermes sont localisées en "adret" en ligne à peu près continue autour de la courbe de niveau de 850 m, mais quelques unes montent jusqu'à plus de 900 m: Kampeseter est à 930 m. En ubac, au contraire, les fermes, beaucoup moins nombreuses que celles exposées au sud, ici ne dépassent pas 700 m (c.t. Vinstra 1/100,000e). Cette situation de vallée affluente se retrouve d'une part à Heidal, drainé par le cours de la Sjoa où dans Skogbygdi, à l'amont de ce voisinage, la ferme de Ellingsbörusti atteint 750 m tandis qu'en exposition sud Nyborg est à 807 m et d'autre part dans une vallée adjacente de celle de l'Otta, dans la commune de Vågå, à Skårvangen où sur l'éperon entre les versants de la Finna et de la Skjerva qui confluent au sud, la ferme de Skjerstulen est à 814 m et celle de Preststulen à 822 m, la plus élevée d'un groupe où toutes les exploitations sont au-dessus de 700 m (c.t. Vågå).[26]

Dans la vallée de l'Otta, le peuplement n'a pas tendance à monter très haut mais dans Nordherads-solsida ("l'adret" de Nordherad) la ferme de Vistdal atteint 701 m au-dessus de l'extrémité orientale du lac de Vågå. Au sud de ce dernier, Bakken qui appartient à un groupe de fermes qui sont en revers du versant à l'ombre atteint 856 m. Il en va de même à Lom qu'à Vågå, c'est-à-dire qu'au-dessus des rives du lac de Vaga, les fermes ne s'élèvent pas très haut, 655 m en adret à Brenna. Dans une vallée affluente de l'Otta au Sud du lac de Vågå, Bøverdalen, l'habitat agricole monte jusqu'à 805 m à Storli dans la vallée adjacente de Leirdalen mais ne dépasse pas 720 m dans la vallée "principale" de la Bövra à Rudningar (c.t. Lom 1/100,000e).[27]

Le fond de la vallée du Lågen, dans les deux communes de Dovre et Lesja est nettement plus élevée que celui de la vallée de l'Otta, mais pourtant l'habitat permanent ne monte guère ici à plus de 650 m et paradoxalement c'est à l'aval dans les limites de Dovre où le plancher de la vallée est plus bas que l'habitat monte environ 100 m plus haut qu'à Lesja.[28]

Mais il n'y a pas que dans la partie supérieure du Gudbrandsdal et dans les vallées adjacentes que l'habitat permanent peut s'élever assez haut sur les versants; il en va localement de même dans la vallée principale, celle du Lagen de Gudbrandsdal à l'aval de la confluence avec l'Otta dans la commune de Sel. C'est le cas de Sør Fron où au sud ouest, sur le versant sud de la vallée, au lieu dit Bankåhaugen, Asmundstad se situe à 690 m tandis qu'au nord est, au nord de la vallée, il y a des fermes entre 700 et 750 m.[29]

Dans le grand ensemble régional de l'Østerdal où le fleuve principal artère majeure du réseau hydrographique est la Glomma et où il y a de nombreuses vallées affluentes, les fermes ne montent pas particulièrement haut sur les versants de l'axe principal. Ainsi, à l'aval, dans la commune de Stor Elvdal, la plupart des fermes sont installées au fond de la vallée de la Glomma sur les deux rives du cours d'eau, à une altitude comprise entre 250 m à l'amont et 230 m à l'aval. Les versants ne sont pratiquement pas occupés. Aucune tendance de l'habitat permanent à s'élever ne se décèle ici (c.t. Stor Elvdal 1/100,000e).[30] Un autre exemple sera pris encore dans la vallée principale mais à l'amont, cette fois-ci dans la région de Tynset. L'essentiel du peuplement dans la vallée de la Glomma au nord de l'agglomération de Tynset est lié aux rives du fleuve qui sur un trajet de 50 km environ s'abaisse de 590 m à 478 m. Les fermes ne s'élèvent guère au-dessus de 600 m sauf en deux groupes de fermes situées sur le plateau accidenté encadrant la vallée; celui de Hodal au sud est de Tolga atteint 800 m mais dans celui de Vingelen au nord ouest de Tolga, les fermes, du moins la majorité d'entre elles, ne dépassent 700 m que de peu, sauf celle d'Osteberg qui est à 793 m (c.t. Tynset 1/50,000e). Au nord-ouest d'Os, dans le "canton" de Dalsbygda, la ferme de Holla atteint 833 m[31] et est une des plus élevées de tout l'Østerdal. Comme dans le Gudbrandsdal, c'est dans les vallées affluentes et surtout à l'amont que l'habitat révèle une tendance à s'élever assez haut. Ainsi dans Folldal, vallée qui descend depuis le massif du Dovre de l'ouest vers l'est, les fermes les plus haut à l'amont dépassent 900 m. Entre Folldal et la vallée de l'Atna dans Nordre Atnedalen, des seuils assez bas permettent le passage; sur les seuils qui réunissent la vallée de Folldal à celle d'Atnedal qui descend aussi du massif de Rondane, il y a des fermes haut situées comme Streitlii (865 m), Holen 815 m, Stodsbuøyi, groupe de fermes au-dessus de 790 m et en descendant Atnedal, jusqu'à l'extrémité méridionale du lac Atnasjø, la plupart des fermes sont au-dessus de 700 m, ainsi Solvang à 765 m, Neset à 740 m, Straumbuen à 741 m, etc... (c.t. Rondane 1/100,000e).[32] Entre la vallée de la Glomma qui naît au-delà de la limite administrative du Hedmark avec le Trøndelag, et le grand lac Femund allongé au Nord au Sud, s'étend une haute surface plus ou moins montagneuse, là se trouve le groupe de fermes d'Hodalen déjà cité entre 765 m et 795 m mais il y a aussi d'autres habitats assez élevés dans les vallées entaillées dans ce plateau comme celles de la Nöra, au sud de la limite administrative dans le Hedmark septentrional, de Håelva et de Hitterdal au nord de la limite dans le Sør Trondelag.[33]

A l'est de l'Østerdal, il y a quelques secteurs des rives du lac Femund qui sont peuplés, les fermes ici sont à plus de 660 m.

Une fois franchie au nord la limite entre les deux grands ensembles régionaux, Østlandet en Norvège du Sud Est et Trøndelag, la limite supérieure de l'habitat s'abaisse et dans Oppdal au nord du massif du Dovre, peu de fermes sont situées au-dessus de 600 m.

C) LE VERSANT SUÉDOIS DE LA CHAINÉ DES SCANDES

A ce versant, il est licite de rattacher la petite région de Trysil Engerdalen en Norvège à l'est de l'Østerdal inférieur. Du grand lac Femund sort un émissaire la Klara qui coule à travers le canton norvégien de Trysil d'où il passe au-delà de la frontière avec la Suède pour rejoindre le lac Väner en traversant la grande province du Värmland où il constitue l'artère principale du réseau hydrographique. Engerdal est une petite vallée affluente de celle de Trysil à l'est. Non seulement la région située en-deçà de la frontière norvégienne a des traits hydrographiques communs avec les parties voisines de la Suède mais aussi des caractères orographiques. En effet, il y a là des avants postes isolés de haute montagne (c'est-à-dire un volume de relief suffisamment élevé pour que la limite supérieure de la forêt soit dépassée en plusieurs endroits) dans Transtrand, Idre et Särna, les subdivisions administratives les plus septentrionales de la Dalécarlie et dans Trysil. Ce sont des bastions détachés du domaine continu de la haute montagne qui s'étend à l'ouest et au nord et en Suède ces massifs véritablement montagneux, Transtrandsfjäll, Fulufjäll, Langfjäll sont les plus méridionaux de tout le pays. Il n'est donc pas sans intérêt d'examiner sommairement la limite supérieure atteinte par l'habitat permanent. A Trysil, l'habitat permanent se situe en gros entre 350 et 650 m d'altitude. C'est sur l'exploitation de Sjøli, au nord de Drevdalen qu'il monte le plus haut à 753 m au-dessus du niveau de la mer.[34] D'autres exploitations pionnières installées entre les deux guerres mondiales atteignent 700 m à Flendalen.[35]

Quant au versant suédois proprement dit de la chaîne des Scandes, il est nettement plus septentrional que le versant oriental norvégien et par suite la limite supérieure de l'habitat permanent y est en moyenne plus basse et elle s'affaisse lentement vers le nord dans les parties supérieures des vallées peuplées, mais cependant pas d'une façon continue des vallées les plus méridionales vers les plus septentrionales mais de manière assez irrégulière.[36]

La Dalécarlie septentrionale et le Härjedal peuvent être considérés comme un ensemble où l'habitat permanent se tient en général à des limites supérieures assez élevées. Tous les sites qui, en Suède, sont placés au-dessus de 700 m sont rassemblés dans cet ensemble régional. Il y a même quatre unités de peuplement au-dessus de 800 m. Parmi elles Högvålen, le plus haut village de Suède à 830 m d'altitude situé sur le plateau entre Idre en Dalécarlie supérieure et Tännäs en Härjedal, Backvallen est à 809 m à l'est de Ljusnadal, Gröndalen à 810 m dans la région des sources de la Ljusna et Lövåsen à Idre à 805 m. D'autres hameaux ou même villages avoisinent les 800 m, il y a cinq exemples à Tännäs dont Fjätdalen à 790 m et dans la paroisse de Storsjö, Tossåsen est à 791 m.[37]

En moyenne l'habitat le plus haut placé, entre 600 et 700 m à l'amont et entre 500 et 600 m dans la partie moyenne des grandes vallées parallèles, est situé tantôt dans les parties supérieures de celles-ci qui s'élèvent par paliers successifs vers l'ouest et dans les principales vallées affluentes, tantôt sur les larges interfluves à caractère de plateaux qui les séparent. Du premier type relève l'habitat à Tänndalen et Särvsjö au nord ou à Glöte-Lofsdalen à l'est. Du second celui qui existe par exemple sur la haute surface au sud-ouest d'Idre.

Dans la partie centrale du Jämtland, l'habitat se situe notablement plus bas. Le réseau des vallées est plus profondément incisé qu'au sud. Les groupes de fermes de la vallée d'Åre et du lac de Kall ne dépassent que de peu 400 m. Dans le canton d'Hotagen, au nord du Jämtland, l'habitat n'atteint pas 400 m dans la vallée principale. Même dans les vallées affluentes des artères principales, il ne s'élève pas très haut, à 600 m à peine dans Vålådalen et reste même plus bas dans la vallée d'Ånnsjö et de Medstugu. Le seul autre groupement d'habitat de quelque importance qui dépasse 500 m ici se trouve aux confins des communes d'Offerdal et d'Hotagen, dans le massif d'Ansätten où un village atteint 689 m à savoir Ansätten le lieu-dit habité en permanence le plus haut situé en Suède au nord du Härjedal, et un autre Jänsmässholmen 623 m. Dans le réseau des vallées secondaires d'Åre et de Kall, le village d'Äggsjön à 652 m est un cas unique.

Dans le nord du Jämtland, le peuplement se situe encore plus bas que dans la partie centrale de la province qui vient d'être évoquée. Dans le canton de Frostviken à l'extrémité nord-ouest du Jämtland, l'habitat ne dépasse 600 m et même 500 m que dans quelques cas isolés dans la partie supérieure du réseau des vallées secondaires. Le peuplement dans cet ensemble régional se tient en règle très générale le long des vallées et ne s'aventure que rarement sur les hautes surfaces.

Franchie vers le nord la limite administrative de la province du Västerbotten, la hauteur moyenne de l'habitat augmente sensiblement. Au sud du fleuve Ume(älv), s'étend l'échine montagneux de Stötting(fjäll) où le peuplement dépasse 500 m loin vers le sud est, et même 600 m à Risträsk. Dans le véritable domaine montagneux qui s'étend à l'ouest, dans les limites de la commune de Vilhelmina, l'habitat s'élève aussi à 600 m autour de Storransan. Il en va de même dans la vallée du Vapst(älv) qui s'écoule vers la Norvège.

Dans la vallée de l'Ume(älv) et celles de ses affluents l'habitat reste plus bas que dans les villages de la dorsale de Stöttingfjäll et les fermes ne s'élèvent au-delà de 500 m que dans la partie supérieure de l'artère principale. Quant aux hautes vallées secondaires, elles sont dépourvues de peuplement à deux exceptions près, Bjellojaure dans la commune de Sorsele et Vesken qui atteignent plus de 600 m.

Dans les bassins de Juktå(dal) et de Vindel(älv), les fermes situées au-dessus de 500 m n'apparaissent que dans quelques vallées secondaires assez haut perchées. La petite ferme de Nuolpen sur un versant au-dessus du lac Tjul(träsk) est située à 625 m et rivalise avec Vesken déjà cité pour le titre de l'exploitation agricole la plus élevée de Laponie suédoise.

Au nord du fleuve Vindel(älv), la région qui correspond à la commune de Malå, le peuplement ne s'élève qu'à 400 m peine au nord.

Dans la province la plus septentrionale du pays, le Norrbotten, le district d'Arvidsjaur et d'Arjeplog se distingue par une indéniable originalité. L'habitat n'y dépasse 500 m que dans quelques rares points sur la partie supérieure de la vallée du Skellefte(älv) et aussi en quelques endroits sur le réseau adjacent des vallées latérales. Le plateau des interfluves qui dépasse 500 m ne porte que des habitations très clairsemées. Dans la vallée du Pite(älv), la plus septentrionale de ce secteur, la ferme la plus haut située Västerfjäll atteint à peine 450 m, bien loin de la ligne de partage des eaux et de la frontière avec la Norvège.[38]

Dans les parties moyennes et septentrionales du Norrbotten intérieur, les plus élevées en latitude de toute la Suède, l'habitat permanent n'a que peu d'extension dans le domaine proprement montagneux et s'élève assez peu en altitude, ce qui n'est pas étonnant. Aktse près de Laitaure dépasse 500 m mais le peuplement dans le bassin du Lilla (Petit) Lule(älv) dépasse en général de peu 300 m excepté dans la vallée du Pärl(älv) où il atteint plus de 400 m. Le long de la vallée principale du Stora (Grand) Lule(älv) élargie par tout une système lacustre, l'habitat ne s'élève qu'à 370 m à peine. Dans la vallée de Kaitunį, il se termine à 500 m excepté pour une petite ferme lapone Lietikkåba qui atteint 550 m ce qui en fait sans doute l'habitat permanent de caractère agricole le plus élevé dans tout le Norrbotten. Dans la vallée frontalière du Torne qui est limitrophe de la Finlande, l'habitat s'arrête à 350 m mais dans la vallée de Köngämä, il dépasse à nouveau 400 m. La ferme la plus septentrionale de la Suède, Kummavuopio est à 470 m.[39]

Arrivé à ce point, il paraît intéressant de comparer la limite supérieure de l'habitat permanent sur le versant oriental suédois de la chaîne des Scandes, versant "continental" à ce qu'elle est sur le versant occidental maritime, "atlantique" qui correspond à la Norvège du Nord, au sens large.

D) LA LIMITE SUPÉRIEURE DE L'HABITAT PERMANENT SUR LE VERSANT OCCIDENTAL DE LA CHAINE DES SCANDES AU NORD DE LA DÉPRESSION DE TRONDHEIM

Il paraît licite de rattacher ici à cet ensemble de la Norvège du Nord (les trois provinces du Nordland, du Troms et du Finnmark) le Trøndelag du Nord, au nord de la dépression nidrosienne. Pour cette étude qui n'a guère été entreprise jusqu'ici à notre connaissance, étant donnée l'extension considérable de ce vaste ensemble régional surtout en latitude, qui rendrait fastidieux l'examen systématique des cartes topographiques, un choix de quelques secteurs caractéristiques sera ici proposé.

Les deux premiers exemples seront pris dans le Nord-Trøndelag. Dans la partie méridionale de cette province, l'habitat permanent ne s'élève guère au-dessus de 300 m et ceci dans les parties supérieures des affluents de la vallée principale de Stjørdal (c.t. Stjördal 1/100,000e). Dans la vallée de la Sona, au sud de l'artère principale, la ferme la plus élevée Sundalen est à 335 m et au nord sur le promontoire formé par les deux versants de la Forra et de la Vigra avant leur confluent, la ferme d'Øvre-Baune est à 309 m seulement, ce qui n'est pas très élevé. Plus au nord, encore sur le versant qui s'étend au nord de l'émissaire du lac Store Grönning, l'exploitation de Vinkjoerran est à 310 m et celle de Storbakken à 334 m. Dans le Nord Trøndelag septentrional, sur le cours du Namsen, artère majeure de la région, si l'habitat permanent se tient au-dessous de 200 m, il s'élève notoirement dans les vallées adjacentes et atteint 400 m à Trøndalen sur le versant nord de la vallée du Gröndalselven émissaire du lac Tunn(sjö) et dépasse 350 m sur les rives de cette nappe d'eau (c.t. Trones 1/100,000e). La ferme de Småvandene au milieu d'un complexe lacustre marécageux à l'ouest du lac Tunn(sjö) est à 378 m. L'exemple suivant concerne la province du Nordland. Dans la région de Rana, sous 66° lat. N., l'habitat n'est pas sensiblement plus bas qu'au Nord Trøndelag méridional situé nettement plus au sud sous 63°30'. Dans le groupe de fermes de Dalsbygden, dans une vallée affluente du cours d'eau Ross(åen), la ferme de Inderdalen est à 322 m. Sur le cours de la Bjerka, à peu de distance des rives du grand fjord qui pénètre jusqu'à MoiRana, la ferme de Bjurboekdalen est à 250 m. Sur le Dals(elv), Bjerkniven est à 311 m à l'amont mais sur le versant exposé au midi, plusieurs fermes sont situées plus haut telles Kjeldalsenget à 350 m et Bjerklien à 323 m (c.t. Rana 1/100,000e). Un peu au nord, sur les versants méridionaux des reliefs qui supportent le grand glacier de calotte du Svartisen, dans le groupe de fermes de Ytteren, celle de Sletten est à 260 m au sud du lac Lang(vatn), Steinkjönlien est à 280 m et les exploitations de Ransandakslen sont à près de 300 m ce qui est assez étonnant à quelques kilomètres seulement de la masse d'air froid entretenue par le grand glacier (c.t. Svartisen 1/100,000e). Au sud-est de la calotte glaciaire dans la région de la grande vallée de

Dunderlandsdalen, l'habitat n'a tendance à s'élever qu dans les vallées secondaires. Ainsi, au sud est du confluent entre le Rauen(elv) et le Langvassågå, au flanc de l'échine allongée de Langfjell, l'habitat s'élève au-dessus de 300 m à Svarttjenneng (ou Tjernskog) 311 m, Solstad, Raurabben (338 m), Bygdas; dans la petite vallée du Grönfjelldågå ici aussi l'habitat dépasse 300 m, Gronli et Kvandalen sont ainsi à 320 m; au nord du grand coude décrit par le Ranenelv avant le confluent avec le Bjellågå et le Stormdals(elv) la ferme de Grauneset est à 285 m sur une terrasse et celles de Bredek sont à 320 m en position analogue. Mais les altitudes maxima sont atteintes ici précisément dans la vallée principale qui s'enfonce dans la masse montagneuse. La ferme d'Andfjeldneset est à 390 m (c.t. Dunderlandsdalen 1/100,000e). Au nord-ouest du glacier Svartisen dans les régions côtières, même avec des reliefs élevés (plus de 1,000 m) proches de la surface d'eau des fjords maritimes, l'habitat ne s'élève guère. La courbe de niveau de 150 m n'est guère dépassée: 170 m à Skaugvolddalen (c.t. Meløy 1/100,000e); au nord est du même glacier dans la vallée de Beiardalen qui est drainée par le Storågå lequel coule du sud vers le nord l'habitat permanent tout-à-fait à l'amont dépasse 200 m d'altitude (218 m à Leiramoen). Dans Tollådalen, vallée adjacente de celle du Storågå (ou Beiar(elv)), l'habitat s'élève même à presque 400 m à Skolneset et à Boeverholmen (c.t. Beiardalen 1/100,000e).

Dans la région du Skjerstadfjord au nord de 67° lat. N. l'habitat s'élève peu au-dessus de la limite supérieure de la transgression marine quaternaire qui a déposé des sables et des argiles fertiles; cette limite atteint 120 m environ. Et il y a peu d'exceptions à cette règle dans cette région très ouverte vers l'ouest aux influences maritimes. Cependant il y a trois fermes situées à 226 m sur Skipmandvik, un cap peu accusé entre le Skjerstadfjord et le Saltdalsfjord. Et il y a même sur le versant oriental de la vallée de Salt(dal), à Båtskar, une ferme à 415 m (c.t. Saltdal).[40] Dans la région de Narvik sous 68°30' lat. N environ, l'habitat peut atteindre 300 m comme à øvre Marken qui se situe à 322 m à l'amont de la vallée de l'Håkvik(elv) au sud du Bersfjord. Au nord du Rombaken fjord, quelques fermes atteignent 300 m comme Nygårdsmarken. Dans la vallée de Gratangseidet qui prolonge Herjangsfjord, une ferme se situe à 311 m à Vasose au-dessus du lac Osc(vatn). Mais ici (c.t. Narvik 1/100,000e), la moyenne des fermes les plus élevées ne dépasse guère 200 m.

A Bardu, belle vallée du Troms, c'est dans les vallées affluentes que le peuplement s'élève; il atteint 249 m à Bjørkås la ferme la plus haut située dans la vallée de Kirkes(dal), il dépasse même 300 m à Hanstad la ferme la plus à l'amont dans Østerdalen (c.t. Bardu 1/100,000e), mais la plupart des fermes n'en sont pas moins situées sur des terrasses fluviales au-dessous de la limite supérieure de la transgression marine quaternaire.[41] Au-delà de la latitude de la région de Bardu soit 68°40' N., il paraît superflu de déterminer la limite supérieure d'un habitat permanent qui ne dépasse guère l'extension des dépôts marins quaternaires et s'établit donc à basse altitude (75 m dans la région de Polmak au Finmark) que ce soit dans les régions d'Alta[42] ou celle de Polmak par exemple.[43]

II. LA LIMITE SUPÉRIEURE DE L'HABITAT TEMPORAIRE PASTORAL DANS LA PÉNINSULE SCANDINAVE

L'habitat temporaire en question est fondamentalement constitué par les chalets pastoraux occupés naguère pendant quelques mois ou même seulement quelques semaines chaque été par le personnel chargé de la garde et du soin des troupeaux aux estives. En règle générale mais pas absolue, les chalets étaient toujours situés plus haut que les exploitations agricoles, habitat permanent dont ils dépendaient. Mais il y a eu d'assez nombreuses dérogations à cette constatation d'ensemble étant donné la variété des situations topographiques respectives des fermes et des chalets et aussi la diversité des types de vie pastorale dans la péninsule scandinave.

L'état de la documention est tel en Norvège à propos des chalets pastoraux qu'il est possible de donner quelques indications assez précises sur la limite supérieure de l'habitat temporaire en ce qui concerne trois types différents de chalets pastoraux, les demi-chalets ou chalets "laitiers" d'abord: la vachère passait la plus grande partie de la journée à la ferme et ne montait qu'en fin de journée au chalet pour la traite du soir; elle y passait la nuit et redescendait au matin après la première traite de la journée, en portant à la ferme tout ou partie du lait recueilli; les chalets "proches" ensuite utilisés à mi-pente ou au sommet des versants des grandes vallées de l'Est (Østland) ou des grands fjords de l'ouest (Vestland) quand ceux-ci sont dégagés de la neige qui les recouvre, donc au printemps et encore à l'automne dans un type de vie pastorale dite "complète" (c'est-à-dire avec préparation sur place des produits laitiers au chalet dit dans ce cas chalet "complet"); enfin les chalets "lointains" (ou d'été, ou encore de montagne) utilisés pour exploiter les hauts pâturages d'altitude au coeur de l'été.[44]

Pour le premier type de chalets, localisé essentiellement dans la partie moyenne des grands fjords du Vestland, à savoir celui des "demi-chalets", cet habitat temporaire pouvait atteindre des altitudes assez fortes: du nord au sud, 730 m dans la partie méridionale du Sunnmøre, 675 m à Eid et Stryn dans le Nordfjord, 830 m dans le Sunnfjord, 550 m à Balestrand et Lavik dans le Sogn moyen, 600 m à Luster dans le Sogn supérieur, 588 m à Voss et Evanger, 700 m dans le Nordhordland, 910 m dans le Hardanger. Dans le Nord de la Norvège au sens large où il y avait aussi dans les régions côtières des demi-chalets, ils atteignaient évidemment des altitudes plus basses: 350 m seulement dans le Snillfjord pourtant localisé dans le Sor Trøndelag sous 63°20' N ce qui n'est pas particulièrement septentrional en Norvège mais 300 m encore pour les demi-chalets les plus haut situés dans Nord Rana dans le Nordland central. Ces altitudes qui ne sont pas considérables mais dont certaines sont assez fortes représentaient le plus souvent dans le Vestland des dénivellationss très marquées car beaucoup de fermes n'y sont pas très élevées, souvent proches du niveau de la mer et d'autre part, les distances qui séparent fermes et chalets sont relativement courtes.

Les chalets proches de printemps et (ou) d'automne en exploitation pastorale "complète" (c'est-à-dire avec traitement des produits laitiers sur place) étaient parfois eux aussi situés à une grande hauteur au-dessus du niveau de la mer notamment dans les parties supérieures des grandes vallées de l'Est (Østland) comme il se doit. Ils étaient placés plus bas dans le Vestland où la limite supérieure de ce type de chalets s'élevait à 579 m dans le Sunnmøre septentrional mais à 270 m seulement à Volda et Dalsfjord toujours dans le Sunnmøre, 450 m dans le Nordfjord à Davik, Eid, Gloppen, 850 m dans le Sogn interne, 1,093 m dans le Hardanger mais seulement 550 m dans le Ryfylke. Cette limite s'élevait parfois notablement dans les grandes vallées du Sørland et de l'Østland: 820 m dans le Setesdal, 800 m dans le Telemark à Sauherad et Seljord et 1,300 m à Vinje et Rauland (ce qui constitue un record); dans la grande vallée de la Glåma, elle varie de 400 m à l'aval en Solør, à 875 m en Sør Østerdal (méridional) partie moyenne de la vallée et à 910 m en Nord-Østerdal à l'amont, 1,112 m en Hallingdal supérieur, 1,036 m en Valdres; elle s'abaisse du Sør Trøndelag: 836 m à Gauldal, et 950 m à Oppdal, au Nord Trøndelag: 600 m à Stjørdal et Verdal, Meråker et Hegra.[45]

Reste la dernière classe de chalets, celle des estives lointaines ou de montagne. Le plus souvent, ils étaient haut placés même dans le Vestland. Ils atteignaient ainsi 1,030 m à Sunndal dans le Nordmøre, 1,100 m dans le Nord-Sunnmøre, 1,000 m à Volda et Dalsfjord dans le Sunnmøre, 650 m dans le Sunnfjord, 900 m dans le Nordfjord, 1,000 m dans le Sogn externe et 1,200 m dans le Sogn interne, 740 m dans le Nordhordland mais plus à l'intérieur du Vestland à la même latitude, 1,200 m dans le Hardanger en général et même 1,300 m à Granvin, 850 m dans le Ryfylke, toutes ces données concernant le Vestland.[46] L'altitude de l'habitat temporaire pastoral s'abaissait dans la Norvège du Sud Ouest à 850 m dans le Ryfylke en Rogaland et à Tonstad et Sirdal en Vest Agder; elle se relevait à 1,100 m à l'amont du Setesdal en Aust Agder.[46 bis]

Dans les cantons situés à basse ou moyenne altitude du Telemark inférieur qui opèrent en quelque sorte une transition entre le Sørland ou Norvège méridionale et les grandes vallées incisées dans les montagnes de l'Østland, les chalets "lointains" ne s'élevaient qu'à 950 m et même 700 m seulement à Drangedal et Sannidal au sud ouest mais dans les communes montagnardes du Telemark occidental qui peuvent être rattachées à l'Østland, ils atteignaient 1,300 m comme en Hallingdal.[47] Ainsi, sur le large plateau qui surmonte le versant septentrional de la vallée de la Votna, commune de Ål en Hallingdal, il y a une concentration de chalets au pied des grands escarpements montagneux car la surface est couverte largement de schistes cambrosiluriens donnant des sols assez fertiles sur lesquels poussent des pâturages réputés qui ont en outre une certaine extension en altitude ce qui favorise les conditions de leur exploitation car toutes les estives ainsi ne sont pas utilisables en même temps. Des groupes de chalets comme Eitrestölen (1,230 m) et Reineseter (1,216 m) sont parmi les plus hauts situés en Norvège (c.t. Hemsedal 1/100,000e).[48] Dans les vallées de Valdres et du Gudbrandsdal, l'altitude des chalets s'élève à 1,350 m mais le record à cet égard serait atteint dans l'Østerdal supérieur et plus particulièrement dans la commune de Sollia où il y avait des chalets qui s'élevaient jusqu'à 1,405 m.[49] Sollia est un canton montagnard qui est situé au sud du grand massif de Rondane où beaucoup des versants les plus haut situés sont exposés au sud sur des schistes sparagmitiques qui donnent de bons sols. L'habitat permanent et la limite de la végétation forestière atteignent à Sollia des altitudes supérieures à celles de la plupart des autres cantons montagnards de la Norvège du Sud. Aussi n'est-il pas étonnant qu'ici les chalets pastoraux atteignent des altitudes assez inhabituelles, plus fortes pour les extrêmes que celles du reste de l'Østerdal supérieur où elles plafonnent à 1,123 m.[50] L'altitude des chalets les plus hauts reste forte encore en Sør Trøndelag (méridional) avec 1,350 m à Oppdal et Orkdal mais n'atteint plus que 920 m à Gauldal toujours en Sør-Trøndelag.

Dans les régions plus basses de la Norvège du Sud-Est et topographiquement moins accidentées, la limite supérieure de l'habitat pastoral temporaire est plus basse que dans la partie méridionale du versant oriental de la chaîne des Scandes. Dans l'Østfold, elle n'atteignait pas 300 m et pas 500 m ni en Romerike inférieur 480 m, ni en Aker et Follo 450 m seulement, ni dans la partie inférieure de la vallée de la Glomma en Odal, avec les cantons d'Eidskog et Vinger, 470 m. L'altitude ne se relève que dans le Romerike supérieur à 645 m, en Solør, partie de la vallée de la Glomma, le grand fleuve de l'Est norvégien, qui débute à Elverum, à 577 m et plus encore en Hedmark dans son ensemble à 900 m.[51]

En Suède, sur le versant oriental de la chaîne des Scandes, l'altitude atteinte par les chalets en Dalécarlie occidentale n'est pas particulièrement élevée; dans la commune de Lima un exemple peut être donné, celui du groupe de chalets de Södra Kallberget composé de trois grands enclos de fauche dont le moins élevé s'élève à 576 m au-dessus du niveau de la mer.[52] L'altitude des chalets s'élève dans les cantons en partie montagneux du Jämtland occidental comme Undersåker. Dans le secteur le plus élevé à l'ouest de cette dernière commune, la majeure partie des chalets se situe entre 500 m et 600 m, entre 600 m et 650 m il y a encore une série d'habitats pastoraux: Prästvallen, Toppvallen, Sunnestvallen à Edsåsen, etc..., les plus haut placés étant Dalsnyvallen et Grönvallen (700 m).[53] Dans la commune voisine d'Åre au nord, le groupe de chalets le plus haut situé est Enesvallen à 740 m mais plusieurs autres atteignent 700 à 710 m dans différents secteurs de cette vaste paroisse, les exemples précis ne manquent pas.[54]

En Jämtland, à l'est du lac de Kall, la limite supérieure des chalets s'élève nettement plus qu'à l'ouest où elle reste dans la majorité des cas inférieure à 550 m. Dans la partie orientale la plus montagneuse de cette commune de Kall, la majorité des chalets sont situés entre 550 et 650 m et plus loin vers l'est dans la commune d'Offerdal, ils dépassaient la courbe de niveau de 650 m et la majeure partie d'entre eux étaient situés entre 670 et 680 m. Ici dans la région de Kall et d'Offerdal, la limite supérieure de l'habitat pastoral temporaire s'élève donc de l'ouest vers l'est en même temps que la limite supérieure de la forêt.[55] Si la comparaison est faite avec l'autre versant, celui de l'ouest de la chaîne des Scandes au nord de la dépression de Trondheim, les chalets les plus hauts atteignaient 660 m dans Namdalen au Nord-Trøndelag, 640 m dans la région d'Inderøy et Snåsa toujours en Nord-Trøndelag, 610 m dans Hattfjelldalen en Nordland, 550 m à Rana en Nordland encore.[56] L'altitude s'abaisse nettement dans la région de Saltdal, à l'est du lac Botn(vatn), où il y a un groupe de quatre chalets à 220 m d'altitude seulement celui de Grønliseter (c.t. Saltdal, 1/100,000e).[57] Dans la région de Bardu, le chalet de Björkåsseter atteint 439 m (c.t. Bardu 1/100,000e). A Steigen dans l'ile d'Engel(øya) au large du Nordland, l'habitat pastoral s'élevait à 270 m et dans l'archipel des Lofoten, à 50 m seulement à Gimsøy et Vågan.[58] Quant au Finnmark, la province la plus septentrionale de Norvège, les déplacements entre fermes et chalets qui en dépendaient avaient une composante presqu'exclusivement horizontale ce qui prouve que l'altitude des chalets était basse, ce qui est somme toute normal sous ces latitudes élevées.[59]

CONCLUSION

Ces données nombreuses et sans doute fastidieuses sur l'altitude des fermes les plus hautes et des chalets les plus élevés en montagne correspondent en gros aux variations de la limite supérieure de la forêt plus élevée sur le versant oriental de la chaîne des Scandes que sur celui de l'ouest. Mais elles montrent aussi qu'en ce qui concerne l'habitat permanent, la limite supérieure de celui-ci ne coïncide pas avec celle de la culture des céréales arrivant à maturité. En effet, bien des fermes de montagne ("fjellgård") dans la péninsule scandinave vivent à peu près exclusivement des produits de l'élevage. Ainsi dans la région de Kall et Offerdal sur le versant suédois en Jämtland, il y a des fermes de montagne habitées en permanence où ne sont cultivés que quelques champs de pommes de terre au-dessus de la limite climatique de la culture de l'orge, la seule céréale venant à maturité ici.[60] Il n'en va pas tout à fait de même dans les cantons supérieurs du Hallingdal et du Numedal où la limite supérieure de la culture de l'orge atteint entre 800 et 900 m d'altitude notamment dans Ustedalen la vallée la plus élevée du Hallingdal supérieur où l'orge était régulièrement cultivée jusqu'à 850 m au moins. Nulle part ailleurs en Norvège, la limite supérieure de la culture de céréales, l'orge en l'occurence ne monte aussi haut.[61] La ferme de Randen à Vass atteint 857 m au-dessus de la vallée de la Votna.[62] C'est une des plus élevées de ce secteur où les fermes de montagne s'essayaient à cultiver de l'orge même si celui-ci ne parvenait pas à maturité car il servait alors de fourrage vert. La culture des céréales ici servait autant dans le passé aux besoins de l'indispensable élevage qu'à celui de la fourniture de farine panifiable pour l'alimentation humaine. Mais il y a aussi en Norvège des fermes de montagne sans culture de céréales comme celles qui sont situées autour du lac Møs(vatn) dont le plan d'eau se situe à 914 m; il n'y a pas ici de grains cultivés et les fermiers vivent exclusivement d'élevage.[63] Il n'y a donc pas partout de liaison absolue entre la limite climatique de la culture des céréales et celle de l'habitat permanent pas plus qu'entre celle de la végétation forestière qui fournissait le bois indispensable à la préparation des fromages au chalet et celle de l'habitat pastoral estival, le bois était transporté sur bien des chalets dépourvus de ressources en combustible aux environs soit en été par cheval de bât, soit en hiver par traineau ainsi dans le Hallingdal supérieur.[64] Mais chaque exemple est un peu un cas d'espèce en fonction de l'exposition du site, de la topographie, de la nature des sols et d'autres conditions physiques locales qui mériteraient chacune un examen particulier.

NOTES

1 d'après NORGE/La Norvège, vol. II, Geografisk leksikon/dictionnaire géographique, J.W. Cappelens forlag/édit., Oslo, 1963, col. 1 et 2. Le chiffre de population estimé pour 1870-1880 est dû à NIELSEN, Yngvar, den faste bebyggelse i Norge fra 600 til 1000 meters hoide over havet og dennes forhold til vore skoves uttynding/L'habitat permanent en Norvège de 600 à 1 000 m au-dessus du niveau de la mer et ses rapports avec le recul de nos forêts, Den norske Turistforenings årbog for 1879/Annuaire de l'Association touristique norvégienne, Kristiania, p. 23 à 47, article essentiel, bibliographie p. 23 et 24.

2 Mention de Maristova sous le nom "Fjeldstuen Maristuen" (l'auberge de montagne Mariestua) dans NIELSEN, Y., op. cit., 1879, p. 30-31; contrôle systématique des lieux cités et de leurs altitudes par l'auteur de ces lignes sur les cartes topographiques modernes (c.t.: contrôle sur carte topographique).

2bis voir SUND, Tore, et SÖMME, Axel, Norge i kart/La Norvège par la carte, N.W. Damm & Søn édit., Oslo, 1947, vol. A, texte, 88 p., vol. B, croquis et photos. 52 p.; voir l'analyse de la carte au 1/50 000e Vest Jotunheimen, vol. A, p. 60, et la c.t. Sygnefjell au 1/100 000e. Dans la vallée d'Utla(dal), la ferme de Vormeli abandonnée depuis le milieu du XIXè s. atteignait 610 m. Un chalet pastoral l'avait remplacé à la fin du XIXè s. et au début du XXè s.

3 pour la situation à la fin du XIXè s., NIELSEN, Y., op. cit., 1879, p. 31 et remise à jour par ASAK, Per-Otto, Hvor høgt går den faste bebyggelse i Norge/Jusqu'où monte l'habitat permanent en Norvège, Geografen/Le géographe, Vol. 7, fasc. 1, Oslo, 1972, p. 49-54. Voir p. 51.

4 SUND, T., et SÖMME, A., op. cit., vol. A, 1947, p. 51.

5 NIELSEN, Y., op. cit., 1879, p. 31.

6 NIELSEN, Y., op. cit., 1879, p. 32.

7 SÖMME, Axel, Suldal, Norsk Geografisk Tidsskrift/revue norvégienne de géographie, vol. III, fasc. 5-8, 1931, p. 406 à 428; voir p. 411.

8 d'après SÖMME, Axel, Øvre Sirdal, en fjellbygd på Sørlandet/un canton de montagne dans le Sorland, Norsk geografisk tidsskrift, vol. III, fasc. 5-8, 1931, p. 429 à 437; voir p. 430. Il n'existe pas de cartes topographiques à l'échelle 1/100 000e pour un vaste ensemble comprenant les régions de Sirdal et du Setesdal et une grande partie de la Hardanger(vidda). Contrôle des lieux dans le Norge Atlas/Atlas de Norvège, J.W. Cappelens forlag/édit., Oslo, 1965.

9 NIELSEN, Y., op. cit., 1879, p. 33.

10 NIELSEN, Y., op. cit., 1879, p. 33 et contrôle sur l'Atlas de Norvège, Cappelen édit.

11 NIELSEN, Y., op. cit., 1879, p. 33 et 34 et contrôle sur l'Atlas de Norvège, Cappelen édit.

12 NIELSEN, Y., op. cit., 1879, p. 34 et 35; voir aussi ISACHSEN, Fridtjov, chapitre Setesdal, Telemark, Numedal, Hallingdal, p. 97 à 188 dans Norge, vårt land/la Norvège, notre pays, 3è édit., sous la direction de WERENSKIOLD, Werner, Volume II, Gyldendal norsk forlag/édit. Oslo, 1950, 816 p.; voir p. 150 et sq. pour les fermes de Møsstrand. Voir aussi HOLT-JENSEN, Arild, Fjellbygda Rauland/le cantone de montagne de Rauland, dans la série Ad Novas, Norwegian geographical studies, n° 6, Universitetsforlaget/Presses Universitaires de Norvège, Oslo, 53 p. + 16 cartes hors-texte, résumé anglais p. 49-51, bibliographie p. 53; voir le chap. 1, p. 9.

13 NIELSEN, Yngve, op. cit., 1879, p. 35.

14 NIELSEN, Y., op. cit., 1879, p. 35.

15 NIELSEN, Y., op. cit., p. 35.

16 ASAK, Per-Otto, op. cit., 1973, p. 51.

17 Il convient de remarquer qu'administrativement Dagali relève de la commune de Hol qui appartient pour l'essentiel au Hallingdal. Pour Nygård, voir den høyest beliggende bondegården/la ferme la plus haut placée, note anonyme dans Den norske turistforening årsbok/annuaire de l'association touristique norvégienne, série Fjell og vidde/montagne et plateau, n° 2, 1971. Thème fjellgårder/fermes de montagne, Oslo, 1971, 239 p., voir p. 199-201.

18 L'existence de plusieurs habitats permanents est signalée ici par NIELSEN, Y., op. cit., p. 36 sans autres précisions.

19 SUND, T. et SÖMME, A., op. cit., vol. A, Oslo, 1947, p. 39.

20 ISACHSEN, F., op. cit., 1950, p. 184.

21 ASAK, P.O., op. cit., 1973, p. 51.

22 HERTZBERG, Ludvig H., Nydyrkingen i tre Østlandsbygder 1918-1939/mise en culture nouvelle dans trois cantons de l'Østland (Norvège orientale), publié par Det Norske Geografiske Selskab/la société norvégienne de géographie, A.W. Brøgger édit., Oslo, 1944, 88 p., Bibliographie p. 87-88, voir p. 25.

23 WERENSKIOLD, W., chapitre Valdres, Gudbrandsdalen, Østerdalen, op. cit., Oslo, 1950, p. 200.

24 ASAK, P.O., op. cit., Oslo, 1973, p. 51.

25 NIELSEN, Y., op. cit., 1879, p. 38.

26 NIELSEN, Y., op. cit., 1879, p. 38 et ASAK, P.O., op. cit., 1973, p. 52.

27 NIELSEN, Y., op. cit., 1879, p. 38.

28 NIELSEN, Y., op. cit., 1879, p. 39.

29 ASAK, P.O., op. cit., 1973, p. 52.

30 SUND, T. et SÖMME, A., op. cit., vol. A, Oslo, 1947, p. 33.

31 pour l'Østerdal, voir NIELSEN, Y. op. cit., 1879, p. 39 et 40 et ASAK, P.O., op. cit., 1973, p. 52.

32 Pour les étapes du peuplement dans Nordre Atnedalen, voir les cartes hors texte V, VI et VII de l'ouvrage de PAY DAHL, Blanche, Utviklingen av bosetning, eiendoms og bruksforhold i Nordre Atnedalen/le développement du peuplement, des conditions de propriété et d'exploitation dans Nordre Atnedal, dans la Série Ad Novas, Skrifter fra det Norske Geografiske Selskap/Publications de la société norvégienne de géographie, Universitetsforlaget/Presse universitaires de Norvège, Oslo, Bergen, 1960, 145 p., résumé anglais, p. 131 à 136, bibliographie et sources p. 137 à 143, XIV cartes en hors texte.

33 ASAK, P.O., op. cit., 1973, p. 52.

34 HERTZBERG, L.H., op. cit., Oslo, 1944, p. 37. Voir aussi du même auteur Trysil, en forpost mot øst/un avant poste vers l'est, dans le recueil Fritid og feriemiljø/Loisir et milieu de vacances, dans la série Ad Novas, Norwegian geographical studies, n° 8, Universitetsforlaget/Presses Universitaires de Norvège, Oslo, Bergen, 1969, 162 p., voir p. 59 à 63, une carte.

35 HERTZBERG, L.H., op. cit., 1944, p. 49.

36 Un travail essentiel sur la question en Suède, RUDBERG, Sten, Ödermarkerna och den perifera bebyggelsen i inre Nordsverige. En diskussion av vissa orsakssamband bakom fördelningen bygd-obygd/L'"anoekoumène" et l'habitat périphérique en Suède septentrionale intérieure. Une discussion de certaines liaisons de causalité en rapport avec la répartition de l'espace habité et de l'espace inhabité, série Geographica, Skrifter från Uppsala Universitets Geografiska Institution/Publications de l'Institut de Géographie de l'Université d'Uppsala, n° 33, Uppsala, 1957, 192 p., résumé anglais p. 174-182, 10 cartes hors texte. Voir notamment la très remarquable carte n° 1, altitude de l'habitat, et le & Bebyggelsens höjde över havet i inre Nordsverige/altitude de l'habitat au-dessus du niveau de la mer en Suède septentrionale intérieure, p. 51 à 56.

37 d'après RUDBERG, St., op. cit., Uppsala, 1957, p. 52. Pour Högvålen, le village d'habitat permanent le plus haut peuplé en Suède (830 m altitude), voir le commentaire des clichés photographiques fig. 12 et 13, p. 82 et 83 de la contribution de FRÖDIN, John, den nord- och mellansvenska byns organisationsformer och upplösning/Les formes d'organisation et la dissolution du village en Suède septentrionale et centrale, 108 p. du recueil Bidrag til bondesamfundets-historie/ Contribution à l'histoire de la société paysanne, Tome II, Instituttet for sammenlignende kulturforskning/ Institut pour la recherche en civilisation comparée, série A, vol. XV, Oslo, 1933, 358 p. Pour Tossåsen, voir ÖSTER, Johannes, une confrontation des relations qui existent entre

certains facteurs d'ordre naturel et la culture dominante de l'habitat rural dans une région montagneuse de la province de Härjedalen (Suède), Folkliv/Vie populaire, Revue d'ethnologie nordique, 1937, p. 217 à 223, une carte, photos.

38 ce qui précède d'après RUDBERG, St., op. cit., Uppsala, 1957, p. 52 à 55; pour le cas particulier d'Arvidsjaur et Arjeplog, voir la thèse magistrale de BYLUND, Erik, Koloniseringen av Pite Lappmark till och med 1867/La colonisation du district lapon de Piteå jusqu'en 1867 dans la série Geographica, Skrifter från Uppsala Universitets Geografiska Institution/ Publications de l'Institut de Géographie de l'Université d'Uppsala, n° 30, Uppsala, 1956, 448 p., résumé anglais p. 414 à 425, sources et bibliographie p. 434 à 447, deux cartes hors texte; pour la limite supérieure de l'habitat permanent avant 1867, voir la p. 266.

39 d'après RUDBERG, St., op. cit., Uppsala, 1957, p. 55.

40 d'après SUND, T. et SÖMME, A., op. cit., vol. A, Oslo, 1947, p. 73.

41 d'après SUND, T. et SÖMME, A., op. cit., vol. A, Oslo, 1947, p. 80.

42 SLURÖD, SO. et TRØMBORG, D., op. cit., Oslo, 1976, p. 36.

43 SUND, T. et SÖMME, A., op. cit., vol. A, Oslo, 1947, p. 86 et sq.

44 d'après les travaux désormais classiques de REINTON, Lars, Soeterbruket i Norge/la vie pastorale en Norvège, Tome I, Soetertypar og driftsformer/Types de chalets pastoraux et formes d'exploitation, Instituttet for Sammenlignende Kulturforskning/Institut pour la recherche en civilisation comparée, Série B, XLVIII, H. Aschehoug & Co. (W. Nygaard), Oslo, 1955, XXVII + 481 p., 3 cartes hors texte; voir le chapitre I, Soeterbrukstypar og soetertypar/types de vie pastorale et types de chalets p. 18 à 50. L'établissement au cours des années trente de listes de chalets avec toute une série de renseignements recueillis par des collaborateurs locaux sur les chalets connus et utilisés de mémoire d'homme (nom, altitude, nature et type, distance de la ferme, etc...) sous l'égide de Instituttet for sammenlignende Kultursforskning permet de donner un choix de renseignements précis car ces listes sont censées être exhaustives.

45 d'après REINTON, L., op. cit., Oslo, 1955, p. 216 et 217 qui s'appuie exclusivement les listes de chalets ayant été utilisés depuis le milieu du XIXè s. selon la tradition locale. Il n'est bien entendu pas question de pouvoir opérer un contrôle quelconque d'après la carte topographique quelle que soit la date d'établissement de celle-ci car les signes de la légende ne distinguent pas entre les types de chalets qu'il s'agisse de demi-chalets et de chalets "complets", "proches" ou "lointains". C'est seulement pour ces derniers qu'un certain contrôle cartographique est possible puisque ce type de chalet est par définition le plus haut situé de toutes les catégories. Dans une région donnée, il faut admettre que la

limite supérieure de l'habitat temporaire est celle des chalets de montagne ou d'été du moins là où il est notoire qu'un système pastoral à deux chalets, l'un utilisé au printemps et (ou) en automne, l'autre en été, était en pratique.

46 REINTON, L., op. cit., Oslo, 1955, p. 217.

46bis REINTON, L., op. cit., Oslo, 1955, p. 217.

47 REINTON, L., op. cit., Oslo, 1955, p. 217.

48 SUND, T. et SÖMME, A., op. cit., Oslo, 1947, vol. A texte, p. 39.

49 REINTON, L., op. cit., Oslo, 1955, p. 217 et ISACHSEN, Fr., op. cit., Tome II, Oslo, 1950, p. 184.

50 SÖMME, A., op. cit., Bergen, 1954, vol. A., p. 125 et 256.

51 REINTON, L., op. cit., Oslo, 1955, p. 217 et 218.

52 VEIRULF, Olle, Södra Kallberget, en byfäbod i Västerdalarna/un chalet village en Dalécarlie Occidentale, dans Geographica, n° 15, Geografiska studier tillägnade John Frödin den 16 april 1944/études géographiques dédiées à J. Frödin..., Uppsala, 1944, p. 174 à 191; voir p. 176.

53 KILANDER, Sven, Fäbodväsendet i Undersåkers socken/La vie pastorale dans la paroisse d'Undersåker/Jämtland occidental, mémoire dactylographié, déposé en archives à l'Institut de géographie humaine de l'Université d'Uppsala, 31 p., mars 1940; voir p. 6.

54 KILANDER, Sven, Fäbodväsendet i Åre socken/La vie pastorale dans la paroisse d'Åre (Jämtland occidental), mémoire dactylographié, déposé en archives à l'Institut de géographie humaine de l'Université d'Uppsala, novembre 1942, 33 p.; voir p. 7 à 9.

55 FRÖDIN, John, Fäbodbebyggelsen i Kall och Offerdal/L'habitat pastoral à Kall et Offerdal, Geografisker Annaler, 1919, p. 353-386, voir p. 363.

56 REINTON, L., op. cit., tome I, Oslo, 1955, p. 218.

57 SUND, T., et SÖMME, A., op. cit., vol. A, texte, Oslo, 1947, p. 73.

58 REINTON, L., op. cit., Tome I, Oslo, 1955, p. 218.

59 WESTRHEIM, Sonja, Seterdrift i Finnmark/l'exploitation pastorale au Finnmark, mémoire de maîtrise, Institutt for Folkelivsgransking/Institut de recherche sur la vie populaire, Université d'Oslo, automne 1978, 240 p., bibliographie, cartes, photos, ronéoté. Voir p. 62.

60 FRÖDIN, J., op. cit., 1919, p. 362.

61 ISACHSEN, F., Numedal et Hallingdal dans l'ouvrage collectif sous la direction de WERENSKIOLD, W., op. cit., Tome II, Oslo, 1950, p. 178-179 et p. 185.

62 SUND, T. et SÖMME, A., op. cit., vol. A, texte, Oslo, 1947, p. 38 et c.t. Hemsedal, 1/100 000e.

63 ISACHSEN, F., dans WERENSKIOLD, W., op. cit., Tome II, Oslo, 1950, p. 151 et sq.

64 ISACHSEN, F., dans WERENSKIOLD, W., op. cit., Tome II, Oslo, 1950, p.184.

EDITOR'S NOTE

M. Cabouret was unable to attend the conference but his paper was accepted for publication: for technical reasons it was not possible to include the two illustrations provided with this paper.

CHAPTER 23

THE CHANGING IDENTITY OF THE FRONTIER IN MEDIEVAL NOTTINGHAMSHIRE AND DERBYSHIRE.

P.T.H.Unwin
(Bedford College, London)

There are two distinct types of settlement frontier and agricultural margin in both upland and woodland areas. The first is encompassed by the traditional concept of the frontier as the line dividing the civilised and settled world from the wild unknown. The image it presents is of a growing population pushing back the forest frontier and extending the settled, agricultural, area far into the uplands. However, behind this broad line of expansion a second frontier emerges, as within each parish or township small areas of woodland are cleared, trees are uprooted, and the plough is turned to the earth. This division of the frontier into its external and internal aspects is of considerable use in interpreting the pattern of occupation and colonisation that occurred in Nottinghamshire and Derbyshire prior to the middle of the 14th century.

This paper begins with a general description of the physical environment of these two English Midland counties. It then uses archaeological and place-name evidence to evaluate the extent of Anglo-Saxon and Scandinavian settlement in the region before the Norman Conquest of the 11th century. Domesday Book provides an abundance of evidence to suggest that, following an early period of widespread external frontier expansion, the 10th and 11th centuries were characterised by internal colonisation within most townships. The study encounters two critical problems. Firstly, it is clear that during the period before the Norman Conquest it was normal for a township to have had several distinct settlement foci within it, but it is unknown whether these each had their own separate field systems, or whether instead the inhabitants of each settlement participated in a unified field system operating throughout the township. Secondly, and associated with the first problem, is the uncertainty over the date of foundation of the townships and their boundaries, and whether or not these relate to an earlier estate structure (Jones, 1976, 1978).

THE PHYSICAL ENVIRONMENT

The land surface of Derbyshire and Nottinghamshire slopes eastwards from an altitude of over 2,000 ft. (610m), in the Peak District in the west, to the east of Nottinghamshire, where most of the land is less than 100 ft (30m) above sea level (Fig 23:1). This part of England therefore provides a good environment in which to consider the effects of altitude on the margins of settlement. The River Trent cuts across the region from the south-west to the north-east, and in its course

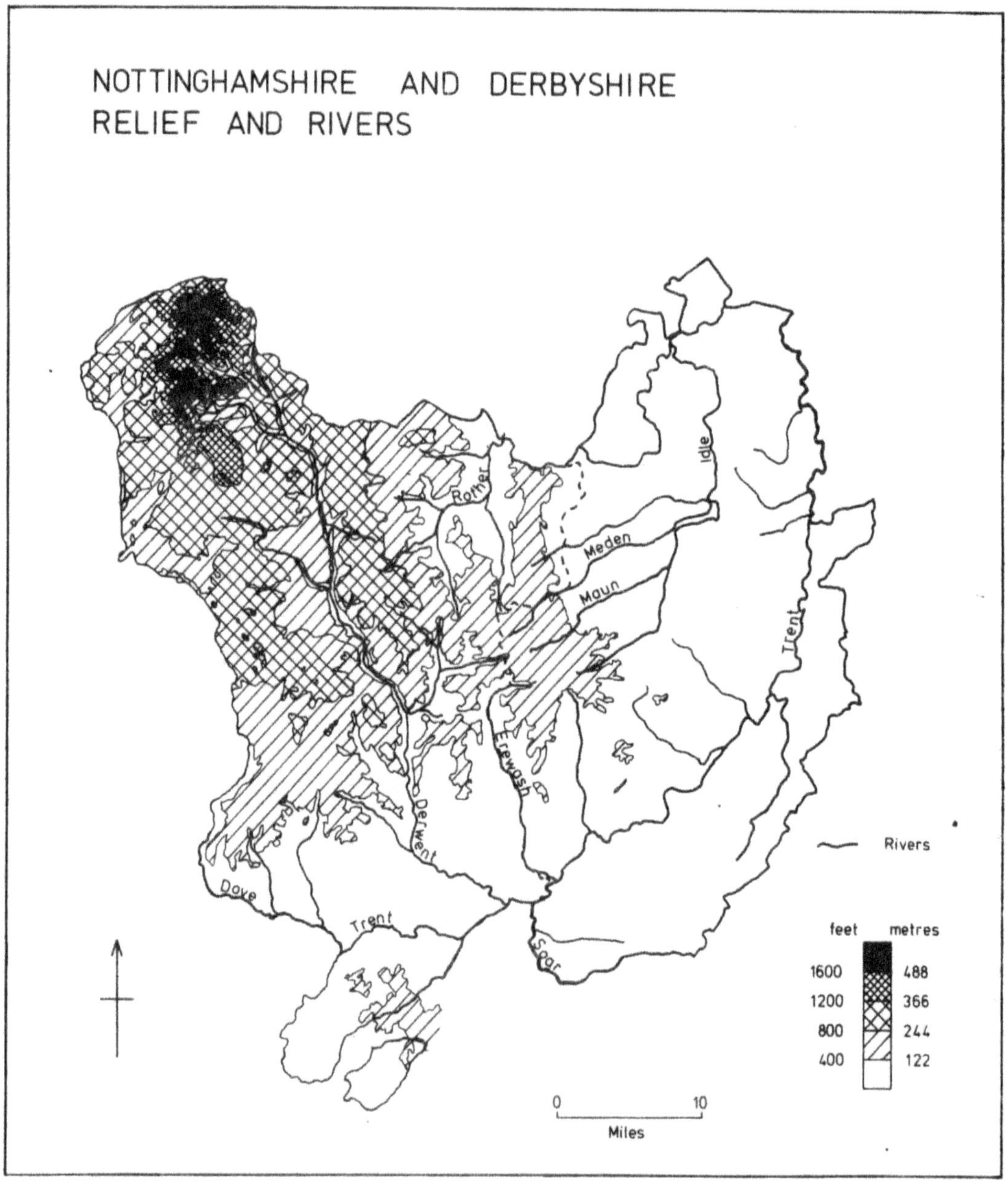

Fig. 23:1 Nottinghamshire and Derbyshire: relief and rivers.

it is joined by several rivers flowing south from the Peak District, notably the Dove and the Derwent. The rivers Rother, Idle, Maun and Meden all flow north or east towards the River Humber and the North Sea. In general the Marls and Clays around the Trent valley provide the more fertile soils, and the Bunter Sandstones in the west of Nottinghamshire, together with the Limestones and Grits of central and northern Derbyshire, remain areas of poor fertility.

POST-ROMAN SETTLEMENT AND AGRICULTURE

The traditional picture of the colonisation of this region, based on early archaeological and place-name work, agrees well with the concept of a receding tide of forested lands on the Bunter Sandstones and the hills of Derbyshire (Cameron 1959: Page, 1906, 1905; Stenton, 1943). The Romans are seen as occupying the Trent valley and the south-east of Nottinghamshire, where there were several villas and wide estates on easily cultivated land. Derbyshire, on the other hand, remained an area of sparse settlement, crossed by several roads, and which included forts such as those at Brough, Little Chester and Melandra; its main economic significance appears to have lain in its lead mines. The Anglo-Saxons between the 6th and 8th centuries, are then seen as colonising the remainder of the lower-lying region, pushing the forest frontier back and expanding the area of settlement. The final stages of this process of early colonisation were completed in the 9th century by the Scandinavians who, in addition to occupying areas of Anglo-Saxon settlement, also pushed further up the river valleys and settled in areas of poorer soils; many Scandinavian place-names are indeed found on Bunter Sandstones in the valleys leading away from the Trent (Cameron, 1965).

Recent archaeological finds and a re-evaluation of the place-name evidence have, however, necessitated a new interpretation of this frontier expansion. It is extremely difficult to estimate the extent of woodland in these two counties at the end of the Roman period. Two main sources suggest that, whereas much of Derbyshire was indeed still wooded in the 8th century, most parts of Nottinghamshire had, in contrast, already been cultivated in the Roman period.

Archaeological evidence, in the form of aerial photography and field walking, has revealed that there was a far greater density of occupation in the Trent valley during the Roman period than had previously been expected. Field walking has thus produced Roman pottery from most parts of south-east Nottinghamshire. Of more importance, though, is the discovery by Riley of a vast network of fields in the north of the county, stretching from Worksop and Retford to Doncaster in Yorkshire (Riley, 1978,1979). This is precisely the area in which large numbers of Scandinavian -by place-names, such as Barnby, Ranby, Serlby and Scrooby, are to be found. Trial excavations

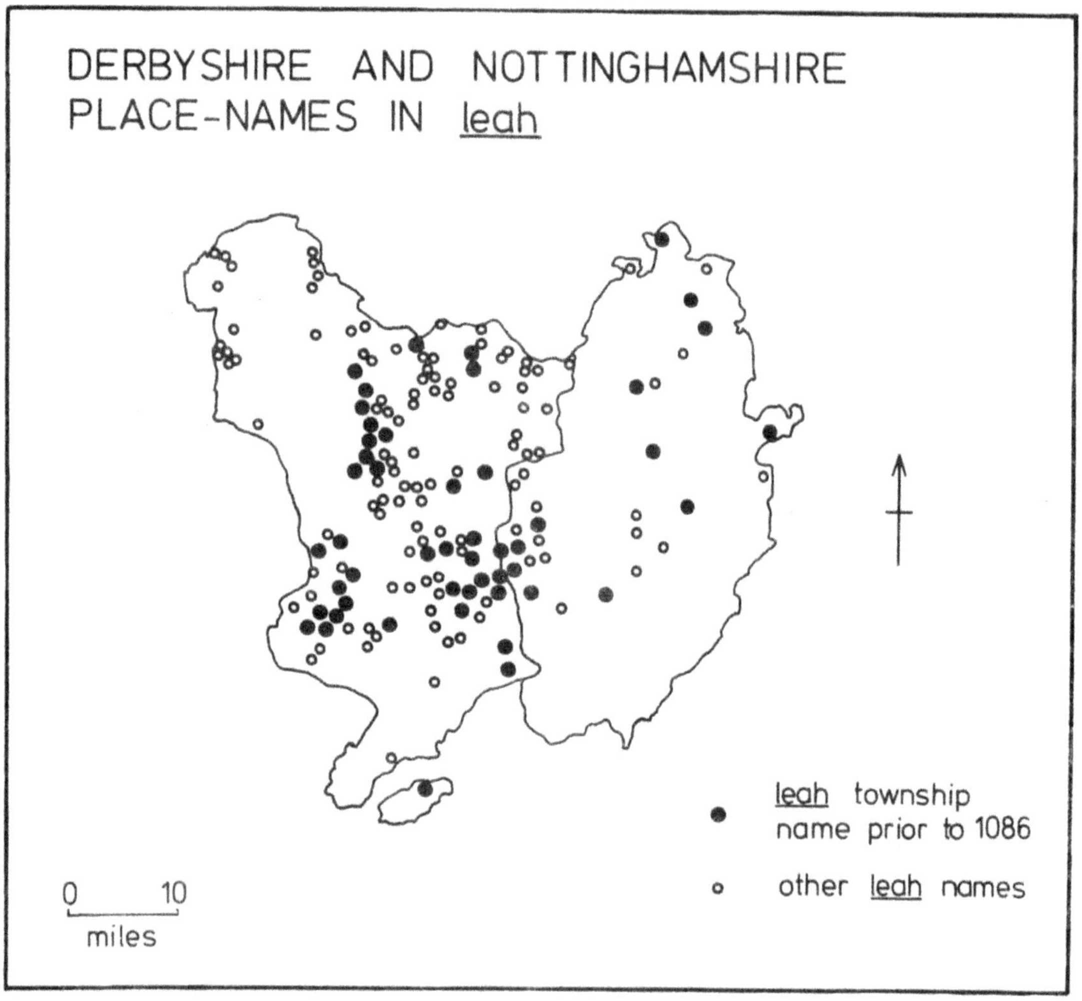

Fig. 23:2 Derbyshire and Nottinghamshire: place-names in leah.

have only revealed Roman material from the ditches of these field systems, and Riley tentatively suggests that they are Roman in date. This region can not therefore have initially been cleared and colonised by immigrants of Scandinavian stock. The only part of Nottinghamshire apparently not to have experienced any major Roman occupation is the upland west between the villa near Mansfield and Nottingham itself. It is perhaps significant that five -leah township place-names, Brinsley, Annesley, Strelley, Greasley and Kimberley, indicating woodland clearings, are to be found precisely in this part of the county.

This introduces the second type of evidence, namely that of place-names, that suggest that Nottinghamshire was far less wooded than its western neighbour. Nottinghamshire has very few -leah township place-names dating from Domesday Book or earlier, with, apart from those already mentioned, only six other names in the north, and two others, Bathley and Lambley near the River Trent in the centre of the county (Fig 23:2). Only one of the townships to the south and east of the Trent, Wigsley, has a -leah name. This lack of woodland township names, associated with extensive archaeological evidence of Roman occupation, continues along the Trent valley into southern Derbyshire. However, in the centre of Derbyshire along the River Derwent there are at least 11 -leah township place-names mentioned in Domesday Book, and in the east, between Alfreton and Longeaton, there are a further 9 -leah township names. In addition 3 -leah township names are found in the north of the county and 7 in the west. The majority of these place-names remain in valleys close to rivers, which further adds to the impression that many areas of relatively low land in north and central Derbyshire remained unoccupied at the end of the Roman period. Extensive post-Norman wood clearance in Derbyshire is attested by the total of 149 -leah place-names noted in the county by Cameron (1959), which can be compared with the total of only 28 in Nottinghamshire.

The archaeological evidence for the early Anglo-Saxon occupation of these two counties is far less extensive than for the Roman period, and both settlements and cemeteries tend to be found mainly in close proximity to the River Trent. This pattern is closely supported by the distribution of -ham and -inga-/-ingas place-names, with the majority of these also lying close to the River Trent in Nottinghamshire; there are only 4 -ham names throughout Derbyshire.

This evidence suggests that there was a recession of the settlement frontier within these two counties at some time between the 5th and the 7th centuries. Although much of the land was probably clear of woodland, both the margins of agriculture and the frontiers of settlement appear to have contracted into the Trent valley. The simplest explanation for this contraction is the existence of a much reduced population in the early Anglo-Saxon period, which permitted for example the earlier cultivated fields in the north of Nottinghamshire to be left to return to scrubland and secondary wood. This would agree closely with Cunliffe's (1978) and Fowler's (1978) arguments for a major population decline in the late-Roman period. The cause of this decline is uncertain, and although Shrewsbury (1970) has argued that Bubonic plague carried by the rat-flea could not have entered Britain at this time, the accounts by Gildas and Bede of famine and disease do give some credence to the suggestion that some major epidemic may have raged in the 5th or 6th centuries.

There is little evidence from either of these two counties that the actual Anglo-Saxon settlements were located on the same sites as, or even in close proximity to, previous Roman settlements; -ham place-names are not to be found in association with villas or Romano-British settlements. Of more interest, though, is the fact that in the north-west of Nottinghamshire the township boundaries bear no relation to the underlying pattern of fields of possibly Roman date (Riley, 1978, 1979). This again suggests a discontinuity between the Roman and later occupancy patterns in this part of the region. However, the township boundaries in the east and south of Nottinghamshire are themselves of particular interest. Throughout the course of the River Trent, from the north-east of the county as far as Nottingham, almost all of the parishes that adjoin it do so with their long axes at right angles to the river. It is as though a swath of land on either side of the Trent has been laid out and subdivided into townships of a similar size, each with a river frontage of between 1 and 2 kms.. These stand in marked contrast to the remaining townships in the north and west of the county in the area that became Sherwood Forest, and also in the south-west. 8 of the 13 -ham and -ingaham place-names in the county are found within these townships adjacent to the River Trent. It thus seems likely that following the late Roman population decline the earliest Anglo-Saxon settlement in the 7th century did indeed lie within the narrow confines of the Trent valley, and it is possible that it was at this date that the boundaries of these townships were first delimited. Most of the actual sites chosen for settlements at this date appear to have been on river terrace gravels.

The subsequent settlement frontier would seem to have expanded quite rapidly westwards following the lines of the major rivers; -tun place-names are thus to be found scattered throughout both counties, although there is a notable absence of them in the Sherwood Forest area and the adjacent eastern parts of Derbyshire. It is perhaps significant that this area has a number of the -leah place-names within it. The precise nature of Sherwood Forest presents a problem. Crook (1979) comments that most authorities accept that all of Nottinghamshire north of the River Trent was considered to be Royal Forest after the Norman Conquest and prior to the reign of Henry III. However, the term Forest did not necessarily mean that the area was totally wooded. By the mid-13th century perambulations indicate that the legal area of the Forest lay between Nottingham and Caythorpe in the south, and Mansfield and Perlethorpe in the north. There is considerable evidence through rights of assart that much of this smaller area was indeed wooded at this time. It is, however, difficult to assess to what extent this Forest had remained woodland, and thus outside the margins of agriculture, since the 5th century. What is certain, though, is that the imposition of Forest Law greatly curtailed the expansion of both settlement and agriculture in this part of the county, and it was only after the relaxation of the laws in the 17th century that the margin of agriculture can be said to have included Sherwood Forest.

It seems probable that it was between the 8th and 10th centuries that major settlement, other than that of the Romans, for the first time extended up the River Derwent and into the Peak District. The juxtaposition of the -tun and -leah place-names also suggests that it was at this time that woodland clearance took place permitting the extension of agriculture into this region.

By the time of the Danish invasions of the 9th century most of lowland Nottinghamshire and Derbyshire had been resettled, and the margins of agriculture pushed out into the hills of Derbyshire and the relatively infertile Bunter Sandstones of west Nottinghamshire. Both counties lay within the land of the five boroughs of the Danelaw (Lincoln, Derby, Nottingham, Stamford and Leicester), but Scandinavian influence, in terms of both place- and personal-names, was far greater in Nottinghamshire than in Derbyshire. The expansion of Anglo-Saxon settlement prior to the Scandinavian immigration implies an increasing population, either through natural increase alone, or more likely also through immigration, and the further Scandinavian immigration necessitated a yet greater expansion of the settled area. While it is clear that the Danes settled in areas of Anglo-Saxon occupation, and in the instance of the so-called Grimston hybrid place-names, probably renamed many earlier Anglo-Saxon sites, it is also evident that they created new settlements. It is particularly interesting to note that most of the -by settlement names, most probably indicating new Scandinavian settlements, are located in areas of poor soil and away from the main regions of Anglo-Saxon population (Cameron, 1965). Indeed several of these names lie within the region of Roman fields in the north of Nottinghamshire, and perhaps represent the final recolonisation of land that had been farmed at least five centuries previously.

By the mid-11th century nearly all of these two counties apart from the Sherwood Forest region and the high Peak District of north-west Derbyshire, with its moors and peat hags, had been settled. The external margins of both settlements and cultivation had been extended beyond the limits that they had reached in the Roman period. However, the estate structure of these two counties provides further information concerning the relative dates of settlement colonisation and also indicates further features of the process of internal colonisation. A comparison of the manorial linkages in Nottinghamshire and Derbyshire derived from Domesday Book, 1086, (Figs. 26:3 and 26:4) reveals that the pattern is much simpler in the west than in the east. In Nottinghamshire there tend to be more manors per township, and estates tend to be more dispersed and scattered than in Derbyshire. If, as seems likely, the Anglo-Scandinavian estate structure became more complex through time, this would add further support to the relatively late settlement of Derbyshire, with its simpler estate structure.

The royal berewick estates in the Derwent valley, based on the manors of Metesforde (Matlock Bridge), Werchesuorde (Wirksworth), Aisseford (Ashford) and Hope (Hope), provide the clearest examples of compact estates consisting of a number of complete and adjacent townships in the region. Prior to the Norman Conquest they were held by King Edward and, while it is possible that they had been formed from an amalgamation of earlier units, their relatively small size suggests that they had in fact remained as compact estates since their foundation, or that they had at some time been carved out of a single larger estate. The simple manorial structure in the Derwent valley provides further support for the evidence of the -leah place-names that this part of the county was settled relatively late in the Anglo-Saxon or Scandinavian periods; there is as yet no evidence that these estates had Roman precursors.

The distribution of manors within individual townships at the time of the Norman Conquest, 1066, provides further information on the subject of the margins of agriculture in the region. This again brings to the forefront the problematic question of the date of the delineation of township boundaries. The evidence already cited suggests that many townships in this part of the Midlands were first laid out during the Anglo-Saxon period, quite possibly making use of some sections of earlier Roman boundaries. Thus in the south-east of Nottinghamshire in Bingham wapentake the Fosse Way, a Roman road, is used as a township boundary throughout its length, but in the neighbouring wapentake of Newark the road is never used as a boundary. If the townships were created at the time of the Anglo-Saxon settlement it is then possible to envisage either that there was one settlement per township, or that a single township was made to include the area of several settlements. If townships were a slightly later creation they could have been laid out deliberately to include several pre-existing villages, but this scenario seems improbable. The regularity of the townships in the south and east of Nottinghamshire, their small size, and the fact that this area was densely populated during the Roman period, suggest that

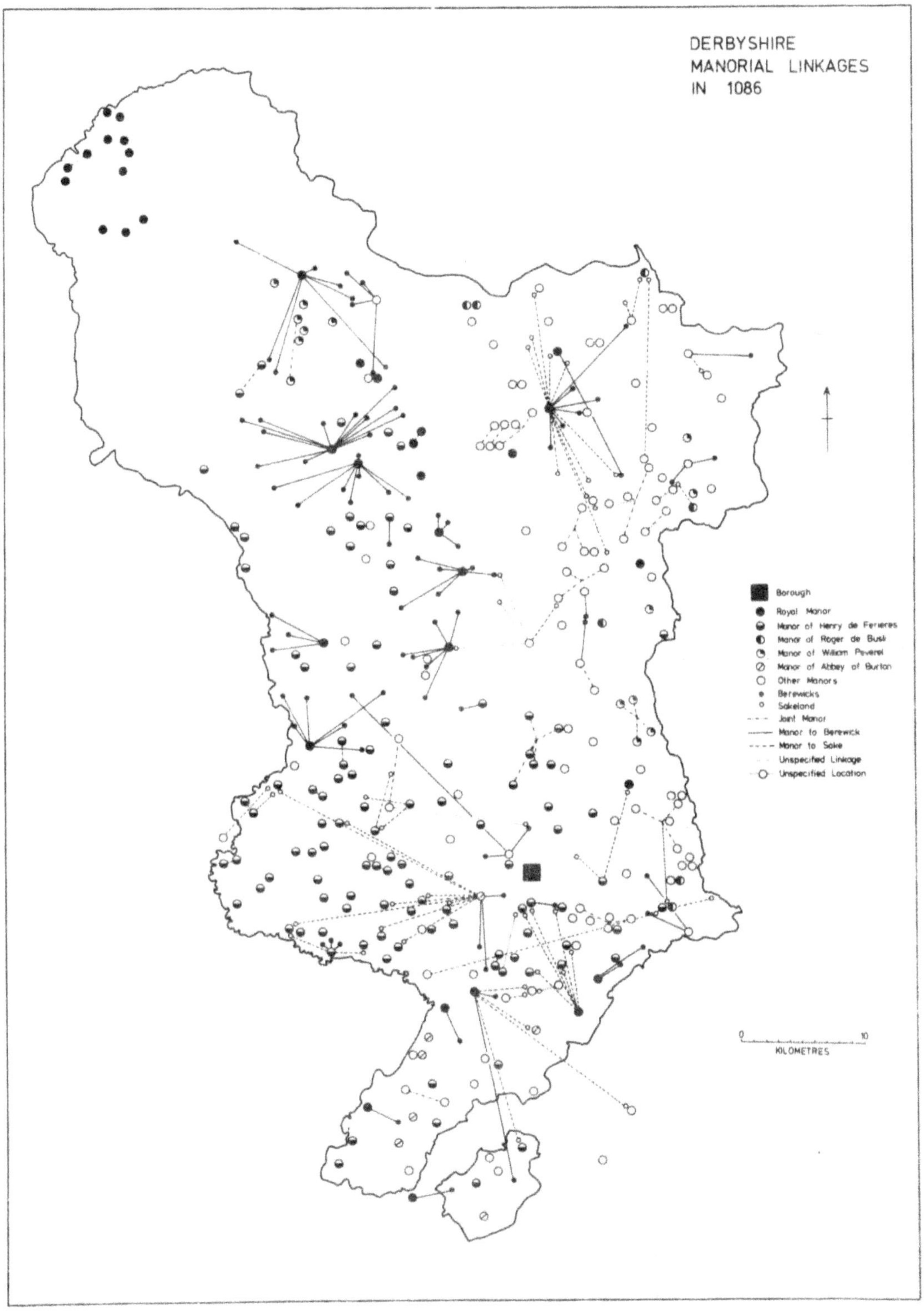

Fig. 23:3 Derbyshire: manorial linkages in 1086.

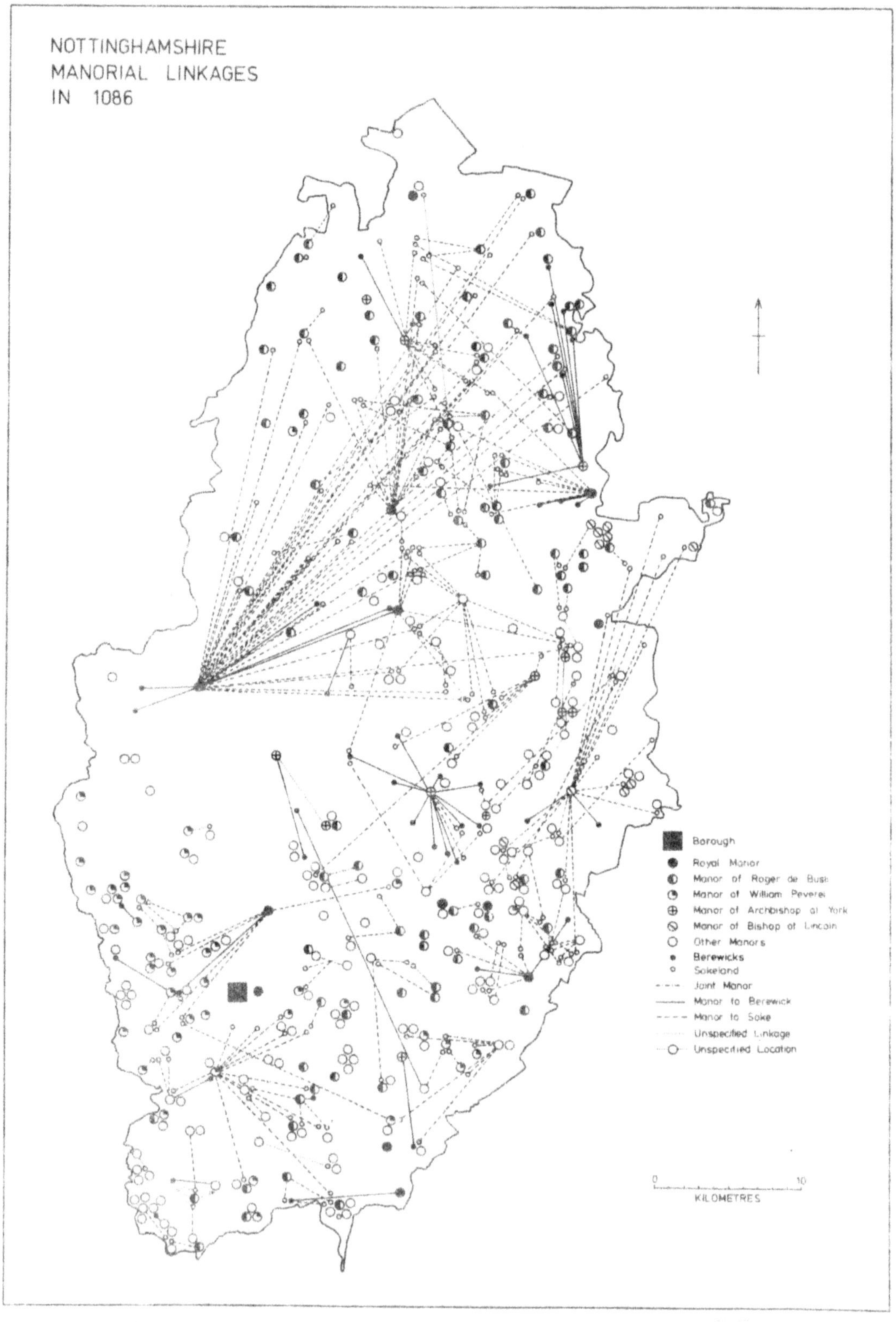

Fig. 23:4 Nottinghamshire: manorial linkages in 1086.

township boundaries here did not evolve gradually through the expansion of agriculture into woodland and waste around each settlement, although it is possible that this occurred in the uplands of Derbyshire.

Approximately 40% of the townships in Derbyshire at the time of the Norman Conquest had only a single manor within them, whereas in the county of Nottinghamshire the corresponding figure was as low as 24%. Thus the vast majority of Nottinghamshire's townships had two or three manors within them at the end of the Anglo-Scandinavian period. If Derbyshire was the later settled area this evidence would seem to suggest that townships normally initially possessed single manors, and that townships with more than one manor represent situations of internal settlement colonisation, partible inheritance giving rise to several settlements within one township, or the granting away of land in one part of a township where an additional settlement was then created. The township name could then represent either the initial settlement in the township or the settlement that became dominant.

CONCLUSIONS

Whatever the details may be of the processes involved it seems clear that, once the external margins of woodland had been pushed back, there followed a period of internal settlement expansion and agricultural colonisation. This caused increasing complexity in the settlement and estate structure, especially within south and east Nottinghamshire. Although it is difficult to estimate population densities at this time it seems likely that the Trent valley by the mid-11th century had regained population levels similar to, or greater than, those of the 4th century; there was little room for further internal colonisation at the given level of agrarian technology. Most of eastern Nottinghamshire records Domesday "population" levels of approximately 10 people (heads of household) per square mile, whereas figures for the Sherwood forest region were as low as 1.5 per square mile and all of north-west Derbyshire had densities of less than 2 recorded people per square mile. To convert these figures to actual populations requires the use of multipliers and is fraught with difficulty. A clearer example of the levels of population in south-east Nottinghamshire can be obtained by looking at one particular township within this region. Staunton, with one manor and a berewick within its bounds, had a recorded population of 30 in Domesday Book, 1086, in an area of 1371 acres. If this represents 30 households, apart from that of the manorial lord, and if the yardland of each family included approximately 30 acres of arable (Adams, 1977), then 900 acres, or 66% of the total area of the township, would have had to be arable land. Part of this township was marshland well into the 18th century, and land for pasture and wood would also have been required for livestock. It thus seems clear that internal colonisation by the 11th century in this part of the region had almost reached its capacity.

In contrast most of Derbyshire remained sparsely settled. The frontiers of settlement had moved up to around the 800 ft. (244m) contour in the uplands, with under 20 settlements in Derbyshire being recorded higher than this in Domesday Book, 1086. Domesday Book indicates that to the north and west of the River Trent much woodland still remained, particularly in western Derbyshire. While Sherwood Forest on the Bunter Sandstones remained largely closed for agriculture during the later Middle Ages, internal colonisation of woodland continued within Derbyshire and some northern and western parts of Nottinghamshire. As Dodgshon (1978) has commented, the history of the period 1066-1350 in England "would appear as a play on themes already established by the time it began". By the mid-14th century south-east Nottinghamshire was no longer the most wealthy and populous part of the region. Settlement development and the expansion of the margins of agricultural land had led to a far more even distribution of population throughout the area, although the uplands of north-west Derbyshire remained as bleak and unoccupied as they are today, except perhaps for the omnipresent sheep.

REFERENCES

ADAMS, I.H., 1978: Agrarian landscape terms: a glossary for Historical Geographers, Institute of British Geographers Special Publication, no.9.

CAMERON, K., 1959: The place-names of Derbyshire vol.1 Cambridge University Press for the English Place-Name Society vol. XXVII.

CAMERON, K., 1965: Scandinavian settlement in the territory of the Five Boroughs: the place-name evidence, Inaugural lecture, University of Nottingham.

CROOK, D., 1979: The struggle over forest boundaries in Nottinghamshire, 1218-1227, Transactions of the Thoroton Society of Nottinghamshire, LXXXIII, 35-45

CUNLIFFE, B., 1978: Settlement and population in the British Iron Age: some facts, figures and fantasies, p.3-24 of Lowland Iron Age communities in Europe, Edited by Barry Cunliffe and Trevor Rowley, B.A.R. International Series (Supplementary), 48.

DODGSHON, R.A. 1978: The early Middle Ages, 1066-1350, Chapter 4, p.81-117 of An Historical Geography of England and Wales, Edited by R.A. Dodgshon and R.A. Butlin, London, New York and San Francisco: Academic Press.

FOWLER, P.J., 1978: Lowland landscapes: culture, time and personality, p.1-12 of The effects of man on the landscape: the lowland zone, Edited by S. Limbrey and J.C. Evans, C.B.A. Research Report, no.21.

JONES, G.R.J., 1976: Multiple estates and early settlement, Chapter 3, p.15-40 of Medieval Settlement, Edited by P.H. Sawyer, London: Edward Arnold.

JONES, G.R.J., 1978: Celts, Saxons and Scandinavians, Chapter 3, p.57-79 of An Historical Geography of England and Wales, Edited by R.A. Dodgshon and R.A. Butlin, London, New York and San Francisco: Academic Press.

PAGE, W.(Ed.), 1905: The Victoria History of the County of Derbyshire vol.1, London: James Street.

PAGE, W.(Ed.), 1906: The Victoria History of the County of Nottinghamshire vol.1, London: James Street.

RILEY, D. 1978: An early system of land division in south Yorkshire and north Nottinghamshire, p.103-8 of Early land allotment, Edited by H.C. Bowen and P.J. Fowler, B.A.R. 48.

RILEY, D. 1979: Air photography based on Sheffield, Current Archaeology VI(7), 219-22.

SHREWSBURY, J.F.D. 1970 A history of bubonic plague in the British Isles, Cambridge University Press.

STENTON, F.M., 1943: Anglo-Saxon England, Oxford University Press.

CHAPTER 24

DEVELOPMENT AND CHANGE IN THE FUNCTION OF CENTRAL PLACES IN AN ISOLATED MINING REGION AT THE SETTLEMENT FRONTIER OF THE BOREAL FOREST: THE EXAMPLE OF THE KLONDIKE GOLD FIELD IN THE NORTHWESTERN CANADIAN MINING FRONTIER

H. BECKER
(Bamberg University)

At the turn of this century a great number of rich placer gold deposits were discovered in the Canadian Yukon Territory and in Alaska. After the discovery of a new gold field thousands of people poured into such an area, often they were more than ten thousand in number.(1) Thus within a very short period of time several important isolated concentrations of population sprang into existence amidst the wilderness of the boreal forest. In each gold field numerous settlements quickly developed; they ranged from urban centres to small goldmining camps.

As it is shown by contemporaneous archival materials the settlements of such a gold field formed a rank size pattern of central places serving the great number of people living in the goldbearing creeks as well as in the urban centres of the region. The degree of centrality of these places normally passed from a dominant regional centre to subregional and local centres down to the dispersed goldmining-camps at the lowest hierarchical level.

In the course of a research project I was able to investigate the development, change and decay of these central systems in several isolated settlement regions on the mining frontier in the boreal forest. In this paper some special aspects of the problem will be demonstrated using the example of the Canadian Klondike gold field.(2)

The first rich placer gold deposits in that district were discovered in small creeks near the Klondike River in August 1896. In the same month Dawson City was founded by a trader as a supply settlement on a small flat at the confluence of the Klondike and Yukon Rivers.(3) Only a few months later, during 1897, smaller towns grew up in the hinterland of Dawson City. Normally they were founded by small businessmen (BECKER 1980, pp. 339 sq.). Thus two years after the first discovery of gold deposits the settlement pattern of the Klondike gold field was fully established. A census taken by the North West Mounted Police in July 1898 documented the number of people living in the hinterland at about 13,000 (Table 1); Dawson City itself had a population of roughly 16,500 persons.

Table 2: The business inventory of settlements of different centrality in the Klondike goldfield (1901)

(Sources: FERGUSON 1901; The Semi-Weekly Klondike Nugget, vol. 6, Nr. 43, May 5, 1901.
Legend: 1 = Dawson, 2 = Grand Forks, 3 = Caribou, 4 = Gold Run Village)

retail trade	1	2	3	4
bicycle store	+			
shoe-store	+	+		
tobacco-shop	+			
clothing store	+	?		
drugstore	+	+		
dry goods (store)	+			
fancy goods (store)	+			
furniture-store	+			
general merchandise (store)	+	+	+	+
gents' furnishing goods (store)	+	+		
grocery	+			
hardware store	+	+		
jewelery	+	+		
ladies' furnishing goods (store)	+			
lumber shop	+			
meat market	+	+	+	
millinery	+			
news-stand	+	+	+	
stationer's shop	+			
tin shop	+			
wholesale trade				
butchers' wholesale	+			
hardware wholesale	+			
liquor's wholesale	+			
craft				
bakery	+	+	+	+
barber-shop	+	+	+	+
blacksmith-shop	+	+	+	+
carpenter' shop	+	?		
confectionery's shop	+	?		
dressmaker's workshop	+	+		
shoemaker shop	+	?	+	
tailor's workshop	+			

service trade	1	2	3	4
accountant	+			
advocat, barrister, solicitor	+			
assayer	+			
broker and agent				
merchandise broker	+			
mining broker	+			
real estate agent	+	+		
bank	+			
bath house	+		+	+
dentist	+	+	+	
freighter	+	+		
hotel, roadhouse & lodging house	+	+	+	+
hospital	+			
insurance agency	+			
laundry	+			+
pawnbroker	+			
photographer	+			
physician	+	+	+	
printing shop	+			
restaurant	+	+	+	+
saloon	+	+	?	?
stage line	+			
transportation company	+			
miscellaneous				
furrier	+			
livery stable	+	+		
machinery	+	+		
newspaper (publishing house)	+			
power company	+			
sawmill	+			
telephon company	+			
theatre	+	+		
thawing machinery	+			

Table 1: The population of the Klondike district in 1898.

Dawson City	16,560(4)		
Bonanza Creek	3,540(5)	Too Much Gold Creek	54
Eldorado Creek	945	Leota Creek	36
Hunker Creek	1,400	Eureka Creek	36
Dominion Creek	1,917	Quartz Creek	260
Sulphur Creek	680	Swede Creek	25
Bear Creek	540	Stewart River district	3,500
All Gold Creek	132		

Besides Dawson City, the dominant urban centre, there existed quite a number of smaller towns in the hinterland for supplying this population with goods and services. In addition there were single businesses and service establishments lying dispersed in the creeks (Fig.24:1). A business inventory of Dawson City, Grand Forks (Bonanza Creek), Caribou (Dominion Creek) and Gold Run Village (Gold Run Creek) demonstrates the central installations in 1901 (Table 2). It also shows the generally well known phenomenon that the number and the quality of central institutions depends on the size of a settlement(6).

In the course of further development the importance of the different settlements as central places changed. Normally it happened in only a few years and connected with it was nearly always a decrease of centrality. Table 3 illustrates the process by the example of two selected settlements in the Klondike gold field as it is shown by the decline in the number of business installations between 1901 and 1917. Settlements in the hinterland of Dawson City, originally centres of some importance, became simple dwelling places; later on they were abandoned. Of the numerous settlements in the gold field only Dawson City stayed alive. But today it is just home to some hundred people and its economy is of a seasonal character.

In order to make a systematic survey of the change regarding centrality in the settlements of the Klondike gold field we need homogeneous sources. It is fortunate that there exists the Alaska - Yukon Gazetteers and Business Directories from 1901 to 1923 (FERGUSON 1901, POLK 1902 sq.). For the thirties and forties we have the "British Columbia and Yukon Directories" and for more recent data we are able to use telephone directories. The earliest of them (FERGUSON, POLK) seems at first to be an excellent source. The residents of the settlements are listed by name and usually by profession. The existing business and service establishments are given in local registers as well as in a separately classified business directory. But the first view deceives. The sources possess considerable deficiencies. The most important of them are:-

— In the volumes of the Yukon directories many of the settlements existing at that time are missing. Ferguson (1901) for instance only mentioned Dawson City and Grand Forks. There is no mention of the other centres (e.g.

Table 3: The decreasing importance of creek-centres in the Klondike goldfield as reflected in the decline of the business inventory (the examples Grand Forks and Caribou)

Grand Forks

	1901	1917
hotels / roadhouses	12	2
restaurants	12	--
general merchandise	5	1
hardware stores	1	--
machine shops	1	--
drugstores	1	--
news-stands	2	--
boots and shoes (stores)	1	--
gents furnishing goods (stores)	5	--
dressmaker's workshops	1	--
jeweleries	1	--
blacksmith-shops	3	1
meat markets	2	--
bakeries	2	--
barber-shops	2	--
physicians / dentists	4	--
real estate agencies	1	--
freigthers	1	--
stage lines	--	1
livery stables	2	1
theatres	1	--

Caribou

	1901	1917
hotels / roadhouses	7	--
restaurants	2	--
bath houses	2	--
general merchandise	4	--
news-stands	1	--
blacksmiths-shops	1	--
shoemaker-shops	1	--
meat-markets	3	--
bakeries	1	--
barber-shops	1	--
news agencies	1	--
physicans / dentists	3	--
mining companies	--	1

Sources: FERGUSON (1901) POLK (1917);
The Semi-Weekly Klondike Nugget vol. 6, Nr. 43
May 5, 1901, p. 4.

Caribou, Bear Creek, Gold Run Village, Sulphur etc.).

- In the several volumes of Polk's directories the listed settlements are different in number.
- Numerous places are just named without any further information about residents' professions or business establishments.
- The information in the directories for those settlements with particulars are based on statements of the then postmasters. This means that the information is different in quantity and quality for, without doubt, not all the postmasters answered in the same way.

On the whole the said directories have considerable deficiencies. Thus they can be used only with extreme caution and blanks as well as inaccuracies must be accepted. In spite of this we cannot do without them because the data are not available elsewhere. In some cases, however, the information of the directories can be supplemented by other contemporaneous sources such as newspaper-advertisements, descriptions or reports.(7)

The well-known methods of determining centrality are confined by the special kind of historical data. The tried method of interviewing residents living in the surroundings is of course excluded, since such interviews are only possible for the present time. As is well-known, central places are marked by their special facilities and by their functions which reach beyond the settlement itself. Their role as supply and distribution centres correspond to institutions within the settlements. Therefore the number of central institutions was often used as an indicator of the degree of centrality. It is not sufficient, however, to put up simply a catalogue of representative central installations for making a comparison. Therefore Boustedt's method (1962) was reverted to. Boustedt was able to show empirically the multiplicity of existing institutions as a measure to determine centrality. He determines this multiplicity of existing institutions in a settlement by means of a catalogue of selected trade branches and general service institutions. The more trade and service branches of his catalogue which exist in a settlement the higher is its central importance.

Table 2 clearly shows that the principle of this method is applicable to the Klondike settlements in historical time. Important central places possess a great variety of business and service branches of every kind; less important places show a significantly lower variety of such branches.

As Boustedt developed his system on the evidence of Bavaria twenty years ago his catalogue could not be accepted unchanged for the study of the Klondike region. For a region at the northern periphery of the ecumene and for a historical situation other conditions have to be considered. Thus by a complete list of all branches in different places in the

Table 4: Catalogue for determining centrality in goldmining towns of the Klondike district

I. retail trade and craft (15)

clothing store
drugstore
dry goods store
furniture store
general merchandise store
grocery
hardware store
jewelery
meat market
news-stand
stationary store
bakery or confectioner's shop
barber shop
blacksmith shop
tailor's or dressmaker's workshop

II. general central institutions (12)

advocat, barrister or solicitor
bank
freighter or transportation company
livery stable
machinery
newspaper (publishing house)
physician or dentist
real estate agent
saw mill
stage line
theatre
wholesale

Klondike gold field (8) a catalogue of 15 trade and craft branches and 12 general service brances was singled out. The catalogue was empirically tested (Table 4).(9)

Further evaluation was simple; with the help of the sources the number of catalogue-branches for every place and year was tabulated. The result was used to determine the degree of centrality in Table 5.(10) The degrees of centrality found in table 5 (and using Fig.24:2) established the ranking of settlements in particular years. These rankings were then mapped.

Table 5: The determination of the degree of centrality by number of central institutions (according to BOUSTEDT 1962).

Degree of centrality	Retail trade and craft	General central institutions
1	13-15	10-12
2	8-12	4-9
3	5-8	1-3
4	2-4	0

A comparison of the maps (Figs.24:3-6) shows the following principal features of the development. In spite of the gaps in the source material a fully developed pattern of central places is to be found for the year 1901 (Fig.24:3). Dawson City was the centre with the highest degree of centrality. But at a distance of about 20 km Grand Forks, a 2nd category place - that means of high centrality, too - was situated. Between them, about 3½ km downcreek of Grand Forks, was Magnet City, a local place of 4th category. (11) The pattern of central places in the western part of the gold field with the highest density of population at that time was, as we see, fine-meshed. In the peripheral parts the net was clearly more wide meshed. Although information for Hunker and Sulphur Creeks is missing we still have the data for Dominion and Gold Run Creeks (12) (see Fig.24:1). The settlement of Caribou was a 3rd category central place; Gold Run Village had a more local importance (4th category). The other places in Dominion Creek are not specifically mentioned in the sources; with the exception of Granville they are shown as small building-concentrations in Fig.24:1. In the hierarchy of central places they were of no importance at that time. This also applies to Granville which grew up at that time. (13)

As Fig. 24:4 shows the situation had changed six years later: Dawson City still belonged to the highest category of centrality. Grand Forks, however, had lost some of its importance; it dropped from 2nd to 3rd category. On the other hand, some settlements in the eastern creeks had increased in central importance, especially Granville which rose to 3rd category. The decline of Caribou from 3rd to 4th category can

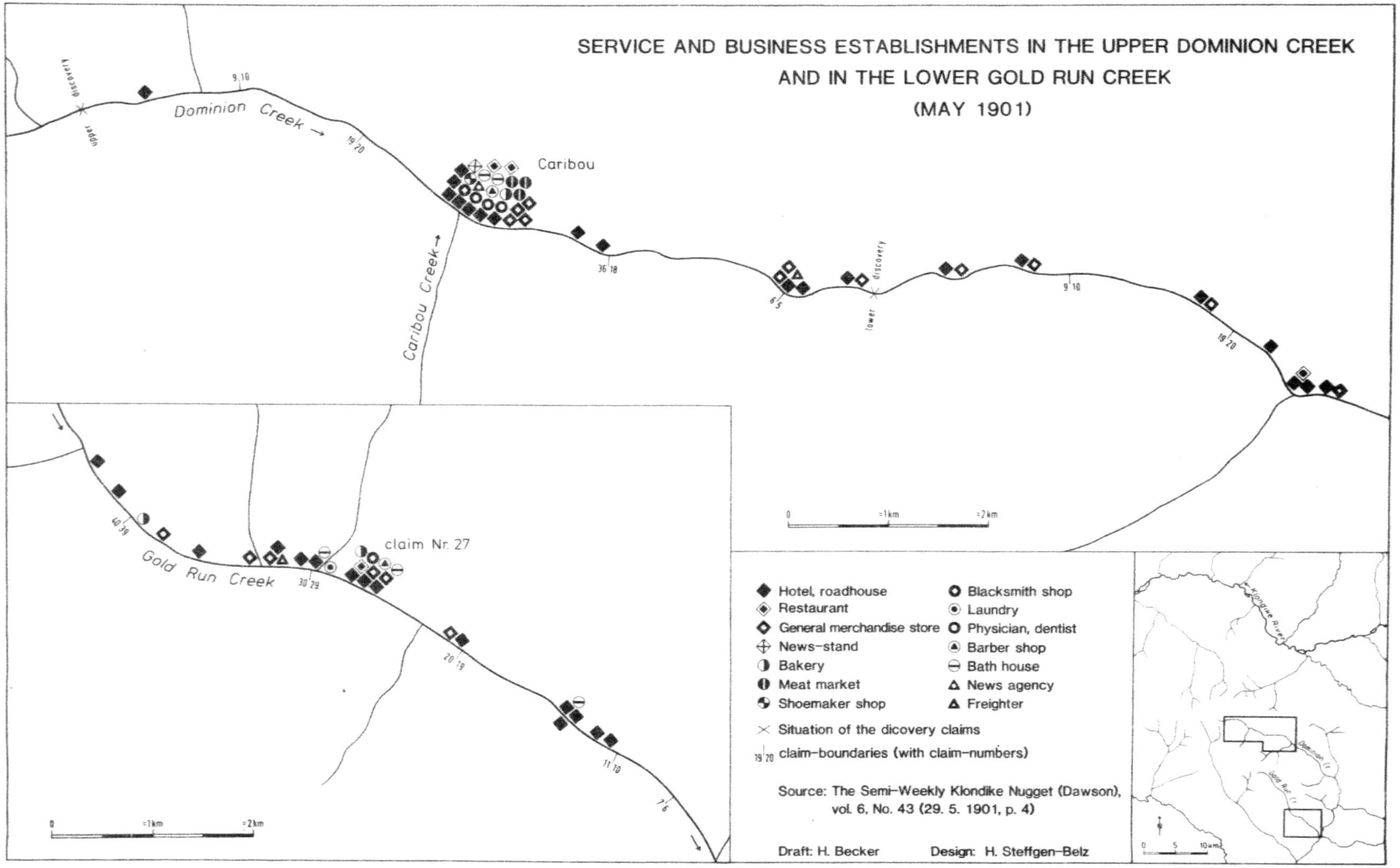

Fig. 24:1 Service and business establishments in the Upper Dominion Creek and in the Lower Gold Run Creek (May 1901).

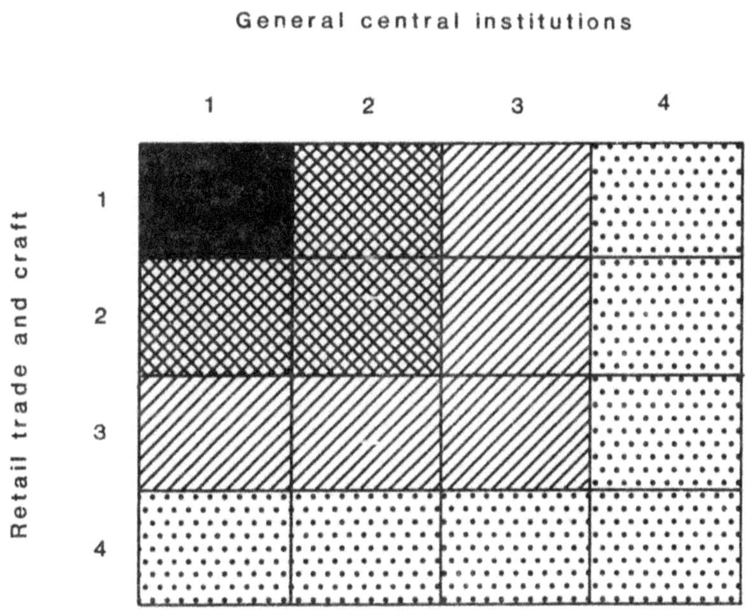

Fig. 24:2

The determination of centrality (according to Boustedt 1962).

CENTRAL PLACES IN THE KLONDIKE GOLD-FIELD (1901)

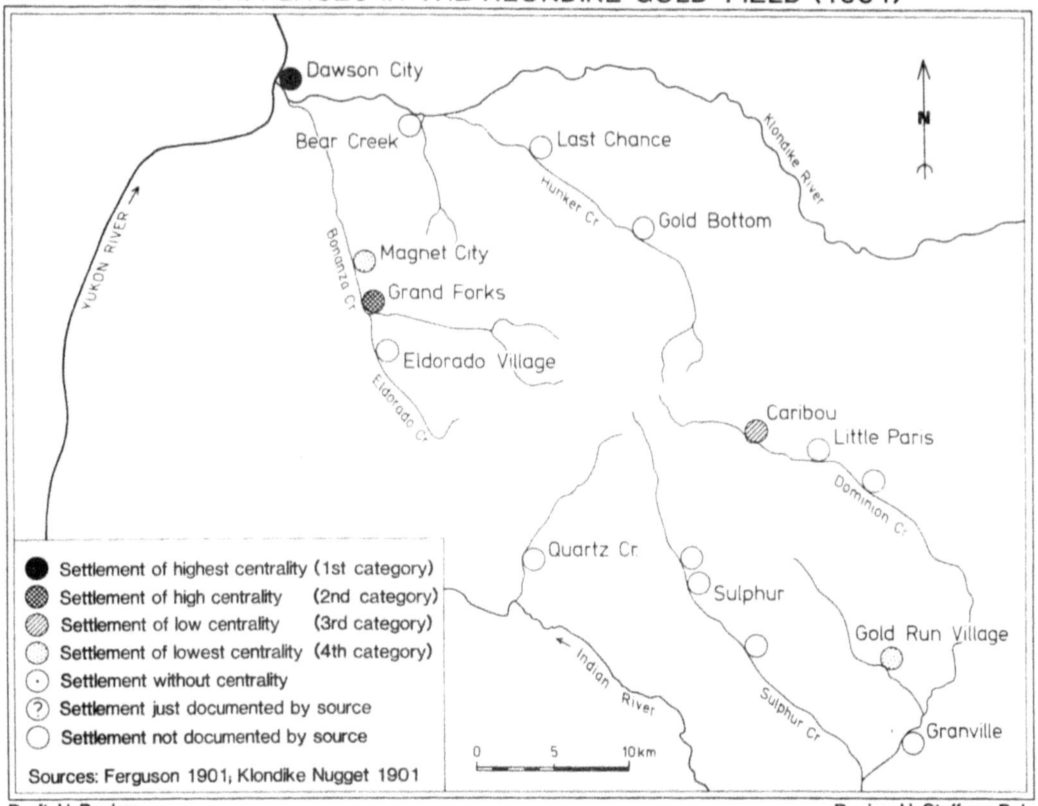

Draft: H. Becker

Design: H. Steffgen-Belz

Fig. 24:3

CENTRAL PLACES IN THE KLONDIKE GOLD-FIELD (1907)

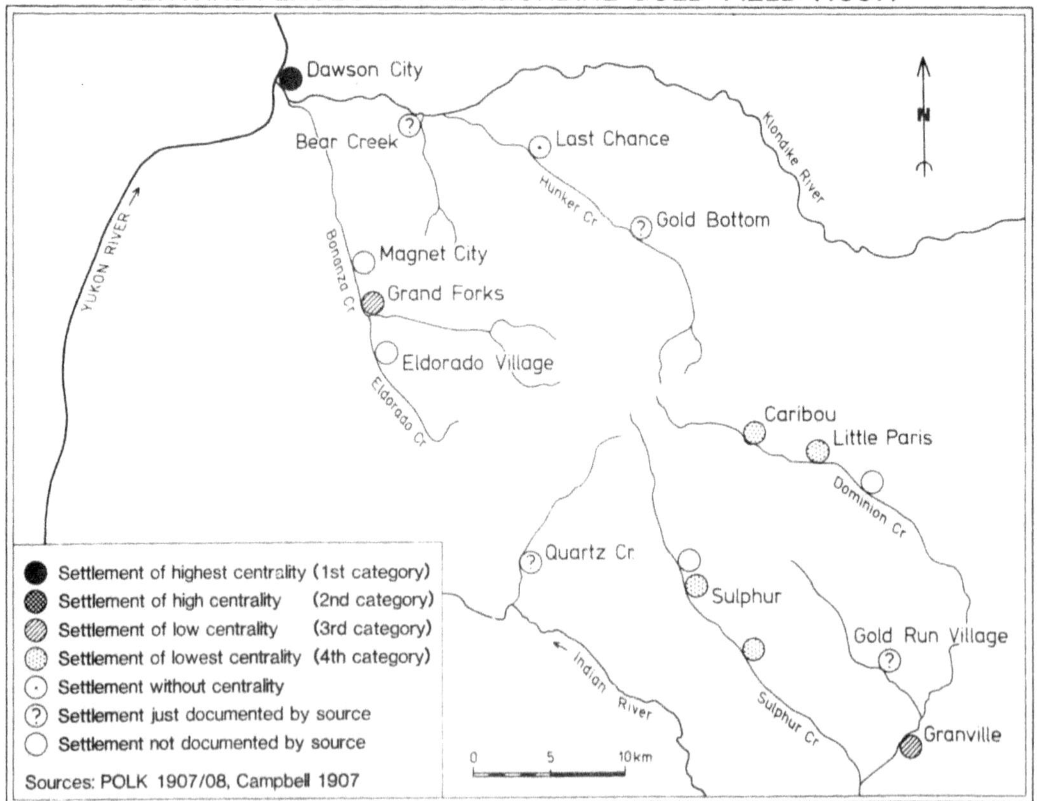

Draft: H. Becker

Design: H. Steffgen-Belz

Fig. 24:4

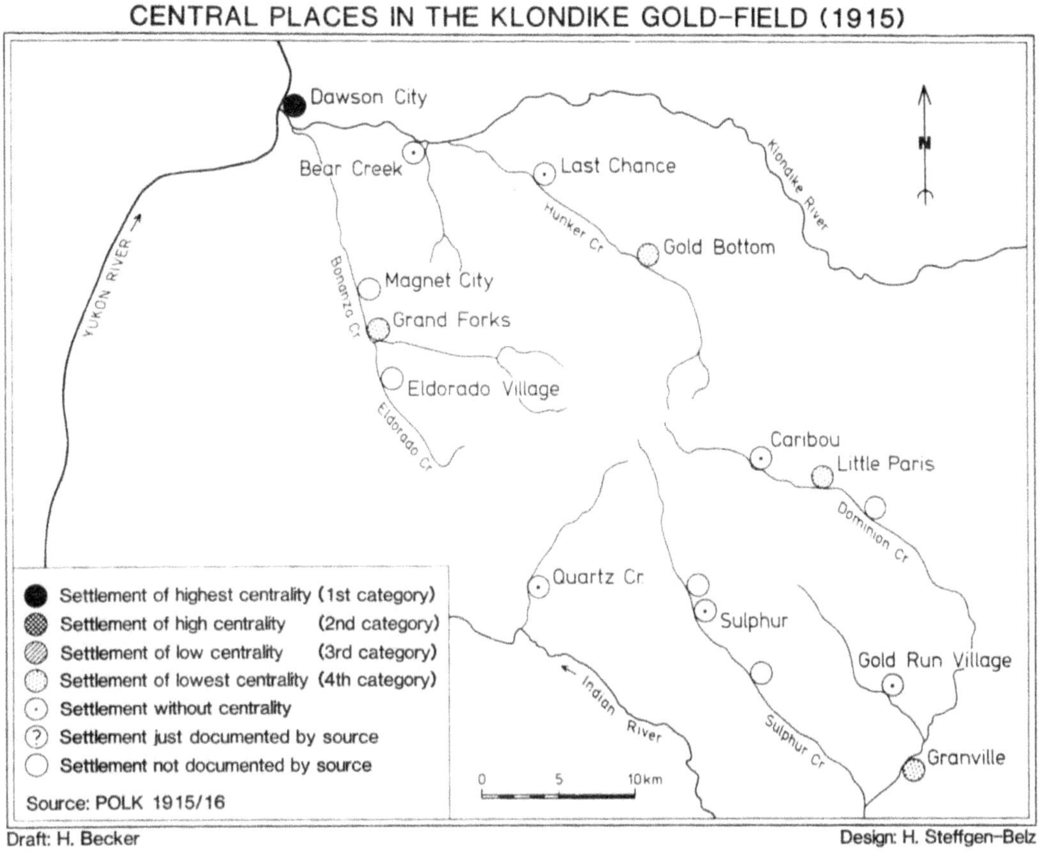

Fig. 24:5

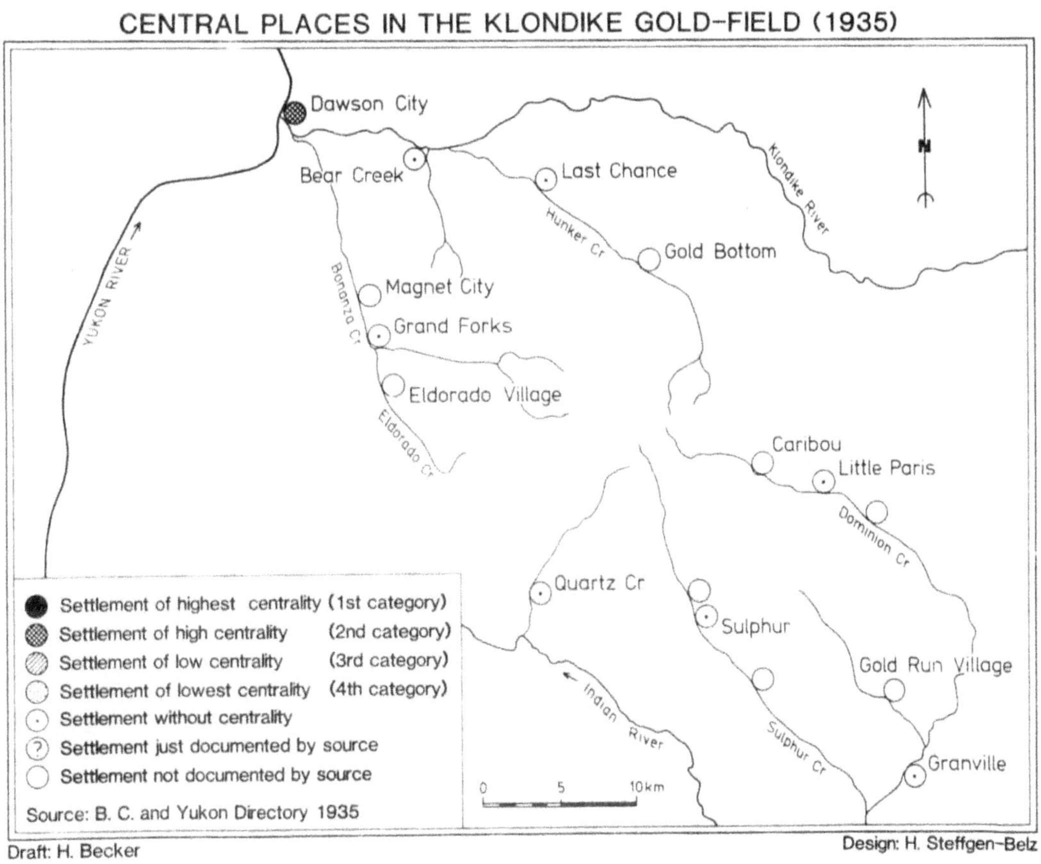

Fig. 24:6

be explained by the growth of the nearby settlement of Little Paris.

With the exception of Dawson City there was a distinct loss of centrality in all parts of the gold field by the year 1915 (Fig.24:5). There are only some places of the lowest centrality, (14) but most of the existing settlements were just dwelling places without any central functions.

Fig.24:6 shows that by 1935 only Dawson City remained as a central place. All the other existing settlements lost whatever importance they had had as places of centrality. Figures 24:3-6 document an almost entire loss of central places in the Klondike gold field. The process passed in phases. At the beginning the settlements in the western part lost their importance, while those settlements in the southeastern periphery of the gold field increased in centrality (see Figs. 24:3 and 24:4). This different development in several parts of the district is to be explained by gradual and selective use of new mechanical means of handling the gold-bearing gravel, and by the partial introduction of mining companies. Pick, pan and shovel were replaced by dredges and hydraulic giants. Campbell (1907) records that in those days dredges and hydraulic mining were to be found in the Klondike Valley, in the Bonanza Creek; and in the lower Hunker Creek; in the more peripheral parts of the gold field, however, individual mining went on. Gold mining by companies using mechanical means needed far fewer workers compared to individual mining.(15) Later on, stock companies and mechanical mining-methods were also introduced into the peripheral parts. This caused a decline of population in the south eastern part of the gold field, and central places for supplying larger numbers of people were no longer needed. In the process of decline of central places transportation changes also played an important part. With the use of trucks, snow-tractors etc. supplies for the remaining population could be carried over greater distances.

NOTES

1. The North West Mounted Police for example reported for 1898: "It is estimated that 28,000 people passed Tagish Post during that year on their way to the Klondike District..." (Public Archives Canada, Ottawa: R.G. 85, vol. 657, file 3242). This estimated number of people is the figure for one year only and just for one of the seven main points on several ways to the Klondike; the number of people passing the other points in that year is unknown.
2. The research project is supported by the Volkswagen Foundation (Hannover, Germany). I would like to thank the foundation very much for the generous financial assistance.
3. It was a planned and regulated foundation made with the intention to speculate in real estate (BECKER 1980, pp. 334 sq.).
4. Public Archives Canada, Ottawa: R.G. 85, vol. 655, file 3008.
5. All numbers in this table are from report of Superintendent Steele, Commanding North West Mounted Police in the Yukon, 1898. Ottawa 1899, p.22.
6. With the exception of Dawson City the numbers of residents are unknown. The scale of magnitude, however, can be roughly estimated by contemporaneous photographs. On this evidence the biggest town after Dawson City was Grand Forks with a population of approximately 1,500-2,000 persons. Caribou may have had about 300 or 400 inhabitants: Gold Run Village was just a small centre with perhaps 100 or 150 persons.
7. In the paper presented the figure of 1901 was done with the additional help of a report about two creeks in the "Semi-Weekly Klondike Nugget" of May 5, 1901 and that of 1907 with the help of a detailed description in the "Alaska-Yukon Magazin" of 1907.
8. The number of the existing branches was even larger than in Table 2.
9. The development in the long period from 1901 up to the present time was considered by changing some institutions (e.g. from about 1925 "stage line" was replaced by "bus line" or "livery stable" by "car rental or taxi").
10. Details of Table 5 were slightly altered from Boustedt's original by empirical tests.
11. Information about Magnet City was taken from "The Semi-Weekly Klondike Nugget", vol 6, No. 41 (May 23, 1901), p.8.
12. The source is: "The Semi-Weekly Klondike Nugget", vol 6, No.43 (May 29, 1901), p.4.
13. Yukon Archives, Whitehorse: R.G. 91, vol. 27, file 9825.
14. Grand Forks, Gold Bottom, Little Paris, Granville.
15. In the period of individual mining it was no rarity that a crew of from 10 to 25 men were at work on one claim having an extension of only 152 m (BECKER 1980, p. 344).

REFERENCES

ARCHIBALD, M.E. 1972 — Grubstake to Grocery Store: The Klondike Emporium, 1897-1907. (Parks Canada, Dept. of Indian and Northern Affairs, National Historic Parks and Sites Branch) Manuscript Report 178.

BECKER, H.& R. LOTTES-GRÖTSCH 1979 — Luftbild Dawson City (Yukon Territorium, Kanada). Die Erde 110, pp. 265 - 274.

BECKER, H. 1980 — Siedlungsgründungen des ausgehenden 19. Jahrhunderts an der kanadischen Bergbaufrontier im Klondike-Goldfeld. Die Erde 111, pp. 329 - 352.

BENNETT, G. 1978 — Yukon Transportation: A History. Hull. Canadian Historic Sites, Occasional Papers in Archaeology and History 19.

BOUSTEDT, O. 1962 — Die zentralen Orte und ihre Einflussbereiche, eine empirische Untersuchung über die Grösse und Struktur der zentralörtlichen Einflussbereiche. Proceedings of the IGU Symposium in Urban Geography Lund 1960. Lund Studies in Geography, Ser. B, No. 24, pp. 201 - 226.

THE BRITISH COLUMBIA AND YUKON DIRECTORY (1935): Vancouver.

THE BRITISH COLUMBIA AND YUKON DIRECTORY (1942): Vancouver.

THE BRITISH COLUMBIA AND YUKON DIRECTORY (1947): Vancouver.

CAMBELL LOWDEN (Mrs.) W. 1907 — Mines of the Klondike. Alaska-Yukon-Magazin IV (3), pp. 206 - 212.

CANADIAN NATIONAL TELECOMMUNICATIONS 1968 — Official Telephone Directory Northern British Columbia and Yukon. Dawson Creek.

ELLIS, H.I. 1915 — Dawson, Nome and Fairbanks. Mining Journal 99, pp. 517 - 520.

FERGUSON, M.L. 1901 — Dawson City, Yukon Territory and Alaska Directory and Gazetter.

NORTHWEST TELECOMMUNICATIONS CO. 1979 — Official Telephone Directory Northern British Columbia and Yukon. Whitehorse.POLK 1902 Alaska-Yukon Directory and Gazetteer. Seattle.

POLK 1903 — R.L. Polk and Co's Alaska-Yukon Gazetteer and Business Directory. Seattle.

POLK 1905 — R.L. Polk and Co's Alaska-Yukon Gazetteer and Business Directory 1905-06. Seattle.

POLK 1907 — Polk's Alaska-Yukon Gazetteer and Business Directory 1907-08. Seattle.

POLK 1909 — Polk's Alaska-Yukon Gazetteer and Business Directory 1909-10. Seattle.

POLK 1911 — R.L.Polk and Co's 1911-12 Alaska-Yukon Gazetteer and Business Directory. Seattle

POLK 1915 — R.L.Polk and Co's 1915-16 Alaska-Yukon Gazetteer and Business Directory. Seattle

POLK 1917 — R.L.Polk and Co's 1917-18 Alaska-Yukon Gazetteer and Business Directory. Seattle

POLK 1923 — R.L.Polk and Co's 1923-24 Alaska-Yukon Gazetteer and Business Directory. Seattle

CHAPTER 25

RURAL INDUSTRIES IN HESSE IN THE NINETEENTH CENTURY: VILLAGE STRUCTURE AND "RE-AGRARIZATION"

I. Leister
(University of Marburg/Lahn)

An historical demography of Germany can be summarised in a swinging curve: a positive swing in high-medieval times followed by the late-medieval negative one, with a new positive one commencing in the late 15th century and levelling off by about 1580/1610 to become positive, once more, after 1700. Of these, the overall-trends, only the high-medieval one was borne equally by urban and rural sectors. During the late-medieval urbanization phase rural depopulation contrasted with the continued growth of German towns; by early-modern times the towns went into decline or - at best - stagnation, leaving the overall population increases to be dealt with by the countryside.

Germany had no colonies, no navy or sea trade to help in maintaining a balanced population. Also, with international trade shifting to the Atlantic, German towns lost most of their international business, and their trade was further hampered by the many territorial dukes each erecting customs barriers.

The era of the merchant prince was past. Scaled down to mostly the middle and lower brackets, urban incomes suffered all the more heavily from the Early-Modern Inflation and from a Thirty Years' War. Their growth potential was too low for the towns to syphon off much surplus population and was lower still after the war; indeed, many towns in Electoral Hesse had by 1747 not yet regained their pre-war size (Dascher 1968 p. 161).

Protestant towns in particular were barely capable of looking after their poor, usually a third of an urban population, yet gained no help to compensate for the loss of urban monastic houses. In self-defence a town would tend to apply the rules of admission to both citizenship and craft guild rather strictly and could in Electoral Hesse be selective, keeping the less remunerative trades out with the help of a sham guild organisation, called Ämterzunfte. Once such a guild which had reference to an administrative district (Amt) was established, rural or urban location of its members became irrelevant. They are on record since 1564 (Zimmermann 1933 p. 291).

1. Ruralization of surplus population

For accommodating a growing population the late 15th century countryside still held some openings in farming: lost villages might be cleared for re-occupation, the more numerous lost individual farms be tenanted. Thirdly, as grain prices recovered, substantial farms of two Hufen or more might be split, split to more than one advantage where foreland and assarts were still available for extending the permanent arable.

Mainly, however, the population surplus became the rural substratum, recte substrata in NW-Germany which, as time went on, developed a neat ranking of cottars and of lesser men, admitted only the 1st class cottars to the farmers' settlement and left the others to form new settlements out on the commonage. In Hesse, by contrast, as in other Haufendorf-regions, all socio-economic groups were united in one village. In the 16th century there had been mainly three groups, namely farmer, cottar and the new cottager, but the distinction between cottar and cottager became so blurred subsequently, that the one term can here cover all grades of the substratum.

The new cottager signified the ruralization of surplus population. If the town was to be relieved and not cause overcrowding of villages close to an attractive town like Kassel, the burden had to be spread evenly and was in Electoral Hesse spread to more or less all villages by means of the number of village rights.

One village right entitled the holder to one share of baking time, of water, brushwood (firewood and timber if available), of commonage and of a plot whenever a piece of commonage was privatised. The village total thus reflected the capacity of local communal institutions and served now for a ceiling on the number of houses.

Most villages, in fact, had already reached saturation point, with the two numbers being identical by 1585; physical stagnation ensued until about the mid-18th century when population pressure caused both a more intensive use of extant buildings and the erection of some new houses on condition of the householder in question being able to either plan life without a village right or to persuade a rich farmer to part with his.

Reaction of the villages varied. Broadly two modes can be established, related to upland and lowland respectively.

Hesse encompasses several structural basins of varying size, separated by uplands of, be it metamorphic rock or sandstones (Bunter) or igneous rocks. The lowlands, favoured by free-draining soils and a dry-warm local climate, have of old formed core areas of settlement and have proven their flexibility by adapting to ever-changing agrarian systems. The uplands, on the other hand, only began to be colonized in early-medieval times; development was probably retarded by law and constitution as much as by ecology. As upland colonisation progressed, a range of sites came into being which from the lowlands' flexibility ranged to the point of inflexibility as became evident after the 13th-century reform of agrarian organisation. A rôle-thinking moreover, required the uplands to support the same peasant economy as the lowlands: it demanded too much from the habitat and the catena became one of disadvantages increasing with altitude. It is just possible that in Hesse too these were compensated for by the introduction of industry, mainly linen. (1) Certainly the early-modern upland village where tillage was not really a paying proposition saw little cause to resist gavelkind; and gavelkind's downward-equalizing effect on the social pattern caused ruralization to result in social uniformity at the smallholder level. A lowland village, on the other hand, would by about 1500 be inhabited by a pastor having the use of a substantial farm and by a limited number of farmers on medium to well-sized holdings. They formed the village elite, and farm sizes remained fairly stable, on the whole, until the late 17th century. Early modern ruralization, therefore, could not but augment an extant cottar-substratum by means of the new cottagers. The village developed a marked socio-economic dichotomy, was composed of a landholding group and those with little or no land who, therefore, were both socially inferior and economically alien, and who represented a money economy in the village. After the upheaval of a Thirty Years' War, however, it became more difficult for the village to retain a more or less strictly dual economy(2); though not in all villages to the same degree, but in general cottagers were now gaining access to more than fringe plots and so to the mixed economy favoured by them.

II Employment

A cottager's household was based on a multiple income, to which his wife and children contributed. Girls in particular would relieve the household by going into service when still quite young; farmers usually had one or two female servants living in ,(less often a male servant), who presumably earned their keep by spinning during the slack season. The wife, besides looking after cattle and land, if any, and gleaning the harvested fields, did many of the hand labour dues, worked as a harvester and further produced for sale, either handknitted woollen goods, or, more regularly, linen yarn. The husband as main provider might also have to follow several trades in the course of the year. It was an additive household budget, pieced together from several sources and as such quite beyond a straight-forward record(3). Here it must suffice to briefly list the main sources of employment:-

a) FARMING in Hesse meant a grain-sheep economy, and farmers met peak demands for labour by agreements with cottagers (male and female) paid in kind and also by engaging day labour; the latter did not create neighbourly obligations. A cottager might further be employed in substituting a farmer's labour dues.

b) Villagers in the VICINITY OF TOWNS, even lesser towns,

lived mainly by commuting. Urban landowners hired village farmers for working the town fields and for the occasional carting job; thereby the farmer might earn sufficient to meet his taxes and other money dues. All or nearly all of the town's general labourers were village-based and so were a good many of the building crafts. As to other crafts, relations were more complex, as the village would transfer some custom to the town, whilst the town might order low-priced goods in the village.

c) NON-AGRICULTURAL ENTERPRISES based on localised raw materials, such as a pottery, a coal mine, a quarry, an ironstone mine and washery, etc., were but a mixed blessing to the village(s) concerned. While working, their wages might help a thrifty cottager to enter the land market, but all too often they proved rather short-lived and left their men to fend for themselves, i.e. to share the remainder of labouring jobs within reach. Real unemployment was alien to a system that relied on several part-time jobs to provide one living and therefore gave no room for out-migration as a built-in relief of the village.

d) Because of their wide distribution in Hesse WOODLANDS formed a much more important and also a more stable source of both raw material and employment. Forestry work including the felling, cutting and sawing of timber, and the transport of lumber offered much sought-after winter jobs. Wood-based industries such as the production of charcoal, tar, oak bark, potash etc. also provided fairly regular and organized work on money wages by contrast to, thirdly, the non-organized wood-working sector. The upland-smallholder who availed himself of the less accessible woodlands for making brooms or wooden slats or farm and kitchen utensils was self-employed like a craftsman; huckstering his wares, he made little enough profit, if any. These poverty trades were the last resort of upland locations too marginal for even linen.

e) TEXTILES were easily the most important sector in the Hessian economy. Wool had for long ranked first, with Cologne as trading partner in both raw wool and worsted yarn to the detriment of the urban woollen and worsted weavers(4). As flax gained upon wood in the course of the 16th century and was to hold first place during the following two centuries, the emphasis shifted from the urban to the rural producer.

The village was the fibre producer. Flax as well as wool were cash crops mainly in the lowlands; home farms with labour services owing to them used these to best advantage in the production of raw flax. The village also did nearly all the spinning of both wool and flax, spinning being the most regular penny earner, universal in lowland villages no less than in the upland vicinity of weavers (Fig.1). Weavers organizing their own spinners bought some of the raw flax but most went to yarn hucksters to be put out to the females in a village. For spinning was more than just a feeder to weaving; it was a trade in its own right carried on by the village-based yarn hucksters, Christian or Jew, as petty-business export traders. Their dealings aggregated to quite a sizeable export mainly to Elberfeld (today in Wuppertal) and worth one third of the tax collector's intake from linen (Dascher 1968 pp. 29f., 145 ff).

Linen weaving was a male occupation and in Hesse subject to a division of labour as between town and country. Rural producers absolutely predominated, but on them fell the less remunerative medium to heavy linens wanted for household and sail cloth, for sacking and packing. Unlike neighbouring Thuringia, Hesse was set against the more effective organization by either merchant or master weaver; it rather instituted the Ämterzunfte, which exerted even less control than had their late-medieval predecessors. The weaver was a self-employed craftsman, free to sell his ware to this merchant or that, in one place or the other - an odd system considering that linen accounted for two thirds of Kur-Hesse's export by value.

Commercial linen weaving was broadly regionalised: western Hesse continued to market at the Frankfurt fair which demanded medium to fine qualities, whilst eastern Hesse, the export earner, produced what in English is termed "Hessian", i.e. medium to heavy wares. Eastern Hesse held about 80 per cent of all the country's linenweavers and was further subregionalised as lowland cottagers with better-paid alternatives open to them did not take greatly to linen weaving.(5) Weaving villages were found in those upland locations which supported no more than subsistence farming yet were not too far removed from one of the trading centres(6). Backed by the land, the smallholder-weavers were able to weather short interruptions of the trade by one war or the other; but if the crop ever failed, as it did in the early 1770s, weaving, though for export, would not buy their food and its transport.

Export of "Hessian" had originated with the Netherland's rise as a maritime power. No longer able to produce all the linen required, the Dutch had naturally turned to Westphalia; very soon the Westphalian trading centres, in turn, faced overdemand and began looking to Hesse for the heavier wares. For the Dutch linen shop supplied England as well until the first half of the 18th century when England established direct trade with Hesse via Bremen. Output continued to rise and reached about 200,000 pieces in 1805 - just before Napoleon declared the Blockade of England. Trade in "Hessian", for which England was the one market(7), suffered complete stoppage. Yet in 1815 it immediately resumed as, indeed, on the supply side it was bound to do, considering the situation in upland weaving villages and population pressure which made cottagers in general more ready to take to the loom. Recovery was but short-lived, however; for now the market began changing and was irrevocably lost to "Hessian" by the mid-1830s.

III Rural de-industrialization and the term "re-agrarization"

In fact, whether it was the long-distance effect of England's industrialization or the growth of Germany's railway network favouring the nascent factory industry, all the village-based craft-industries mentioned were doomed by the mid-19th century, priced out of the market by mechanical mass-production. From the factory and from the new concentration of production followed a "de-industrialization" of the villages. The village, fast losing all production except farming, was, so to speak, being 'purified' to become the mono-functional settlement of the later 19th century. For this process German economic historians use the term "re-agrarization", a concept which seems worthy of discussion.

Early-modern ruralization had created villages in Hesse of a size which presupposed non-agricultural industries to deal with underemployment and to keep the many upland sites in being. De-industrialization, therefore meant village-overpopulation aggravated by the general population growth lasting here into the 1840s(8). Emigration, on record since 1837, was relatively heavy during the 1840s/50s and continued until the turn of the century (MÖKER 1977 ch.C). Nearly all of it was to the United States, where the Hessian timber-frame house has since become part of the cultural heritage as far south as Texas (JORDAN 1964). Very few farmers emigrated and few miners or iron workers; mostly the emigrants were craftsmen labourers and shopkeepers, both urban and rural, who had some savings to invest in the new start. These same groups also sent off permanent migrants to the Ruhr, to which, of course, miners and iron workers were also attracted.

With hectic urbanization dis-phased, rural overpopulation in retarded Germany lacked the most effective, because readily accepted demographic relief by the towns. Kur-Hesse in particular offered little scope for short-distance rural-urban migration within the region; its many small country towns had no growth potential whatsoever, and the overall urban rate grew but slowly from 28 per cent in 1819 to 33 per cent in 1885. (MÖKER 1977 p. 25) Weakness of the towns clearly had a repressing effect. The one road still open to the villagers was long-distance seasonal migration. Seasonal migrants went to the Ruhr, went as harvesters as far north as Jutland, followed road and rail construction as navvies whose wages could be up to four times the rate obtaining in Hesse.(9) These men as well as their households would accept a coolie-like existence of semi-starvation for the savings with which to buy land and make land their economic solution.

By contrast to NW-Germany, Hesse held next to no reserves of land. Not even poor land was available in quantity for a sizeable expansion of the arable to support the primary agrarization of cottager households and viable holdings.

Subdivision, i.e. a more intensive sharing of the extant cultivated land, could bring no solution to the uplands or any village dominated by smallholders; it was resorted to in the lowlands but not indiscriminately so. The land market was mobilized to some extent by the land property reforms demanding payment of money, by some holdings subjected to partible inheritance, some even disintegrating and by the exceptional home farm being cut up into some six holdings - yet, on the whole, new units were of considerably less importance than the extant smallholdings/small properties.

Smallholdings should be graded according to both size and aim. Of those which were small farming units, most followed the same routine as the regular farms and, at best, secured subsistence. In the Eschwege basin, though, they could be considered an enriching element, because an imaginative people, sure to find a market niche, proved able by the 18th century to specialise and to select types of production in keeping with the size of a holding.(10)

Small properties were more of a life insurance policy than a farm unit. In fact, the actual farmers in a village can be ascertained only from the old cadastres which note the equipment and draught animals a man owned. Land-holding non-farmers could always rely on hiring a farmer, be it for money or in exchange of labour, or on some share-cropping arrangement. When rural industries closed down, however, these properties became the last ray of hope and highly problematical.

Because re-agrarization was confined in Hesse to the extant cultivated land, in theory it presupposed an intensification of farming. As far as the fallow field was concerned, however, there had been a creeping intensification throughout the 18th century, with plots increasingly used for growing cabbages, flax, tobacco etc; by the late 18th century foddercrops compensated for the loss of fallow commonage and, moreover, gained the farmer a better quality manure by stall-feeding. Soon money was to be earned with Napoleon's sugar-beet, a cash crop-cum-foddercrop, where it was backed by an early factory as at Eschwege and also at Wabern (1835). Sugar-beet, however, was a crop of the good-sized farm and highly selective in respect of soils. Smaller farmers and those in the uplands only had the potato for intensifying land use and had to adapt their diet so as to enable the holding to sustain the household.

The place where re-agrarization happened was the individual village. Of necessity its structural pattern as evolved by the 1830s must condition the local mode of change. In the lowlands four main variants were met with, scattered irregularly.

Many lowland villages, which had not drifted too far away

from agriculture as their primary raison d'être, were clearly stratified; if within reach of Eschwege and Wabern, moreover, their respective elites were further improving their economic strength thanks to the sugar-beet and to an overabundance of labour at extremely low wages. De-industrialization was very much a problem of the substratum, yet only the exceptional minor settlement was ultimately to lose the few cottagers it had had. Since then the first variant forms a hamlet of substantial farms. Agriculture also had the upper hand in those villages, secondly which included a small country-seat and had little room for manoeuvre, as the home farm took up between a third and a half of the village territory. Home farms were often ready to experiment, and to their respective villages they were now, after the abolition of labour dues, becoming a real asset as a local employer of labour. Also, a responsible owner could be expected to assist the redundant household to make a new start elsewhere.

With the more regular variant of stratified village much depended on its size and on the relative weight of its social groups in respect of land. The balanced village, thirdly, still was a valid location for farms; its substratum was augmented but not inordinately so and included landless labourers as well as cottagers owning a few odd acres or a unit like half a Hufe. Re-agrarization (of the village) might involve householders' primary agrarization, with the resultant severe competition for land putting farm sizes in jeopardy. Land was the one secure thing, and cottagers went slaving in order to buy more land; one day it might also form the base of vertical mobility aiming for farmer status. However, unless a marriage turned the laws of inheritance in their favour, cottagers rarely made the grade, and on the railway opening village stations, they began commuting and so reverted to a mixed budget. Problems and poverty increased sharply in, fourthly, the disbalanced villages.(11) These might still boast one good-sized and some medium farms, but the major share of the land had passed to the many smallholders. Practically every household owned land and now looked to the land for survival. Property's immobilizing effect extended to the entire village. The village, whatever the quality of its land, was inflexibilized by social factors, with little prospect of farming once more becoming an economic proposition, until the social constraints were removed. In fact, the smallholder-pattern persisted, was later cemented by commuting and having withstood the remembrements of the late 19th century was finally broken up in the 1960s.

Worse still was the case of upland villages, where de-industrialization was tantamount to the collapse of their economy. They were uniformly (or nearly so) composed of smallholders (or less), who had gained a living from their respective trade and were unable to stand on one leg only. When linen failed or any other trade except iron, villagers had nothing to fall back on and were badly located for competing in the general labour market. Emigration, migration and starvation would reduce numbers, but the remaining villagers still have no more than poverty farming; an upland habitat required more drastic reductions for amalgamated holdings to become viable. Villages had few fully-fledged farmers, if any, and, indeed, where surplus farming had been absent for centuries, agriculture could not now become an economic proposition.

It did not do so even in upland villages connected with the iron trades, although in their case population pressure was greatly lessened by redeployment. Iron was the exceptional upland trade, because its new mode of production did not debase traditional skills. Instead of becoming redundant, skills were transferred to the new mills, be it nearer home or in the Ruhr, and the home village was thus relieved of surplus households. This together with favourable national conditions and a strong market would in some uplands permit farming to be rejuvenated. A habitat as disfavourable to a meaningful re-agrarization as the Westerwald uplands, on the other hand, left no choice; its villages, though reduced in size, had to continue a mixed economy, with iron ore mines, limestone quarries and local transport for non-agricultural employments. The exceptional village, suitably located, might even by-pass re-agrarization and as a commuter settlement be de-agrarized instead.

Rural industry, i.e. the production of non-agricultural goods in the countryside, was in Hesse not the chance result of fecundity producing Haufendorfer. Several industries, e.g. iron, salt, glass, wood-based chemicals, were forced by the demands of their trade to a rural location and often a marginal one. Such was not the case with textiles, however, and, moreover, a rural location does not invariably mean a mixed economy.

The Haufendorf of early-modern times was a matter of policy, both demographic and economic, a policy shared possibly by France but not by England. The village-weaver, so it was thought, produced more cheaply and ensured the trade that flexibility which eased its adaptions to an ever-varying market. Flexibility without responsibility: for technically the Hessian weaver was self-employed, was nobody's Poor Law responsibility and was expected, if the linen market was at a low ebb, to fall back on his other jobs if not his land. Accordingly, Kur-Hesse invited French Huguenots solely for particular skills and settled them, to their dismay, in newly founded villages - where promptly their respective crafts died (ZÖGNER 1966). The mixed economy of farming-cum-craft may well have lost the country local skills as well.

Re-agrarization did not similarly result from a new thinking, a reversal of policy. It resulted from market forces bringing de-industrialization on the village. Government in

Table 1: A Statistical Profile of Electoral Hesse and "Hessian"

	1	2	3	4	5	6	7
Bründersen	55	57	20	57	2	340	411
Altenstädt	56	53	21	37	3	651	649
NAUMBURG						1678	1432
Elberberg	42	30	15	30	1	350	322
Altendorf	46	28	21	24	3	330	341
Wehren	66	30	21	34	4	279	278
Werkel	57	26	47	57	5	524	500
Obervorschütz	54	53	86	94	6	804	870
GUDENSBERG						1927	1875
Maden	59	30	37	55	3	458	483
Böddiger	57	31	37	42	10	473	498
FELSBERG						1136	1026
Hesserode	57	42	26	20	3	193	212
Mosheim	64	22	30	51	2	395	419
Elfershausen	44	23	19	31	2	258	244
MELSUNGEN						3780	3538
Adelshausen	42	15	18	23	12+	262	261
Mörshausen	38	21	40	48	31	376	323
Bergheim	53	38	22	36	28	358	239
SPANGENBERG						1952	1650
Pfieffe	39	37	68	69	48	547	486
Bischofferode	38	35	27	35	28	240	201
Hetzerode	35	40	30	44	33	292	248
Mäckelsdorf	35	20	18	28	21	173	127
Friemen	40	11	26	25	10	252	204
WALDKAPPEL						1096	1021
Bernsdorf	54	62	22	18	12	132	153
Wipperode	54	62	32	44	17	311	257
Niddawitzhausen	45	37	44	41	n.d.	327	376
Eltmannshausen	48	30	52	67	36	488	551
Oberhone	46	19	60	75	8	479	475
ESCHWEGE						5694	8181
Niederdünzebach	50	34	35	55	1	475	455
Aue	56	26	36	49	11	422	468

column 1:	Ertragsmesszahl/assessment of natural productivity
column 2:	Percentage of cultivated land owned by holdings of less than 10 hectares.
column 3:	Number of houses in 1585
column 4:	Number of houses in c1750
column 5:	Number of linen weavers in or after c1750
column 6:	Population in 1835
column 7:	Population in 1871

Sources: Die hessischen Gemeinden, ed. by Hess. Stat. Landesamt, Wiesbaden 1966 (columns 1,2,7); Reimer, 1926 columns 3,4,) StAM Bestand Kataster und Bestand 16 II Kl.Ic Nr.3.

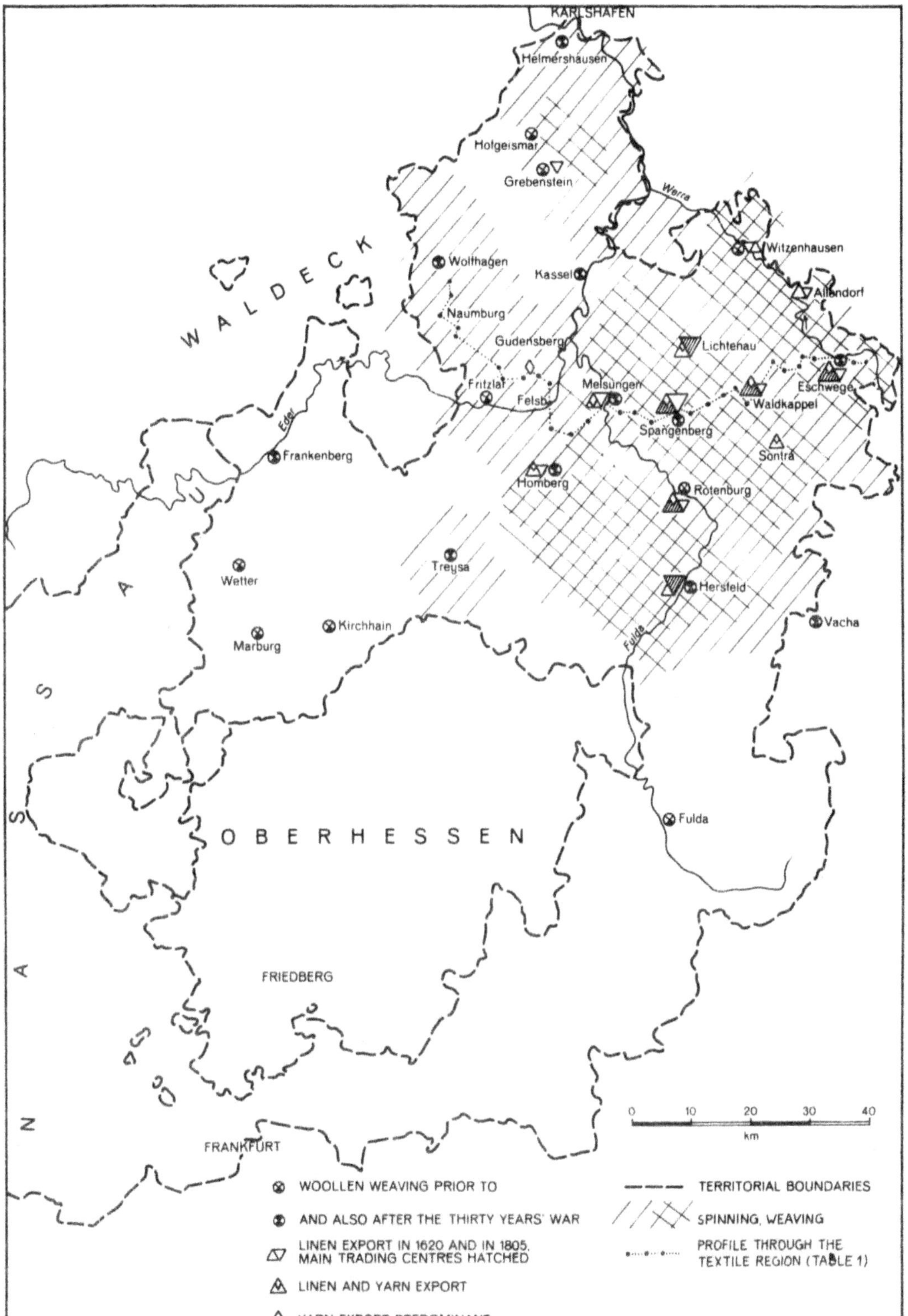

Fig. 25:1 A Profile across the Textile Country in Electoral Hesse (based on Andrees Handatlas, 4th ed. 1901 map 51-2, Dascher 1968).

Hesse remained passive. The village was not restructured, for instance, for the new commercial farming to gain a sound footing; but it also had no guidance or assistance in respect of its no longer employable surplus population. Every villager had to find his own poverty line. Whether or not in these circumstances agriculture really did come into its own again, as the term implies, would depend entirely on the social structure of a village and its habitat. Much of it was agrarization by distress, certainly in Hesse.

Within retarded Germany Hesse lagged far behind, and government appeared to want it so. For centuries it had relied on Ämterzunfte rather than on urban entrepreneurship to organize a trade and had closed its eyes to neighbouring Thuringia's superior competitiveness. Neither was Kur-Hesse to join those German states actively promoting factory-industry. Its guild system was annulled but in 1869, i.e. after Prussia had taken over, and by itself the measure proved insufficient for Hesse to repeat the performance of SW-Germany. Village history in the two regions had been fairly similar, but SW-Germany went the path of re-industrialization which had innate logic in its favour; development here clearly reflected upon the region's mentality and aptitudes, and as, after an interval, the new non-agricultural sector grew stronger, the Southwest became equally well-known for its smallholder-worker villages.

De-industrialization (Entgewerblichung) is a term applicable throughout Germany, whilst the sequel is evidently less easily comprised in just one term. Re-agrarization, no doubt, fits NW-Germany well. In respect of Haufendorf-regions a term like smallholder-agrarization (in a situation of distress) would, it seems, be more appropriate and also name the starting point of further development in Hesse and SW-Germany respectively. For a more provocative term Haufendorf-agrarization would be fully justified in Hesse during the second half of the 19th century.

NOTES

1. For Hesse, unlike Silesia, we lack a record to give hard proof. The assumption rests on, firstly, the location of towns known to have had a medieval linen weavers' guild and Rotenburg's pioneering them; secondly, it rests on the wording of the Treysa guild rules of 1368, the earliest extant, which enforce membership but not the urban location (ZIMMERMANN 1933 pp. 291, 313 ff). By the 15th century Hesse marketed bleached linen yarn and linen cloth at the Frankfurt fair (DASCHER 1968 p.11), and whatever the proportion of cloth, yarn can certaiinly be taken for a rural product.
2. Linen had been gaining on wool throughout the 16th century, and with two commercial fibre crops the lowland economy must have been rather strong. After the Thirty

Years' War had denuded the country of sheep, however, flocks were remarkably slow to recover, and the good-sized farm, its economy thus weakened, apparently gave in more readily to a demand for subdivision.

3. The 18th-century cadastres render the occupation of heads of household only, often naming just one occupation but occasionally adding a qualifying remark in respect of a person or the village in general. For a better insight recourse can sometimes be made to a supplementary register, if still extant (StAM Bestand 49d), which subjects each head of household to a qualitative economic assessment.
4. In a few places, notably Hersfeld and Eschwege, the woollen industries weathered all vicissitudes and were strong enough for worsted to introduce mechanical spinning (Dascher 1968 p.76).
5. See table 1 which, it should be firmly understood, would give the reverse pattern if instead of linen weavers the general labourers were singled out. (see also Leister 1980).
6. The number of linen weavers was recorded for each village by the 18th century cadastres. They provide the best available data but saddle a comparative table with non-uniform dates (after 1740) and also with under-representation. The loom tax tended to enhance a pattern of flux as between commercial and non-commercial looms. This bears even more so on the broad figures of looms surviving. Both sets of figures, however, offer clear proof of eastern Hesse's subregionalization into upland and lowland.
7. Some 18th century 'attempts' at diversifying the market (Dascher 1968 p. 146) indicate no more than a temporary awareness of the problem in governmental circles.
8. Möker 1977 Tabelle 1. Whilst aggregate figures for the Kreise began declining slowly at some date in the 1840s, it will be noted from table 1 that, as usual, local population changes were more complex than the aggregates suggest.
9. Möker (1977 p.39 f.) points out that as early as 1857 farming interests, spoiled by the consequences of overpopulation and sensitive to the least changes, persuaded government into a less lenient attitude towards seasonal migration.
10. Leister 1980 in respect of the 18th century; for a mid-20th century assessment see Meimberg 1951 p. 136 ff. Significant too were the survival of vineyards and viticulture up to the 1740s (severe frost) and the trust placed by landlords in the local people for making tree orchards a success. These orchards on sunny slopes along the Werra, which had been planted piecemeal since the mid-18th century, were a god-send during the problem period.
11. These were the villages of which Pfister, writing in 1840, observed that small towns and big villages as a rule suffered the worst poverty in Hesse (quoted by Höck 1978 p. 77). In fact, the Kammerbericht of 1775, taking stock

after the famine years of the early 1770s, had stressed already that upland villages on the one hand and on the other those lowland villages with an inordinately big substratum could no longer be considered healthy economically (StAM Bestand 5 Nr. 3194).

REFERENCES

BREPOHL, W. 1957 — Vom Industrievolk an der Ruhr. Ruhr und Rhein, H.4, Essen.

DASCHER, O. 1968 — Das Textilgewerbe in Hessen-Kassel vom 16. bis 19. Jahrhundert. Veröff. Hist. Komm. Hessen u. Waldeck, Bd. 28, Marburg.

GREES, H. 1961 — Das Seldnertum im östlichen Schwaben und sein Einfluss auf die Entwicklung der ländlichen Siedlungen. in: Ber.z.dt.Landeskunde 31.

GREES, H. 1971 — Das Kleinbauerntum in Ostschwaben und sein Einfluss auf die Entwicklung von Siedlung und Wirtschaft. in: L'Habitat et les paysages ruraux d'Europe, Liège.

GREES, H. 1975 — Ländliche Unterschichten und ländliche Siedlungen in Schwaben. Tübinger Geogr.Studien, H.58.

GREES, H. 1976 — Unterschichten mit Grundbesitz in ländlichen Siedlungen Mitteleuropas. in: Verh.Dt.Geogr.tag Innsbruck, Wiesbaden.

GRIGG, D.B. 1980 — Population growth and agrarian change. An historical perspective. Cambridge Geog.Studies, Cambridge UP.

HARSCHE, E. 1966 — Die Nebenerwerbslandwirtschaft in Hessen. AVA, Sonderheft Nr. 20, Wiesbaden.

HEITZ, G. 1961 — Ländliche Leinenproduktion in Sachsen 1470-1555. Dt.Akademie d.Wiss.Berlin, Schr. Inst.f.Geschicte, Reihe II,4.

HÖCK, A. 1978 — Witwen und ihre "Hantierung" zu Neukirchen (Am Knüll)im Jahre 1775. in Hess. Blätter f.Volks- u.Kulturforschung Bd.6, Giessen.

JÄGER, H. 1961 — Die Allmendteilungen in Nordwestdeutschland in ihrer Bedeutung fur die Genese der gegenwärtigen Landschaften. in: Geografiska Annaler, 43.

JANSEN, H. 1957 — Uber die sozialgeographischen Entwicklungstendenzen bäuerlicher Dorfer im westlichen Jülicher Land. in: Mitt. Geogr. Ges. München Bd. XXXXII.

JORDAN, T.G. 1964 — German Houses in Texas. in: Landscape, Autumn, 1964.

KELLENBENZ, H. ed. 1975 — Agrarisches Nebengewerbe und Formen der Reagrarisierung im Spätmittelalter und 19./20. Jahrhundert. Stuttgart. Darin: Troitsch, U.: Staatliche Bemuhüngen um die Einfürhrung der Strohflechterei in Kurhessen in der Mitte des 19. Jahrhunderts - ein Beispiel verfehlter Nebenerwerbsförderung.

LEISTER, I. 1980 — Zur Sozialstruktur osthessischer Dörfer im 18. Jahrhundert. in: Dussart-Festchrift, Liège.

MEIMBERG, P. 1951 — Die Landbaugebiete Hessens. Frankfurt/Main.

MÖKER, U. 1977 — Nordhessen im Zeitalter der Industriellen Revolution. Neue Wirtschaftsgeschichte Bd. 13, Wien-Köln.

NUHN, H. 1965 — Industrie im Hessischen Hinterland. Entwicklung, Standortprobleme und Auswirkungen der jüngsten Industrialisierung im ländlichen Mittelgebirgsraum. Marburger Geogr. Schriften, H. 23.

REIMER, H. 1926 — Historisches Ortslexikon für Kurhessen. Veröff. Hist. Komm Hessen u. Waldeck, XIV, Marburg.

ZIMMERMANN, L. 1933 — Der Ökonomische Staat Landgraf Wilhelms IV. Bd. 1: Der hessische Territorialstaat im Jahrhundert der Reformation. Veröff. Hist. Komm. Hessen u. Waldeck Bd. XVII, 1, Marburg.

ZÖGNER, L. 1966 — Hugenottendörfer in Nordhessen. Planung, Aufbau und Entwicklung von siebzehn französischen Emigrantenkolonien. Marburger Geogr. Schriften, H. 28.

StAM = abbreviation for State Record Office Marburg.

CHAPTER 26

THE DANISH HOSIERY INDUSTRY; A SPECIFIC RURAL INDUSTRY IN CENTRAL JUTLAND.

Viggo Hansen
(University of Copenhagen)

The Danish knitwear industry is today concentrated in a very small area of Central Jutland where the two municipalities of Herning and Ikast together employ a labour force of over 8000 in textile and clothing industries. Some 150 factories produce all kinds of articles of clothing from woollen and artificial threads, men's and women's underwear, socks, shirts, and dresses to blankets and carpets: much of this is now exported to other countries of Europe and to the United States. Most of the medium-sized factories (30-80 workers) were established during the economic boom of the 1950's and 1960's, yet some of them date from the 1930's, and a few have existed since before the turn of the century. The traditional raw material in the early period was wool from the local flocks of sheep, and as such this industry has roots far back in time, when Central Jutland was to a great extent heathland, and when only 10-20% was under cultivation and had a very sparse population of peasants.

For a book published in 1947 named Spind og Bind (Spin and Bind), H.P Hansen, the force behind, and later director of, an open-air museum in Herning, collected a large amount of material on old customs and folklore as well as records from public and private archives dealing with the history of the hosiery industry and the hosier tradition in Central Jutland. It is the intention of this paper to extend his research along geographical lines in order to uncover how and when the production of knitwear for home consumption became an industry and a trade beside traditional farming, and also why the old district of Hammerum Herred became the production and trading centre from which the hosiery industry spread to several neighbouring herreds.

The following research will concentrate on the four herreds below, but also take some account of several neighbouring herreds in the two counties of Ringkobing and Viborg (see fig.26:2). According to the Danish occupational statistics from 1845 the number of those gainfully employed in the knitwear industry was distributed as follows:-

Table 1.

Herred	Binders	Weavers	Hosiers
Hammerum	221	71	24
Lysgård	150	56	13
Middelsom	49	85	1
Bølling	72	65	2
Total	492	277	40
The two counties	759	1007	42

The figures refer to the heads of those families whose main occupation was in these industries, but they do not take account of all those for whom knitting was a supplementary income to farming. In Den danske Statistik (1847) A.F. Bergsøe estimated that 25,000 persons were occupied in the production of knitwear, and that they manufactured 220,000 lbs. of goods valued at 230,000 Rigsbankdaler (Rbd). This production consumed 300,000 lbs. of raw wool from 160,000 sheep. Between 40 and 50% of the total production of knitwear was thought to derive from Hammerum Herred and 20% from Lysgård Herred.

There is little doubt that the Danish centre for the knitwear industry in the middle of the 19th century was Hammerum Herred together with the neighbouring Lysgård Herred, even if the industry at that time was spread over 60 parishes in Central Jutland. The hosiery industry was partly organised in small work-shops, where a master employed a few men as well as wives and children, but to a much greater extent by the peasants themselves in home knitting. A water-powered mill at Brunshåb near Viborg in the 1830's employed 180 workers, mostly women and children, but this was an exception.

Men and women, young and old, knitted whenever there was an opportunity: while walking or driving to and from the fields, while rearing the flocks of sheep, and in the house, as is immortalized in drawings and paintings by artists who were charmed by the nature of the open spaces, and in literature by poets who were drawn to the heathland to get new inspiration for their art. Topographers and learned men of that time, whose ideas were to find ways to improve the poor peasants' situation, have described many features connected with the hosiery industry. One of the best connoisseurs of the heathland was Steen Steensen Blicher, a local vicar, highly celebrated for his many novels and short stories, which found much of their inspiration in his hunting tours across the heath. One of his short stories is "E Bindstouw" (The Knitting Parlour), in which in the local dialect he refers to an old custom of social entertainment, that was already on the wane. An earlier description of the knitting parlour is found in the writings of Fr. Ludwig Moltke from 1759:

> "In the winter dusk young and old met village by village and from the nearest surroundings, whence to knit and henceforth to vie to see who can knit the fastest. Each one is meted out 240 yards of yarn - which is enough for six hours at the rate of 40 yards to the hour - by which they reckon their time. ..... When they gather in the evening in the knitting parlour in order to save lighting and pass the time with chatter, they behave quite properly without consuming any kind of drink. They do, however, give the tongue freedom to invent the most untruthful stories they can think of".

The basis of the knitting industry was the wool from the big flocks of sheep that roamed for most of the year through the extensive heathlands under the surveillance of a shepherd, often a young boy. Only after the last sheaves had been removed were they let loose to graze the stubble of the empty fields. Some village flocks numbered a thousand or more and a specialization in sheep rearing was often taken up by smaller manors and granges. Table 2 gives an idea of the number of sheep; a herred included 12 to 15 parishes.

Inventories of cattle had taken place occasionally for the purpose of taxation since the late 17th century, to which reference is made later. At this moment we are only interested in the time when sheep were the basis of an important supplementary income to the normal low income from grain production, and so important that few families in the heathland really were in need of poor-relief.

In order to present the general level of the number of sheep in Denmark a summary of the sheep population is given in the following graph on the basis of the livestock enumerations from 1837 to 1909 (Fig 26:1). From this it can be seen that the number of sheep in Jutland generally peaked in 1871, after which there was a steady decrease except for a shorter period: by 1909 the number had halved. This is the period when Danish farming changed from mainly grain production to intensive animal production, and in which there was no longer room or market for extensive sheep farming. It is also the period of the industrial revolution and the introduction of a market economy in agriculture, which meant import of cheap raw materials, e.g. wool, from overseas. If one compares Jutland with the Danish islands, the main development traits are evidently the same, except that the islands did not experience much growth between 1837 and 1861. In 1871 the sheep population was twice as big in Jutland as on the islands compared with the human population, but equal in relation to the acreage. At the first enumeration in 1837 the sheep population on the islands was greater than in Jutland.

Totals for the most important sheep counties, Viborg and Ringkøbing, are shown on the diagram, and the following points should be emphasized:-

1. The sheep population more than doubled between 1837 and 1871.
2. The decrease after 1871 was much slower here than in Jutland as a whole.
3. There were more sheep in 1909 than in 1837.

The explanation for this specific development may be summarized as follows:-

1. The heath covered far larger areas in the two counties than elsewhere in Jutland.
2. Sheep grazing was the best use of the vast heathland.
3. The wool market was present from earlier times in the two towns of Holstebro and Ringkøbing, both in Ringkøbing county.
4. The hosiery trade had been a monopoly in the four herreds of Hammerum, Lysgård, Bølling, and Ulfborg since 1741.
5. The taking-in of the heath for cultivation started later here and showed slower progress than anywhere else in Jutland until after the foundation of the Danish Heath Society in 1866.
6. Despite enclosure from the late 18th century much of the heath remained common grazing because fencing between individual farms was not popular.
7. Enclosure assigned each farmer several hundred hectares of uncultivated land but did not give him the power or the resources required for its improvement.

Before going further into this question, it is necessary to give a short account of the physical basis of the area (Fig 26:2). The easternmost herreds (Middelsom, Hids, and most of Lysgård) have a surface moulded by the last glaciation (the Würm glaciation), that left behind a morainic landscape with a soil cover of sandy or clayey tills of middle quality. The Würm glacier penetrated to a north-south line, at the edge of which an old trade road (the Oxen road) and military road leads south from Viborg along the watershed.

West of the line fluvio-glacial deposits, brought here by the meltwater streams, created the extensive outwash plains, (also called heath-plains), that cover most of Hammerum Herred. Further west an older moraine emerges having a rather thin layer of heavily leached sandy or stony till on top of tertiary sands. More loamy tills are confined to a narrow spur leading east through Herning, Gjellerup, and Ikast parishes in Hammerum Herred. The two western soil types are rather infertile (see Table 2), and the spread of agrarian settlements, that under these circumstances depended on the few available resources of surface water, created an open

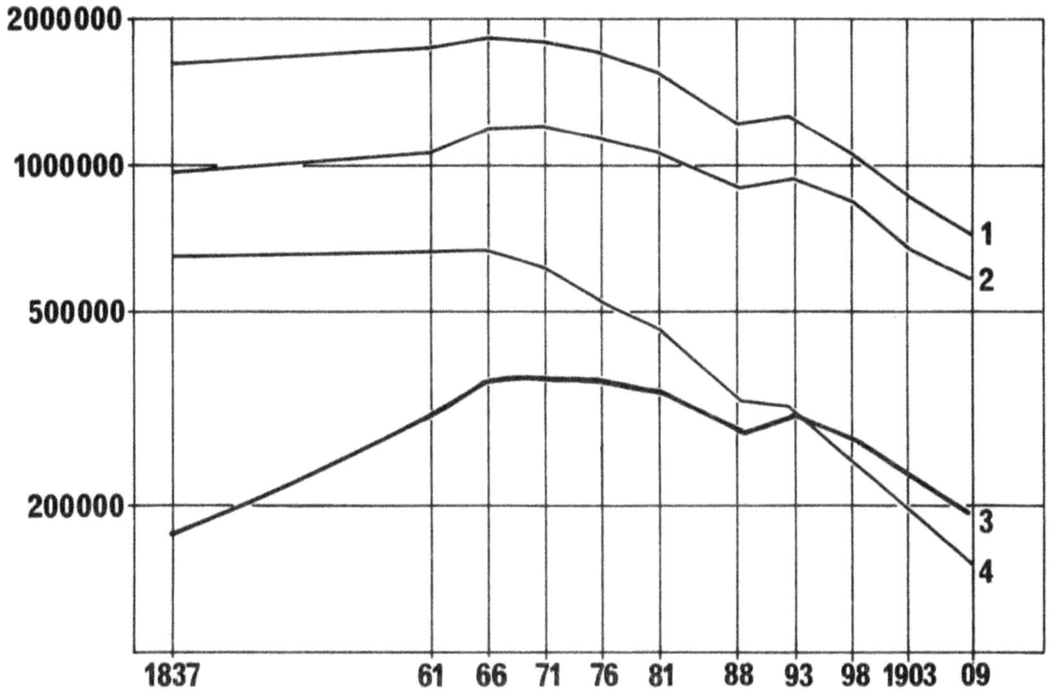

Fig. 26:1 Danish sheep population 1837-1909
1 Total
2 Jutland
3 Viborg and Ringkøbing counties
4 The islands

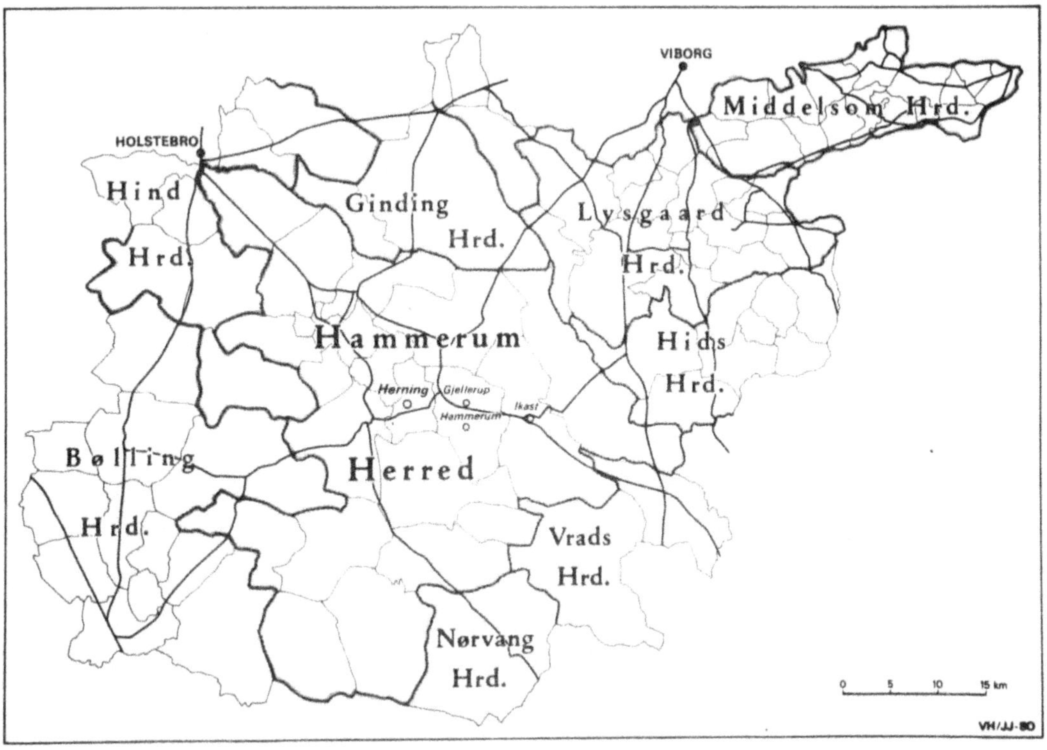

Fig. 26:2 The centre of the hosiery industry in Jutland.

pattern of small hamlets or single farmsteads along the water meadows and local rivulets. As distinct from this settlement type the young moraines to the east had a much denser network of middle-sized villages and hamlets, with a richer and more varied resource base. A population density of 12-15 inhabitants per square kilometre in 1801 in the eastern herreds against 5-7 in the central western parts reflects these differences in resources.

The spatial distribution of the resource base is converted in Table 2 which also elucidates the correlation between this base, the sheep population, and the amount of wool processed for sale.

The population figures of Table 2 are taken from the 1840 census to get as close as possible in time to the acreages in column 2. These are from the land register that came into force in 1844 after parliamentary enclosure, and which replaced the previous register of 1688. From the same source originate the land tax figures in column 3, an early Danish land productivity measure: one tønde hartkorn being the acreage that could sow one tønde (i.e. 3.83 Eng. bushels) of hard corn of barley or rye. Column 4 thus shows the average acreage of land that according to its productivity measure could grow the equivalent of 1 tønde of hard corn. The higher the figure the lower was the productivity as well as the land taxes. The reference figures 1 to 6 give the productivity rates for the whole Herred (in column 4) as well as the rates for those parishes that alone had a production of knitwear for sale (bottom of table). The last figures indicate that only parishes with a very low tax assessment had taken up the knitting business. Column 5 gives the sheep population in 1837 by herred, and by comparison with column 6 (processed wool for sale) it is evident that the two columns do not correlate: 6 herreds in the list processed no wool for sale at all, while some herreds with few wool-producing parishes show rather high figures.

Table 2.

Herred	1	2	3	4	5	6
Hammerum	9,538	110,325	2,068	53	23,600	50,300
Lysgård	6,586	37,977	1,660	23	16,700	22,700
Middelsom	4,939	25,566	1,943	13	15,600	14,600
Fjends	5,679	39,305	1,827	22	18,200	-
Rinds	5,316	39,543	2,218	18	19,500	-
Nørlyng	3,892	25,837	1,359	19	11,000	-
Sønderlyng	4,558	22,916	2,022	11	13,300	-
Skodborg	6,365	41,665	2,519	17	13,700	-
Hjerm	5,711	34,351	2,767	12	15,500	-
Ginding	4,696	58,209	1,783	33[1]	15,400	6,500
Ulfborg	5,009	46,791	1,829	26[2]	13,200	2,100
Hind	6,611	41,711	2,081	20[3]	13,900	300
Bølling	5,875	58,186	2,192	27	14,400	5,200
Hids	3,462	15,809	906	17[4]	9,400	2,800
Nørvang	11,216	77,620	3,242	24[5]	22,700	150
Vrads	5,175	50,597	1,331	38[6]	10,500	1,500

Column 1. Population 1840
Column 2. Area in hectares 1844
Column 3. Tønde hartkorn (see text)
Column 4. Hectares per tønde hartkorn (see text)
Column 5. Sheep population 1837
Column 6. Lbs. wool processed for sale.

1) 3 parishes alone: 87 ha/ tønde hartkorn
2) 3 " " 58 " " "
3) 1 parish " 65 " " "
4) 4 parishes " 40 " " "
5) 1 parish " 60 " " "
6) 2 parishes " 87 " " "

The busiest herreds by far were Hammerum, Lysgård, and Middelsom, of which only Hammerum according to the productivity measure (53 ha on the tønde hartkorn) could be reckoned as having a poor soil, while, on the other hand, Middelsom Herred could boast of a much better land quality (13 ha on the tønde hartkorn). It is evident that herreds with high production figures could not provide sufficient wool from their own stocks of sheep, but were forced to buy from other herreds, e.g. from those that did not process wool for sale but only for home consumption, and this exchange of wool was the hosiers' business.

The inconsistencies here observed may be further

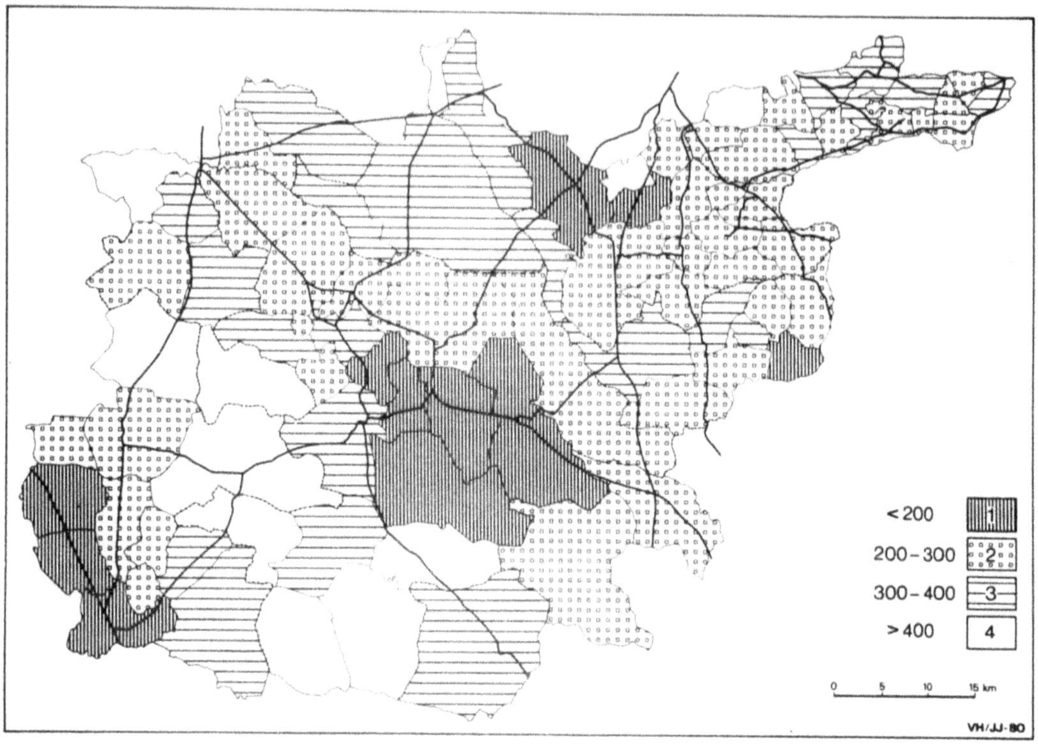

Fig. 26:3 Sheep density per 100 inhabitants (parish basis). The map brings out the fact that the busiest hosiery district had fewer sheep than the neighbouring areas.

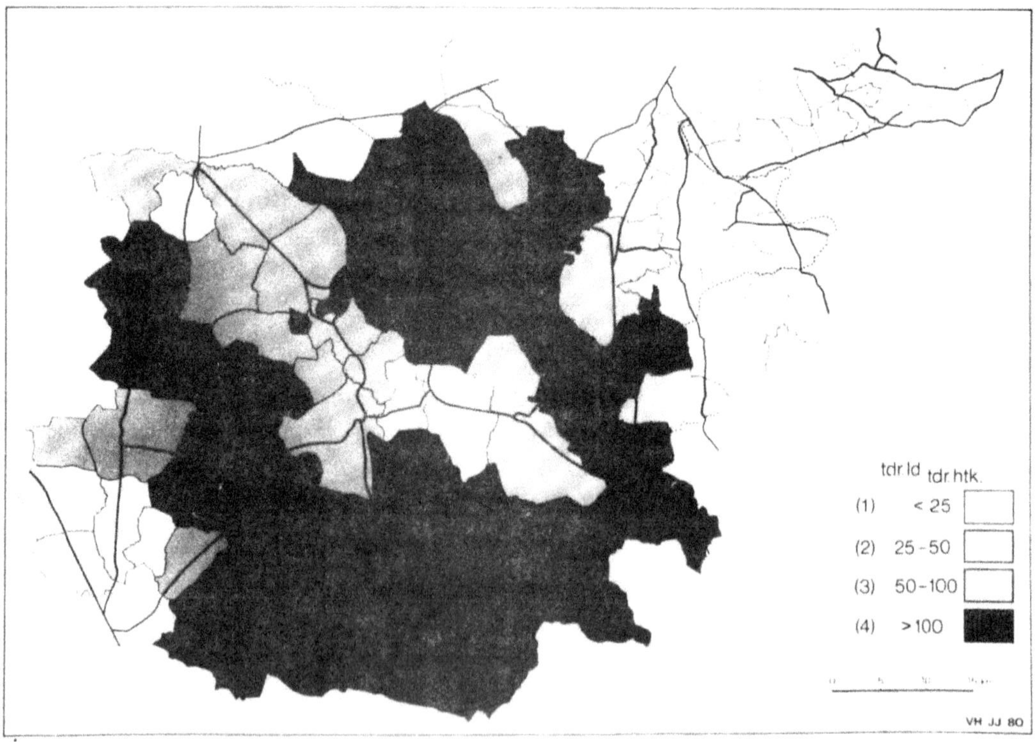

Fig. 26:4 Estimated agricultural productivity in Central Jutland based on the 1844 taxation. The darker the shading the lower the productivity.

illuminated, when one compares the above analysis with Fig 26:3 that shows the sheep density per parish in the central district, covering the same area as in Figure 26:2. Here the lowest density is to be found in the central part of Hammerum Herred, and the further one moves towards the periphery of the area the higher is the density, with only a few exceptions. This situation is further stressed in Figure 26:4 which brings out the differences in relation to soil productivity per parish and once again shows central parishes such as Herning and Gjellerup as having the most productive soils together with the two easternmost herreds Lysgård and Middelsom. From these observations one must conclude that parishes like Herning, Gjellerup, and Ikast, nowadays the centre of the production of knitwear in Denmark, were also the busiest processing areas around 1840 in spite of both a lower sheep density and a higher soil productivity than the surrounding parishes in the same herred. Another observation can be made in the eastern Herreds, where in this case a high degree of correlation exists between a high productivity figure, a large number of sheep and many households in the knitting business, as for instance in the two parishes of Mammen and Vinkel (compare also Table 3 for 1801). It is much more an exception when the few wool processing parishes in Ginding, Ulfborg, and Hids Herreds (see Table 2) show a high degree of positive correlation between a large sheep density, low soil fertility, and production of knitwear for sale. In order to explain several of these inconsistencies the figures from the 1801 population census have been broken down. Many sheep on land of low fertility did, in itself, not create a selling market for knitwear, as has generally been surmised, but other factors must have been at work.

THE KNITWEAR INDUSTRY IN CENTRAL JUTLAND AROUND 1800

In the 1801 census, the first fully reliable enumeration in Denmark, people were asked for their title, profession, business, craft or trade. In order to get a complete account of the spatial distribution and of the local importance of home-knitting in Central Jutland the primary lists have been examined for those families that had this industry as their main occupation as well as a subsidiary income. The total number of those fully occupied in home-knitting was found to be 350, of which well over half (190) lived in Hammerum Herred, 91 in Middelsom, 38 in Lysgård, 27 in Bølling, and 4 in Ulfborg Herred. It was found that the distribution within the herreds was very uneven.

In order to get the best impression of the local importance of this home industry the number of families with a gainful occupation in the trade was compared to the total number of households. The largest percentage was found in Gjellerup parish with 68% of all households, followed by Mammen and Vinkel parishes in Middelsom Herred with respectively 50 and 38%, Rind and Sunds in Hammerum Herred with respectively 27 and 23%, while Torning parish in Lysgård Herred showed 23%.

In fact the industry was of much greater importance to the households than indicated by these figures, if the many remarks by the enumerators on subsidiary incomes are taken into account. In several cases the enumerators added comments such as:-

> "Old and young, farmers as well as small-holders, maids and farmhands, and children too, have made the hosiery into a trade in this parish" (Sønder Felding, Vium, and Lysgård with few alterations).

Or:-

> "Farmers, small-holders, and renters go in for home knitting, regardless of the sexes, as soon as they reach their fifth or sixth year, as the royal taxes and land rents mostly should be payed from this industry, wherefore the hosiery trade must be considered the most important occupation in this parish" (Torning).

In other connections it is noted that:-

> "because of this industry poor-relief is hardly known in this region".

Who were the families that had their main income from the hosiery industry? The answer to this can be deduced from the enumerators' several remarks. In Gjellerup parish (Hammerum Herred) which had the largest percentage of households in the industry, even farmers as well as lower classes took part, and particularly it is noteworthy that in the village of Lund every household had knitting as its main occupation. With decreasing percentage of households that figure in the census as knitters, one sees how farm families "fall out", and the knitting narrows down through a scale from small-holders' families to landless people, to old folks, and to single women and widows. But at the same time it is important not to forget the above quotations on the general importance of the knitting business both as a pastime and as a source of supplementary income in more than 60 parishes in the region.

If one looks at the spatial distribution of knitting as a main occupation, it is more easily seen how poor families had to live on the income from the knitting sold as well as producing for home consumption as everybody did.

It has generally been accepted that the knitting business was a kind of substitution for low farm-income in an area of very low soil productivity. Consequently one would expect to find a correlation between low farm-productivity on the one hand and a large sheep population and a busy knitting industry on the other. Such an analysis has been presented in reference to the 1840's, but previous to that period the sheep enumerations fail as do the figures for wool processed for sale. Table 3 gives the figures for knitting activity and soil productivity on a parish basis for 1801:

Table 3

Parish	per cent households in knitting	soil productivity (ha/td. htk.)
Gjellerup	68	24
Mammen	50	15
Vinkel	38	13
Rind	27	77
Sunds	23	72
Torning	23	41
Aulum	23	43
Bording	20	77
Hjermind	18	14
Torup	16	9

As seen from the figures the above suggestion does not hold true, even if 3 parishes belong to the lowest productivity class, but on the other hand the 3 busiest and further 2 must be classified as having medium soils.

THE MARKETING OF WOOL AND WOOLLEN COMMODITIES

Marketing demands communications, but around 1800 the general miserable condition of the small roads was a major obstacle to communication between parishes in the heathland and to and from the market towns; there were few improved roads at that time. One of these was a road that connected eastern Jutland with the town of Ringkøbing with its rather poor port facilities; it passed through Hammerum Herred and the parishes of Ikast, Gjellerup, and Herning. From Gjellerup a branch road led northwest to Holstebro, where the most important wool market was situated for over a hundred years. But with a distance of 35-40 kilometres from Gjellerup, the peasant going to market at Holstebro was forced to stay overnight. The old ecclesiastical centre of Viborg was even farther away from Hammerum Herred, but on the other hand near enough for a one-day trip for the peasants of Lysgård and Middelsom Herreds. For that reason Viborg was the natural meeting point for that part of the heathland and for the sales of wool and woollen commodities. For the easternmost part of the area considered here, Viborg had a competitor in the port town of Randers where the farm produce could be shipped directly to Copenhagen.

A peasant's diary from the 1780s records how this man

either collected wool after the shearing from several villages on commission, brought it to the market in Viborg, and afterwards settled with the suppliers, or he bought it from other peasants and had a small profit for himself on the sale. Even if we know no further examples of this custom, it may well have been practised by others.

In an area otherwise empty of market towns it is not unusual to meet with periodical markets, and such a one was held twice a year (April and September) in Gjelleruplund (or Lund) at the meeting point of the two improved roads mentioned above (see Fig.26:2). To here came from the east many kinds of articles of wood such as cartwheels and spokes of oak, household articles of beech wood, wooden shoes, etc. besides hardware and garden produce. From the west came cattle, horses, seed-corn of rye, black earthenware, and several home produced articles, but knitwear is not mentioned in the records.

Considering, however, that the nearest parishes were the ones with the best soil in the Herred, had fewer sheep per inhabitant and also had the largest number of active binders, there is reason to believe that to some extent wool was also dealt with at the market. It is well-known that the central knitting area had to be supplied with raw wool from other parts of Jutland and Funen, and when Moltke (1759) reports that wool was carried to the area from Holstebro market and from Tinnet market in eastern Jutland - from where also the wooden articles came - this supports the suggestion that wool was also exchanged at the Gjelleruplund market, and that the binders bought their supply there, as well as selling their knitted goods to buyers. The idea supported by the above evidence, is that Gjelleruplund market acted as a centre of innovation for the knitwear industry. It was at that time said openly that the peasants in Hammerum Herred neglected farming in favour of knitting. That the industry was a good business is confirmed by the fact that maiden servants were reluctant to take service in the summer season, especially in years with low food prices, because they found that they could earn more by knitting for themselves; sometimes they demanded 2 to 4 weeks' holiday, in which period they worked for themselves while the peasant had to feed them (Moltke op.cit.).

It is understandable from this that not only did the knitting generate good earnings, but that the marketing of knitwear was a profitable trade, and this is where the hosier comes in.

THE HOSIER

Hosiers are first heard of in an official document of 1691, when the Copenhagen cloth guild complained of the unbearable competition from Jutlandish hosiers (KUHLE, 1936). That the complaint was well-founded is later shown by the fact that the "invaders" in the capital city in 1694 started to buy houses and to establish shops in the main business district, Amagertorv. A government ban on imported woollen goods, issued in 1698, strengthened the hosiers' position against the cloth guild, and in 1700 nine hosiers, all from Hammerum and neighbouring herreds, got a trading licence.

The Copenhagen hosiers worked through authorized agents in Jutland, and their main districts were the eastern herreds (Lysgård, Middelsom, and Hids) where they collected the knitted goods and brought cartloads with them to Copenhagen. As residents of Central Jutland the hosiers were not only allowed to deal freely in woollen and linen goods, but (in 1741) they got a monopoly as well to do trade all over the country, so long as they possessed a licence giving their home address in one of the four herreds: Hammerum, Bording, Lysgård, or Ulfborg. Some were independent of the Copenhagen companies and worked particularly in Hammerum and the neighbouring herreds to the west. In some instances the hosiers were the selling as well as the buying agents, selling wool to the farmer and buying his knitwear. One report says that "there was hardly any money between the hosier and the peasant". The hosiers were not ordinary tradesmen but had most often passed a training period of 6 years or more, before they were accepted as buying agents, and some of them even had a citizenship in the capital city.

But Denmark (Copenhagen) was not the only market; there was also a significant export to Norway and Sweden, and of even greater importance to Holland, Hamburg and northern Germany directly from Jutland. Knitwear was also exported from Copenhagen to Poland and to the German province of Pomerania. The importance of the wool market in Holstebro, the collecting point for central and northwestern Jutland, attracted agents from Hamburg who arrived in Holstebro up to five times a year, bought up whatever amounts they could of knitwear and wool, and hired peasants to carry it the whole way back to the southern Danish/German border. Here other carriers took over and took it to Hamburg and to other places in northern Germany. According to reports from 1735 and 1746 (Danske Atlas, 1768) other consignments went via the town of Ringkøbing and were shipped to Amsterdam. These reports thus tell us a great deal about a pre-industrial society of poor peasants who nevertheless participated in an extensive trade with one single commodity. Keeping in mind the rather poor earnings the peasants got out of it, it is surprising to find that the total income from knitting in the 1750's amounted to as much as 16-20,000 Rigsdaler (Danske Atlas, 1768). It was a good auxiliary income at that time, but still nothing to compare with 50 years later, when the income from knitting amounted to 70-90,000 Rigsdaler per year in Hammerum Herred alone (G. BEGTRUP, 1812). The same source mentions two hosiers, who each bought wool to the amount of 16-20,000 Rigsdaler yearly.

Prior to 1801 the records do not record details of the knitting business. The censuses of 1769 and 1787 name a few dealers in knitwear in the parish of Rind, and from their names it is obvious that this employment was transmitted from father to son. Due to the small number of hosiers one is inclined to conclude that woollen goods were sold to agents of Copenhagen merchants, or that they were marketed in another way.

Tax lists from 1761-65, 1743-44, and 1719 are not of much further help, as the taxes were on wealthy persons only, and people who had some kind of craft or trade as their main occupation. People having the production of knitwear as a side-line were not included in the tax lists.

THE BEGINNING OF THE HOSIERY INDUSTRY

In 1735 a countrywide royal rescript had asked for reports on the economic conditions of both towns and rural districts. These reports show that sheep breeding at that time was an unstable and unprofitable business because of frequent lung diseases and other epidemic illnesses among the sheep. Also the wool production hardly satisfied the home consumption. Sheep were more or less kept for food and for skin, and were mostly left untended on the common land.

What then caused the progress in sheep rearing after the middle of the 18th century? There is no reason to believe that the peasants began caring more for their animals than hitherto, and the fluctuating market prices would hardly influence the sheep population. The beginning of the 18th century was a period of agricultural crisis; this lasted till after 1735, when a changed political attitude towards a more expansive policy began to show results.

Prior to 1735 malnutrition and diseases among the sheep may well have been influenced by climatic circumstances. Different sources point to the very bad weather conditions that prevailed during most of the 17th century and into the first decades of the 18th, the so-called Little Ice Age, when the weather was cool, windy, and rainy, i.e. the worst possible physical conditions for sheep breeding. The weather seems to have been at its worst in the 1680's according to both the evidence in the Danish land-register of 1688, and several simultaneous tax assessment lists.

An analysis of these tax lists for the number of livestock in the traditional knitting areas shows that less than half the peasant households kept sheep, and these households held no more than 2-4 animals. The result is the same all over, whether in the central area or in the periphery. A comparison of the 1837 enumeration with the corresponding figures for 1684 are given below in rounded figures.

The enumerations were carried out after the lambs were born and before the slaughtering.

Table 4

Herred	1684	1837
Hammerum	1,100	23,600
Lysgård	1,260	16,700
Middelsom	930	16,000
Hids	570	9,400
Sønderlyng	980	4.900

As to the production of wool in the 1680's this would hardly surpass 3 lbs. per animal, even counting on 2 shearings per year from the tough domestic breed. As the amount for home consumption is reported to have been 3-5 lbs per grown member of the family (BEGTRUP, 1812), the small number of sheep would not satisfy the peasant household needs. Thus if the figures are reliable one must conclude that there was no surplus of wool, on which a local knitting industry, meant for sale, could be based. Neither was there any place for the hosier except where manorial estates had large flocks of sheep roaming their extensive and otherwise unproductive heathlands. One possible source of supply for the hosier might have been the eastern herreds like Middelsom and parts of Lysgård, where an industrious population performed several other crafts mainly based on wood, and who were willing to sell wool to the hosier. This was then resold to knitters in the less prosperous areas in return for knitwear.

CONCLUSION

The later and widespread knitwear industry in Central Jutland thus only saw its beginning some decades into the 18th century, after which it spread through the later part of that century to wide areas of the poor parishes as a supplementary means of income to meet the taxes and the rents. As a main occupation the production of knitwear for sale to the hosiers covered only small areas that emerged not so much because of exceptionally poor soils, but because of a better infrastructure including improved roads and the presence of periodic markets.

Documents:

Folketaellingerne 1787, 1801, and 1834. (Rigsarkivet)

Ekstraskatteregnskaber 1682-88, 1719-22, 1743-44. (Rigsarkivet)

Konsumptions, Familie - og Folkeskatregnskaber 1761-65 (").

Lundenaes og Bøvling amtsregnskaber, Konsumptionsregnskaber 1766-73. (Rigsarkivet)

LITERATURE

Faestebonde Christen Andersens Dagbog 1786-97. Landbohistorisk Selskab, 1969.

G. BEGTRUP 1812 — Beskrivelse af Agerdyrkningens Tilstand i Danmark, Bind 7.

A.F. BERGSØE 1847 — Den Danske Stats Statistik, Vol 2.

HANS BJØRN 1976/7 — Staten, Kraemmerne og forlagssystemet 1620-1735. Erhvervshistorisk Årbog.

STEEN STEENSEN BLICHER 1839 — Bidrag til de danske Provindsers naervaerende Tilstand i oekonomisk Henseende, Viborg Amt.

C. DJØRUP 1842 — Bidrag til de danske Provindsers naervaerende Tilstand i oekonomisk Henseende, Thisted Amt.

J.C. HALD 1833 — Bidrag til de danske Provindsers naervaerende Tilstand i oekonomisk Henseende, Ringkøbing Amt.

H.P. HANSEN 1947 — Spind og Bind. Bindehosens - Bindestuens og Hosekraemmerens Saga. København.

ERIK HVIDTFELD 1959 — Praesteindberetningerne fra Viborg Stift 1771. Fra Viborg Amt.

ELI JEPSEN 1971 — Tekstilindustriens udvikling i Hammerum herred. Hardsyssel Årboger.

SEJR KÜHLE 1936 — 'Jyske hosekraemmere i Kobenhavn i det 18. årh'. Historiske Meddelelser om Staden København, 3:11 .

GREVE FRIDERICH LUDVIG MOLTKE 1759 — 'Kort Efterretning om de fordelagtige Manufactur af Strømper, Vanter og Nat-Trøyer i Hamrum Herret og dets Egne.' Danmark og Norges oeconomiske Magazin, Bd. III.

PONTOPPIDANS DANSKE ATLAS 1768 Tome IV.

STATISTISK TABELVAERK, FEMTE RAEKKE, 1842 (Kreaturholdet 1837/38)

STATISTISK TABELVAERK, TREDIE RAEKKE, 1865 (Arealbenyttelse).

CHAPTER 27

UNE INDUSTRIE RURALE ORIGINALE; LES ANCIENNES VERRERIES

J. Pinard
Université de Limoges

Parmi les nombreuses industries qui se sont développées autrefois dans les campagnes (forges, fours à chaux, briqueteries...), il en est une particulièrement originale parce que ses productions ne correspondaient guère à celles que l'on a l'habitude de rencontrer dans les milieux ruraux - il s'agissait souvent de produits de luxe destinés à une clientèle aisée - et que d'autre part les industries actuelles qui en sont les héritières gardent encore de nombreux traits de leurs ancêtres quand ce n'est pas l'implantation originelle: cette industrie est la verrerie.

Si loin que l'on remonte dans le passé, on trouve des verreries installées dans les forêts, alors qu'aux mêmes époques les villes accueillaient déjà d'autres activités manufacturières venues souvent des campagnes; si les raisons en sont aisément décelables à l'origine, la permanence de cette implantation ne peut que surprendre quand les facteurs de localisation ont évolué et on peut se demander comment les campagnes ont pu conserver de telles fabrications quand apparemment tout ce qu'elles en tiraient a disparu et qu'elles vont depuis longtemps le chercher à l'extérieur; et pourtant, quand on consulte un annuaire des verreries françaises, on constate que beaucoup d'usines sont encore dans les campagnes.

I - LES CONDITIONS ANCIENNES D'IMPLANTATION DES VERRERIES

A l'origine les verreries ont trouvé dans les régions forestières, et plus généralement dans les zones rurales, les bases indispensables à leurs productions et à leur développement que ce soient les matières premières, les appuis humains ou les débouchés.

Les premiers verriers installaient leurs fours dans les forêts pour en utiliser le bois comme d'ailleurs les autres activités qui avaient recours à ce combustible. Ils recevaient souvent celui-ci gratuitement des seigneurs du domaine où ils étaient implantés en échange de menus travaux. Tous les bois convenaient en principe pour chauffer les fours de verriers à condition qu'ils ne crépitent pas trop en brûlant pour la fabrication des verres fins car ce phénomène était à l'origine d'irrégularités comme des bulles ou des rayures dues aux flammèches qui étaient alors projetées dans la pâte en fusion. Avec la suppression des privilèges à la Révolution, les verriers durent acheter comme les autres utilisateurs le bois aux enchères publiques, ce qui entraîna de fortes augmentations des prix et de sérieuses difficultés financières pour nombre d'entre eux.

Les campagnes devaient aussi leur fournir les matières premières entrant dans la composition de la pâte: les sables les plus purs extraits de carrières voisines, exceptionnellement de bancs de rivières, la fougère qui en brûlant donnait la soude indispensable à la fabrication du verre et quelquefois le calcaire utilisé comme fondant pour abaisser le point de fusion: même lorsque ces composants seront recherchés dans d'autres matériaux comme la soude dans les plantes marines (varech) au XVIIè siècle, les verreries resteront à leur emplacement primitif. Les verriers ont été aussi considérés comme de gros consommateurs de bois et donc comme des dévastateurs de forêts, au point qu'il a fallu réglementer la profession à la fin de l'Ancien Régime car l'exploitation intensive de celles-ci s'ajoutait à celle des autres possesseurs de fours et à celle des autres consommateurs de bois d'oeuvre (scieries, chantiers navals...).

Les nobles ont très tôt vu l'intérêt qu'ils pouvaient retirer de la production du verre sur leurs domaines, non seulement pour valoriser des terres boisées rapportant peu jusqu'alors, mais par les subsides qu'ils pouvaient prélever sur la vente d'objets fabriqués en définitive à peu de frais: aussi cherchèrent-ils très vite à attirer sur leur manse des verriers qui étaient souvent des immigrés venant d'Italie où l'usage des objets en verre fut précocement répandu, indépendamment des fabrications de luxe qui faisaient le renom de villes comme Venise à la fin du Moyen-Age. Les propriétaires forestiers offraient des avantages financiers à ces artisans-manufacturiers qui avaient ainsi la triple tâche de défricher et de cultiver la forêt ou la lande sur laquelle ils s'installaient, de construire une maison forte et de la défendre avec le reste de leurs installations. Bientôt ce furent les princes qui, pour garder sur leurs terres ces familles où se transmettaient de père en fils des secrets de fabrication, édictèrent des interdictions de se déplacer et surtout leur concédèrent des privilèges dont celui de la noblesse: ainsi des gentilshommes-verriers prirent la direction de petites entreprises employant d'abord quelques ouvriers, puis une main d'oeuvre de plus en plus nombreuse chargée de couper le bois, d'entretenir le feu des fours, de préparer la pâte... Cette main d'oeuvre paysanne continuait aussi à travailler la terre et venait à la verrerie quand la fabrication l'exigeait: cette double activité donnait une grand souplesse à la conduite de ces petites fabriques; elle était beaucoup plus stable que la main d'oeuvre immigrée à l'humeur souvent vagabonde au moins dans les premiers temps.

D'autres facteurs économiques intervinrent aussi très tôt pour retenir les verreries dans les campagnes. Dès l'origine la principale clientèle de ces activités était constituée par la noblesse qui a seule les moyens d'acquérir les produits alors de luxe que sont les gobelets, les flacons ou le verre à vitre lorsqu'il se répand dans la construction. Ce sont les manoirs et les châteaux, à côté des hôtels de ville et des églises utilisant les vitraux, qui sont les premiers à adopter ce mode de fermeture des ouvertures à la fin du Moyen-Age. Puis, lorsqu'au XVIIè siècle la vaisselle en métal (argent ou étain) est progressivement remplacée par une vaisselle plus fragile mais moins onéreuse, c'est surtout le verre qui a la faveur des classes riches avant la découverte du kaolin permettant la confection de la porcelaine: nobles et clercs feront alors appel aux verriers locaux. De même au XVIIIè siècle l'usage de la bouteille de table se répand et les forêts deviennent d'importants centres de production. Les ateliers qui ne se lancent pas dans cette fabrication continuent cependant à produire de la gobeleterie commune ou font des essais dans des compositions plus recherchées comme celles dites à l'italienne ou celle du cristal de Bohême ou d'Angleterre (cristal au plomb): on parle alors de "Petites Verreries" pour les distinguer des "Grandes Verreries", celles-ci désignant les manufactures qui ont été créées à partir du règne de Louis XIV, avec aussi de nombreux privilèges dont celui du monopole et souvent installées également près de forêts réputées comme Saint-Gobain en Laonnais, Bayel en Champagne, Fay-aux-Loges en Orléanais... (Fig.27.1).

La tradition rurale reste donc fortement établie jusqu'à ce que la révolution industrielle vienne modifier quelques méthodes de fabrication.

II - DES PETITES AUX GRANDES VERRERIES

Il existait différents types de verreries à la fin du XVIIIè siècle que l'on pourrait classer selon les productions (verre plat, verre creux et verres spéciaux: verre filé, de laboratoire, porcelaine en verre, émail...), mais le lieu d'implantation et la forme de l'établissement sont plus intéressants à retenir.

Les petites verreries dites parfois itinérantes sont les plus anciennes, c'est pourquoi il en reste peu de traces de nos jours si ce ne sont quelques excavations abandonnées dans des sables, souvent remblayées, quelques tas de scories au fond de vastes forêts, parfois des toponymes sur la carte comme Verrières, Verrines, Verrey... Les installations étaient modestes: une simple maison, un four et un hangar pour abriter les outils. Ces "forestiers" fabriquaient toutes sortes de verroteries, gobelets, flacons, mais aussi colliers, étoffes de perles. Ils pouvaient arriver qu'ils alimentassent avec du verre en feuilles les artisans des villes. "Quelquefois les équipes itinérantes étaient amenées à construire près du monument qui attendait leurs oeuvres, les fours nécessaires à la fabrication des feuilles de verre. S'il existait à proximité une fournaise forestière, il semble qu'elles l'utilisaient pour confectionner leur ouvrage" (R. Chambon). Pour les plus petits établissements deux ou trois personnes suffisaient à la production: le maître-souffleur, un compagnon qui préparait la pâte et un aide qui entretenait le feu; la recherche de la matière première et du combustible avait lieu entre les périodes de fabrication proprement-dite. Comme il fallait environ 200 stères de bois pour obtenir 100 kilos de verre, les parcelles forestières s'épuisaient assez vite et il était nécessaire de changer fréquemment d'emplacement. Les installations étaient facilement transportables car elles étaient constituées de constructions légères et démontables: quelques baraques pour loger les ouvriers, un ou deux hangars en bois pour le matériel et un fourneau très primitif. Chaque région a connu dans ses forêts ces verreries qualifiées parfois de "populaires": elles étaient très nombreuses en Lorraine, en Champagne (l'Argonne), en Laonnais (Saint-Gobain), en Normandie (forêt de Lyons), en Bretagne (forêt de La Guerche), en Poitou, en Nivernais. Dans le Midi de la France les forêts de la Grésigne, de la Montagne Noire, de la Double saintongeaise et périgourdine étaient célèbres pour leurs productions destinées aux

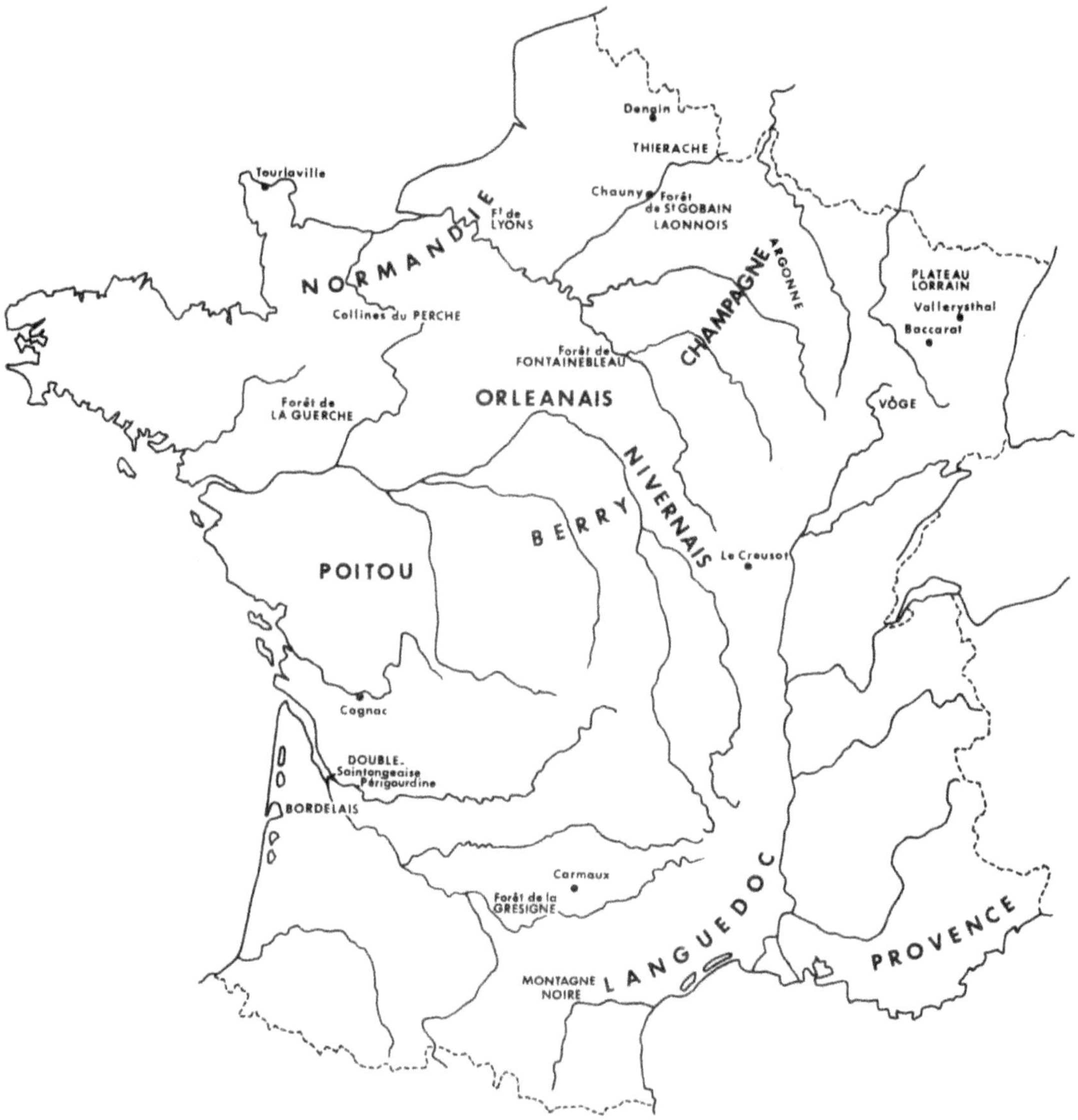

Fig. 27:1 Principaux pays et centres d'anciennes verreries.

pays de vignobles qui les entouraient quand bonbonnes, pipettes et flacons intervinrent dans la vinification avant même que les bouteilles ne fussent utilisées pour le transport ou la vente des vins. En Languedoc et en Provence les fabriques étaient également nombreuses car toutes sortes de flacons, pots, coupes, étaient demandés pour conserver les huiles et les onguents, sans parler de tous les objets d'art en verre fin, en porcelaine, en émail, que pouvaient élaborer certains artisans sous l'influence de maîtres italiens immigrés ou rencontrés lors de voyages.

Lorsque l'usage de la gobeleterie et de la bouteille se répandit et que des verriers voulurent se lancer dans des productions plus fines à l'image de celles de l'étranger comme les cristaux de Bohême et d'Angleterre, des verreries prirent de l'importance et durent adopter un emplacement définitif, recevant alors les soutiens de seigneurs et de financiers. Plusieurs fours étaient désormais nécessaires pour obtenir une production continue, des bâtiments en dur abritaient les moules des modèles dans lesquels la pâte était coulée, des matières premières plus pures et rares (sables très blancs, sels métalliques originaux) devaient être achetées à l'extérieur de la région comme les sables de Fontainebleau très recherchés par toute l'Europe. Une main d'oeuvre stable et spécialisée fut attachée à chaque verrerie, dirigée souvent par un gentilhomme-verrier qui s'entourait d'une équipe importante au sein de laquelle la division du travail devint plus poussée avec les cueilleurs de verre, les souffleurs, les tailleurs à chaud et à froid sans oublier les différents façonneurs de pierres artificielles et la masse des manoeuvres chargés de l'approvisionnement et de l'entretien du four. Alors des établissements vont employer plusieurs dizaines de personnes et fournir une production diversifiée où la création de modèles n'est pas absente comme chez les verriers normands ou lorrains qui acquirent une grand célébrité et qui n'eurent jamais de difficultés pour se faire confirmer leurs privilèges; chez tous l'empirisme joua pourtant encore longtemps un grand rôle.

S'il y avait parfois encore une certaine mobilité du maître-verrier quand il n'était pas propriétaire de son établissement, il s'engageait de toute façon à rester sur place le temps d'une campagne, c'est-à-dire semble-t-il un temps déterminé par la durée des réfractaires du four ou par les possibilités d'approvisionnement en matières premières, essentiellement les cendres végétales entrant dans la préparation de la pâte. Dans les campagnes normandes une période de 9 mois à un an correspond à la durée de chauffe d'un four.

Certaines de ces verreries que l'on peut dire moyennes sont devenues célèbres au moins quelque temps, comme celle du sieur Gaspard Thévenot de Folembray en Laonnais: il fonda une verrerie en 1705 et obtint un tel succès avec la fabrication de bouteilles que son entreprise se développa dans les deux siècles suivants en produisant des cloches de jardiniers et plus tard des isolateurs de verre. Quelques cristalleries du plateau gréseux lorrain eurent ces origines modestes avant de gagner leurs lettres de noblesse ou bien d'autres verreries produisant du verre plat selon la méthode du manchon dite aussi lorraine bien qu'imitée de l'Allemagne ou de la Bohême.

Les grandes verreries furent le plus souvent à cette époque des Manufactures royales qui obtinrent du roi lui-même leurs privilèges et en particulier le monopole de telle ou telle fabrication pour concurrencer une production étrangère que la clientèle aisée faisait venir jusqu'alors du pays d'origine. De grands personnages du Royaume comme les Brossard, les Brucher, les Coqueray, souvent nobles puisqu'ils ne dérogeaient pas en devenant verriers, s'attachaient des maîtres dans cet art prestigieux et fondaient leurs établissements dans la capitale ou à proximité grâce aux grandes forêts qui l'entouraient et qui avaient déjà leur réputation pour les matériaux indispensables qu'elles fournissaient. Par exemple le 14 décembre 1688, le Roi accorda à un avocat "bourgeois de Paris", Thévart, un privilège de trente ans pour faire des grandes glaces en coulant le verre au lieu de le souffler. Ce spéculateur voulait avec des financiers exploiter la découverte faite en 1687 par un verrier orléanais, Bernard Perrot, du procédé de fabrication des glaces coulées. Il demande à Louis Lucas de Nehou, neveu de Richard de Nehou dirigeant la glacerie de Tourlaville près de Cherbourg de prendre la direction de son établissement du faubourg Saint-Germain qui devient en 1693 la "Manufacture royale des Grandes Glaces". A cette date Thévart chercha un nouvel emplacement pour monter un autre établissement destiné au coulage de ses glaces: il jeta son dévolu sur l'ancien château des sires de Coucy en pleine forêt de Saint-Gobain s'étendant sur 5000 ha; le doucissage et le polissage restaient à Paris. Après bien des péripéties et des changements de mains au cours du XVIIIè siècle, la société qui avait été créée maintint après la Révolution non sans difficultés son monopole et le chauffage au bois alors que le charbon avait fait son apparition dans les autres verreries. En 1819 la Compagnie de Saint-Gobain achetait encore des forêts pour être sûr de son approvisionnement en combustible; elle ne vint à la fusion au charbon qu'à partir de 1829, l'affinage se faisant toujours au bois dans un autre four. C'est seulement en 1852 que les deux opérations se firent à l'aide de la houille dans un même four: bel exemple de pérennité des méthodes anciennes.

Deux autres verriers parisiens, Lambert et Voyer, obtenant le privilège exclusif de la fabrication du cristal au plomb (flint glass) sur le modèle anglais, avaient installé en 1783 leur cristallerie dans le parc de Saint-Cloud. Bénéficiant du patronage de Marie-Antoinette, ils donnent à leur établissement en 1784 le nom de "Manufacture des Cristaux et Emaux de la Reine" et font venir d'Angleterre souffleurs et tailleurs de verre. En 1785 des hommes d'affaires, prenant le contrôle de l'entreprise mal gérée, décident de la transférer au Creusot près du charbon de Montcenis jugé indispensable pour chauffer la pâte à l'oxyde de plomb (Plate 27.1). Bien qu'ayant survécu à la Révolution, l'affaire donnant des signes de faiblesse change de mains et est finalement rachetée par des verriers de Saint-Louis et de Baccarat pour être arrêtée en 1832: la concurrence était devenue très vive et la cristallerie ne semble pas s'être intégrée au milieu régional comme dans les cas précédents d'établissements très liés aux ressources du terroir:

Plate 27:1 L'un des fours coniques de la verrerie de la "Manufacture des Cristaux et Emaux de la Reine" transférée au Creusot en 1785. Cliché J. Pinard.

c'est à partir de cette date que les Forges du Creusot prennent l'essor que l'on connaît entre les mains de leurs nouveaux propriétaires, les frères Schneider.

Comme dans d'autres régions européennes (Bohême, Småland en Suède), des entreprises déjà anciennes se sont développées au cours du XIXè siècle et ont essaimé dans plusieurs établissements de quelques dizaines et parfois centaines d'ouvriers chacun, conférant à leur société une taille importante tout en demeurant très fortement implantée dans le monde rural. C'est le cas de la Compagnie de Vallerystal et de la Plaine de Walsch en Lorraine. Dès le XVIè siècle quand le "nomadisme" des verreries cessa dans cette région, les usines devinrent plus grandes, le maître-verrier recruta une main d'oeuvre temporaire de journaliers, de commis des environs pour assurer les travaux annexes comme la recherche du combustible et de la matière première, l'entretien du four, les transports. Ces établissements arrivaient ainsi à faire travailler de nombreux paysans, ne serait-ce que temporairement.

Quelquefois la verrerie-mère créait des filiales placées entre les mains des nombreux enfants qu'ont toujours eus ces familles de verriers très prolifiques et où il fallait subvenir aux besoins de tous. Le maître-verrier devait ainsi se livrer à de nombreuses occupations en dehors du travail du verre proprement-dit sans parler de la culture des terres qui lui étaient concédées; il devait s'occuper de l'organisation des campagnes de chauffe, de la gestion de l'établissement, de la recherche des débouchés et de moyens de transport toujours délicats; il accompagnait parfois ses démarcheurs pour collecter des commandes ou recueillir des idées afin d'orienter sa production selon les désirs de sa clientèle; il traitait aussi de questions financières et cherchait des capitaux pour investir dans ses nouvelles entreprises.

La multiplication des verreries entraîna parfois la constitution de véritables alignements d'usines dans certaines régions comme l'Argonne avec le Val de Biesme en amont des Islettes. Les établissements attiraient les populations par suite des possibilités de travail et des exemptions fiscales qu'ils leur offraient. Dans la Vôge où la dissémination est plus fréquente on parle de groupements de "granges et verreries"; ces fabriques concentrent nombre d'autres activités industrielles: forges, tuileries, scieries, papeteries et toutes sortes de moulins. Il en existe aussi dans l'Ouest comme en Vendée auprès de la mine de charbon de Faymoreau (Plate 27.2).

Dans le sud de la Suède les premiers verreries naquirent à la fin du XVIIè et au début du XVIIIè siècle sur le plateau cristallin du Småland dont l'épaisse couverture forestière et les sables fournissaient respectivement le combustible et la matière première. Si beaucoup devaient aussi leur venue à des seigneurs locaux, ces verriers étaient souvent des immigrés originaires des pays germaniques qui introduisirent dans la péninsule scandinave leurs techniques et leur art, en particulier la taille du cristal dans le style de Bohême. Quelques verreries plus récentes et plus importantes eurent aussi pour origine des forges (Orrefors) qui avaient été créées pour mettre en valeur les ressources minières et les chutes des cours d'eau de la région: ces forges constituaient

Plate 27:2 La verrerie de Faymoreau en Vendée comprenait deux vastes halles: il n'en subsiste plus qu'une. Elle a fonctionné de 1838 à 1880 avec le charbon extrait de la mine voisine. Cliché J. Pinard.

aussi un moyen d'existence pour les habitants de la contrée peu fertile pour les cultures. C'est la concurrence des gisements plus riches du centre (Bergslag) et du nord qui amena l'arrêt des hauts-fourneaux et la transformation de quelques établissements métallurgiques en verreries.

Ainsi des verreries de dimensions moyennes succédèrent-elles fréquemment à de petits ateliers anciens et se maintinrent à côté des grandes verreries dont les produits n'étaient pas concurrents: elles travaillaient conjointement dans un esprit différent mais où l'on retrouve toujours certains traits provenant de leur passé rural.

III - LES SURVIVANCES DU PASSE DANS L'INDUSTRIE MODERNE DU VERRE

Avec les transformations des deux derniers siècles les verreries ont eu tendance, comme les autres établissements industriels, à se concentrer en grandes sociétés mais nombre d'entre elles ont gardé de leurs origines rurales des caractères, ne serait-ce que leur implantation et leur état d'esprit profondément marqué par l'individualisme.

Le recours volontaire ou obligé au charbon à la suite des déboisements intempestifs des siècles précédents a conduit de nombreuses verreries à s'implanter sur les gisements houillers ou près des ports par lesquels était importée la houille anglaise, mais même dans ce cas elles recherchaient des sites isolés comme sur la bordure orientale du Massif Central autour de Rive-de-Gier. Dans la région du Nord des verreries "populaires" subsistèrent très longtemps en Thiérache autour de Sars et les ententes avec les milieux ouvriers furent très tardives au cours du XIXè siècle.

Les progrès techniques dans le coulage des glaces avec la coulée continue ou dans la chimie de la soude accélèrent les absorptions sans toutefois renier les premières localisations: Saint-Gobain créa en 1816 une soudière à ses portes dans l'ancienne verrerie de Charlesfontaine et en 1822 à Chauny sur la voie navigable pour produire de la soude artificielle. L'adoption du four Siemens à production continue et à récupération des gaz permet la fourniture de verre à vitre homogène et de grandes dimensions: les demeures urbaines et rurales vont très vite l'adopter en remplacement du verre en "manchon" ou à cylindre qui avait lui-même succédé au verre dit "à boudine" surtout que l'emploi de la soude artificielle et la fabrication continue en abaissent considérablement le prix de revient tout en augmentant sa qualité. Le nombre des verreries s'accroît au cours du siècle.

La découverte de la machine à souffler le verre et sa mise au point par Claude Boucher à Cognac en 1894 sont suivies de son emploi immédiat dans les grandes verreries de bouteilles (Carmaux, Denain...). Les grandes régions de vignobles comme la Champagne y eurent très vite recours pour développer leurs ventes car de nombreux souffleurs avaient disparu lors du premier conflit mondial; même de petites verreries isolées de campagnes peu industrialisées en firent un grand usage pour les nombreux produits qui commençaient à se vendre en récipients de verre (huiles, parfums...).

De leur passé campagnard les verriers ont gardé des traits psychologiques et sociaux: descendants de gentilshommes-verriers ils ont eu longtemps un fort esprit indépendant et le goût des privilèges: très jaloux des secrets de fabrication qu'ils se transmettaient de père en fils et qui tenaient plus de l'empirisme que d'une véritable science, ils ont toujours voulu se tenir à l'écart du reste de la main d'oeuvre industrielle et leur résidence rurale n'a fait que renforcer cette mentalité et ce désir de toujours se tenir à l'écart des autres salariés.

Les verriers possédaient par ailleurs une petite exploitation sur laquelle ils pouvaient vivre en cultivant quelques légumes et en élevant quelques bêtes et qui les mettait à l'abri des fluctuations de la conjoncture si fréquentes dans une industrie de luxe comme la cristallerie où ils ont été longtemps mieux rémunérés que les ouvriers des autres branches industrielles. De plus les grandes manufactures ont souvent construit dès le XVIIIè siècle des logements pour leur personnel afin d'accueillir des célibataires et des ménages d'immigrés et de les avoir toujours sous la main en particulier pour les coulées de nuit: la main d'oeuvre en a acquis et gardé un certain esprit de l'entreprise lié au paternalisme des dirigeants qui versaient comme à Saint-Gobain des indemnités de maladie et des pensions d'invalidité à leurs salariés dans la nécessité. L'embauche en était facilitée et la stabilité du personnel s'en trouvait améliorée. Celle-ci semble toujours avoir été dans la verrerie, d'après une étude sur l'industrie du verre en Suède, assez forte: sept ans en moyenne pour les ouvriers professionnels et un an et demi à six ans pour les manoeuvres au milieu du siècle dernier tandis que les ouvriers-paysans travaillaient de 150 à 200 jours par an à l'usine et consacraient le reste de l'année aux travaux des champs.

De leurs lointaines origines les verriers ruraux ont conservé aussi des habitudes d'autonomie: isolés dans leurs forêts ils devaient se suffire à eux-mêmes, non seulement pour leurs fabrications trouvant sur place ou dans un périmètre assez restreint matière première, énergie et main d'oeuvre, mais aussi les moyens de transport pour porter sur des marchés parfois éloignés les produits fragiles fabriqués: animaux de bât, chariots bateaux étaient alors utilisés; enfin indépendants ils l'étaient devenus en cultivant les terres que leur avait données leur seigneur à défricher, en élevant quelques bêtes pour les charrois et en se défendant eux-mêmes contre les brigands dans ces régions de forêts frontalières très convoitées par les voisins.

Cet esprit d'autonomie ils le garderont jusqu'à une époque récente en refusant de venir s'installer dans une ville ou même un village quand le pouvoir local - duc, évêque, abbé... - lè leur demandera, en s'efforçant de rompre tout lien avec ceux qui avaient été à l'origine leurs protecteurs et qui leur avaient donné terres, privilèges et parfois argent, et en restant en dehors de tout mouvement corporatif ou syndical, gardant dans les entreprises les plus modestes une mentalité d'artisan et quelquefois d'artiste réalisant de ses propres mains un objet d'art qui saura être apprécié d'une clientèle choisie de connaisseurs.

Ainsi non seulement les verreries qui sont restées dans les campagnes et qui se sont spécialisées dans la production d'objets d'art mais aussi de nombreuses autres devenues urbaines ont-elles conservé presque jusqu'à nos jours des caractères, des aspects de leurs ancêtres vivant et travaillant de façon indépendante et isolée, même si les modes de fabrication ont abandonné tout empirisme et ont considérablement évolué dans les cent dernières années. De profonds atavismes demeurent dans la profession.

De toutes les industries rurales qui animaient dans les siècles passés la vie de nos campagnes les verreries sont les seules qui aient survécu en si grand nombre, certes en se modernisant mais en conservant beaucoup de caractères, de tours de main des activités anciennes. A côté de quelques grosses unités de production fournissant de la verrerie courante ou industrielle qui ont pu aussi bien s'implanter en milieu urbain qu'en milieu rural, des établissements de dimensions moyennes ou petites ont choisi de rester dans le second cadre, en adoptant toutes les techniques du premier, parce qu'ils y retrouvaient les conditions de développement qui avaient favorisé les premières industries du verre (main d'oeuvre stable et consciencieuse, espaces vastes et calmes, activités complémentaires) et qui leur permettent aujourd'hui de conserver une production de très haute qualité indispensable au renom de toute entreprise touchant aux métiers d'art.

BIBLIOGRAPHIE SOMMAIRE

BARRELET J. 1934. La verrerie en France de l'époque gallo-romaine à nos jours. Larousse. Coll. Arts, styles et techniques 1934 208 pp.

BONDOIS P.M. 1936-7. 'Le développement de l'industrie verrière dans la région parisienne de 1515 à 1665.' Revue d'Histoire économique et sociale 1936-7 XXIIIè année, 49-72

BONDOIS P.M. 1936-7. 'Le développement de la verrerie française au XVIIIè siècle.' Revue d'Histoire économique et sociale 1936-7 XXIIIè année, 237-261 et 333-361

CHAMBRON R. La verrerie occidentale du VIIIè au XVè siècle. Annales du premier Congrès des "Journées internationales du Verre" Liège 20-24 août 1968 p. 97-113

NORDSTRÖM Olof 1962. Svensk glasindustri 1550-1960. Lokaliserings- och arbetskraftsproblem (L'industrie du verre suédois 1550-1960. Problème de localisation et de main d'oeuvre). Meddeland från Lunds universitets Geografisk instution. Avhandlingar 41 Lund 1962 176 pp. Résumé en allemand.

PIGANIOL P. 1965. Le verre, son histoire, sa technique. Hachette 1965 261 pp.

ROSE-VILLEQUEY G. 1970. Verre et verriers de Lorraine au début des temps modernes. Paris-Nancy 1970 908 pp. Nombreuses planches.

ABSTRACT

Glassworks are an old rural industry manufacturing luxury products for wealthy customers: window-glass, flasks, goblets...

They were established on the outskirts of forests which provided the firewood; the raw materials, sand, soda, limestone, were fairly common throughout the country. The landowners attracted the glass-makers who thus gave more value to their estates. The lay and ecclesiastical lords granted them privileges, such as that of ennoblement.

Gradually the "large" glassworks developed and were distinct from the "small" glassworks which had limited means and often made more common objects. The large glassworks set up in the seventeenth century were royal factories with a monopoly of one product such as, for example, mirrors. They required considerable investment and resources. Some of them are still famous, for instance those of Saint-Gobain, Sèvres-Le Creusot or those of Lorraine, which copy foreign products such as glass from Bohemia or England.

Today the modern glass industry has retained some features of those ancient activities, such as their rural location and distinctive attitudes: an independent spirit, the feeling of belonging to a special class of workmen and the will to remain aside from other social movements.

RESUME

Les verreries sont une ancienne industrie rurale fournissant des produits de luxe à une clientèle aisée: verre à vitre, flacons, gobelets...

Elles étaient installées en bordure des forêts qui produisaient le bois de chauffage; les matières premières, sable, soude, calcaire, étaient assez répandues dans les campagnes. Les propriétaires fonciers attiraient les maîtres-verriers qui valorisaient ainsi leurs domaines. Les seigneurs laïcs ou ecclésiastiques leur concédaient des privilèges dont celui de la noblesse.

Peu à peu de "grandes verreries" se distinguèrent des "petites verreries" qui étaient au début itinérantes, avaient des moyens limités et fabriquaient souvent des objets plus usuels. Les verreries créées à partir du XVIIème siècle étaient des manufactures royales avec le monopole d'une fabrication, comme par exemple celle des glaces, et exigeaient de gros soutiens financiers. Certaines sont restées célèbres comme celles de Saint Gobain, de Sèvres-Le Creusot ou de Lorraine, imitant des productions étrangères comme les cristaux de Bohême ou d'Angleterre.

Aujourd'hui l'industrie moderne du verre a conservé certains traits de ces anciennes activités tels que leur implantation rurale à côté de grandes créations urbaines, et un certain état d'esprit: une mentalité indépendante, le sentiment d'appartenir à une catégorie particulière d'ouvriers et la volonté de se tenir à l'écart d'autres mouvements sociaux.

CHAPTER 28

LINKS BETWEEN AGRICULTURE, COAL MINING AND POPULATION IN COUNTY DURHAM, WITH PARTICULAR REFERENCE TO THE NORTH DURHAM COALFIELD, c. 1550 - 1810.

R.I.Hodgson
(University of Manchester)

> "... the great obstacle to the improvement of agriculture or the cultivation of the surface of the county...arises from the immense wealth which has been, and still is, derived from the minerals beneath. No other county is so interwoven with a network of public and private railways. In no other is there so large a quantity of land occupied by collieries, manufactories, quarries, waste heaps, &c.; and in none, perhaps, is there a more extensive trade carried on in the working, manufacturing, and sale of coal, limestone, ironstone, and lead, - all inducing a sacrifice of the interests of the surface for the sake of what lies below it."
>
> (T.G. Bell, A Report on the Agriculture of County Durham, 1854).

From the middle of the sixteenth century to the opening decades of the twentieth, a period of almost 400 years, North-east England (essentially the traditional counties of Durham and Northumberland) can be said to have played a prominent role in Britain's economic and social development. At the time of Bell's report in 1854 it stood on the threshold of perhaps its greatest industrial achievements; for during the closing decades of the nineteenth century, the region was better placed than any other to serve our basic needs in coalmining, shipbuilding, engineering and iron and steel manufacture. Sadly, however, the shortcomings of this narrow-based economy became all too evident in the Depression years of the 1930s when no other region suffered greater levels of unemployment and social deprivation. Today, after more than 40 years of government experimentation with regional aid programmes (programmes which have done little more than remove some of the worst excesses of industrial dereliction), many of the scars of depression remain in the economy and landscape and live on in the memory to help fashion current attitudes and preserve what traditionally has been a strong sense of local identity. Little wonder that the North-east has been characterised by the popular press as the most regional of all English regions, inhabited predominantly by cloth-capped coalminers, hard-swearing shipyard workers and ale-drinking football fans[1]

Whether as a spectre to haunt the landscape and mind of twentieth-century man, or as a more tangible experience of the mid-nineteenth century, as depicted by Bell, industrialization is apt to erase from our thoughts all but the most distant echoes of what the process of modernization owed to the physical and human resources of the countryside. Indeed the image of an agrarian economy sacrificed on the altar of industrial capitalism, of a rural society and landscape progressively submerged beneath industrial spoil and the 'barrack rows of industry' is at once compelling and attractively simple. It is, of course, also misleading. As in other areas of early and prolonged industrialization, the industrial growth of North-east England was continuously nourished and sustained by having one of its main tap-roots embedded beneath the fields of the countryside, embedded in a few rich seams of coal that were to be so fundamental to Britain's economic well-being that Nef could argue that if one understood coal, one understood the industrial revolution. This is to suggest that coal mining was not only a rural industry but embraced, within a web of forward, backward and cross linkages, a whole range of variables, that had important implications for a region's population: its size and composition; its means of earning a living, of being housed and fed; its impact on the landscape. Thus, while the actual existence of links between agriculture and industry, countryside and town, may be well known, if seldom remarked upon, the precise nature of these relationships, and how they changed through time and space, was highly complex and is far from being properly understood. It is worthy of further investigation for the very good reason that it touched the lives of many thousands of ordinary families. In the following survey a framework is provided within which some of the relationships between agriculture, industry and population are being explored. Attention focuses upon the main area of County Durham to be affected by industrialization prior to the opening up of the concealed coalfield of east Durham and the widespread growth of a railway network from the 1830s - an area designated 'North Durham coalfield', an area which Defoe, in the 1720s, simply but grandly called 'Coal Country':

> "When we see the colliers crowding into the Thames, we wonder how it is possible for them to be supplied, and that they do not bring the whole Coal Country away; yet when in this Country we see the prodigious Heaps, I might say Mountains of Coals, which are dug up at every Pit, and how many of those Pits there are, we are filled with equal Wonder to consider where the People should live who consume them"

London did indeed consume vast quantities of coal from the middle years of the sixteenth century and most of it came from the Northumberland and Durham coalfield. North-east coal also found its way to other south and east coast markets and, despite frequent disruptions to shipping, some 'sea-sale' coal, often of an inferior quality, was sent abroad. Much poor quality coal was also sold locally in the 'land-sale' trade. The general patterns and processes of industrialization that are thought to underlie these trading relations have been identified elsewhere and are relatively well known. In comparison, the likely chronology of population change and the mechanisms underlying it have been considered only recently and much remains to be done. Exploring the two themes, industry and population, in parallel (Fig 28:1) provides a new way of describing the main trends on the North Durham coalfield, though at the risk of advocating a simple 'cause and effect' explanation of change, and leads inevitably towards a consideration of the role of agriculture.

In 1563 the population of County Durham probably numbered about 35,000 (or 8,500 households), only slightly more than one-quarter the figure recorded at the first official census of 1801. Its distribution still showed a striking relationship to the physical geography of the area, especially its relief. With fewer than ten inhabitants for every 100 acres in the lowlands to the east and south, and fewer than five in the uplands to the north and west, it was still a county which was sparsely settled and markedly rural in character. There were no great towns; only Durham and possibly Bishop Auckland had more than 2000 people. In the Tees Basin and Wear Lowlands and on the East Durham Plateau were nucleated villages with their open fields, often three in number and worked under communal rules. On the High Pennines and Pennine Spurs, unenclosed wastes, fells and forests still dominated the landscape. However, disruptions to this ancient pattern of economic life were afoot; lead mining was already of importance in the Pennines and the 'sea-sale' collieries that were to experience such revolutionary growth in the next few decades were beginning to make an impact on the south bank of the Tyne and, in lesser measure, along the middle Wear valley.

Between 1563 and 1674, the population of the county increased by upward of 70 per cent. However, growth was not spread evenly. In South-east Durham there were some settlements which were already small and which suffered a decline. On the North Durham coalfield, the overall increase in population was probably in the order of 138 per cent. Some parishes, which included small, newly-founded industrial settlements as well as large, old-established centres, witnessed the most dramatic increases of all. Whickham parish, scene of the most intensive mining activity in the seventeenth century, and which was producing 100,000 tons of coal per annum by the 1630s, appears to have trebled its population. Here, as elsewhere, growth appears to have resulted from a combination of high rates of natural increase and in-migration, though precisely what weighting should be attached to each is impossible to say. Folk drawn from Scotland and the Border dales came first as seasonal migrants but later settled in keelmen's colonies along or near the banks of the Tyne at Dunston and Swalwell. South Shields, which also trebled its numbers between 1563 and 1674, became increasingly important

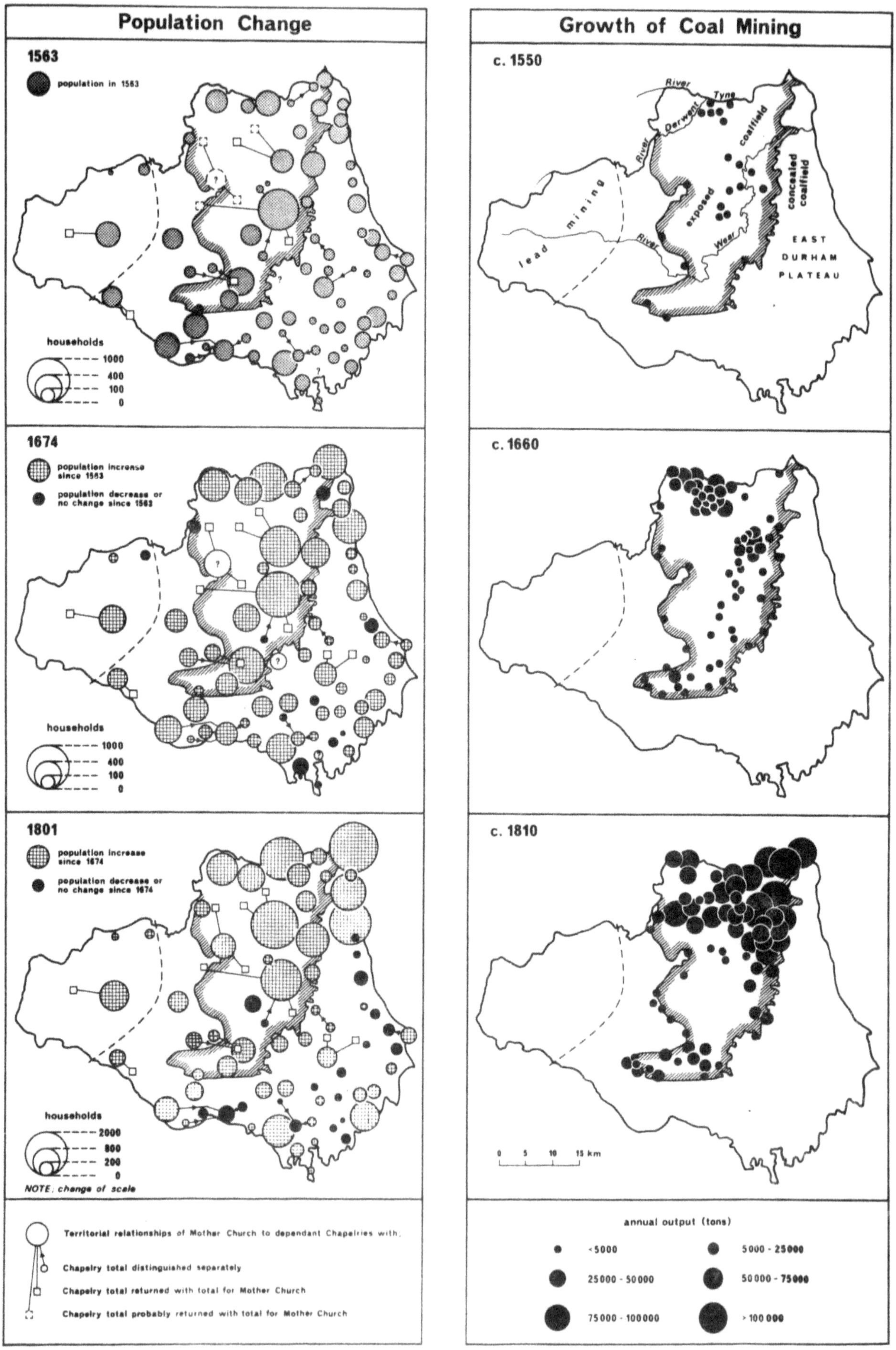

Fig. 28:1 County Durham: Population change and growth of coal mining.

as a shipping haven and centre of salt manufacture. Its experience serves as a reminder that local supplies of cheap coal were a powerful encouragement to the growth of other industrial enterprises. At the mouth of the Wear too, there was salt production and shipbuilding, and in addition, Sunderland and Bishop Wearmouth had a glass and pottery industry. Sunderland is said to have taken advantage of Newcastle's misfortunes in the Civil War to double coal exports from the Wear valley in a matter of 25 years. Certainly the average annual shipments of about 125,000 tons in 1660 were a great advance on the mere 12,000 at the beginning of the century. Although it retained the greatest share of North-east exports, the Tyne valley (drawing upon the collieries of both the Northumberland and Durham shores) only doubled its output over the same 50 years; from about 240,000 tons in 1610 to 480,000 by 1660.

All the industrial changes outlined above were on such a scale and represented such a departure from earlier times that Nef regarded them as constituting an early industrial revolution in North-east England, and indeed in the nation as a whole. Had he been fully aware of the accompanying demographic events that are only now coming to light, he would doubtless have felt even more strengthened in his belief. It is wise perhaps to show some sympathy with those of his critics who maintain that the changes he chronicled for the late sixteenth and seventeenth centuries were far outshone by those of the late eighteenth and nineteenth, but it would be foolish to underestimate the impact of proto-industrialization on the local landscape and the lives of the people. To appreciate this point fully, requires an exercise in imagination as we join Nef's seventeenth-century observer, seated upon a hillside behind Newcastle, surveying the transformed landscape:

> "Picture the mouth of the muddy, narrow river Tyne, jammed with four or five hundred keels and two or three hundred ships, all specially constructed to carry coal; think of the hilly slopes to the north and south covered with hundreds of small carts and wagons, leaving behind them trails of black refuse on the green countryside; and then think of a time when this same countryside was at rest except for the occasional movement of some husbandman driving his mule to Newcastle for corn, when the only evidence of the coal industry was a few pits at the water's edge by the hamlet of Gateshead, and a few ships, whose masters took on board a lading of coals to ballast their cargoes of fish, grindstones, or jersey cloths. In this comparison you have the contrast between an output of about 40,000 tons and one of about 800,000 tons, and a view of the changes wrought

> around the town of Newcastle in the century following the accession of Elizabeth".

A reflective observer two or three generations later might have been equally astonished at the scene confronting him in the North Durham landscape: indeed he may have had difficulty in absorbing all the changes that had occurred in his own lifetime, and he would certainly have needed to cast his eyes far and wide to take in all that had happened. Between 1674 and 1801, the population of County Durham as a whole grew by about 162 per cent, that of North Durham alone by 222 per cent. With the exception of salt manufacture, those industries which had taken advantage of cheap supplies of coal in the seventeenth century (glass, pottery and shipbuiding) continued to expand and flourish. Of particular interest was the founding of ironworks at Winlaton and Swalwell by Ambrose Crowley: the Ryton parish registers show that he employed German swordmakers. Coal output probably trebled over the period despite restrictive practices by combinations of coal-owners and mounting problems of drainage, ventilation and wayleaves; problems in whose solution, some might argue, lay the foundations of Britain's industrial greatness.

By about 1700, the honeycombed landscape by the banks of the Tyne and Wear signalled the end of mining in the most easily accessible coal seams. Thereafter the coal-owners were faced with two choices: to sink deeper or fan out laterally. Neither course of action was easy, both were attempted. The extension of mining activity away from the river banks involved the costly business of wayleaves (charges levied on the carriage of coals or other goods across another person's land) and the introduction of a method of transportation that was superior to cars and wains trundling along rough trackways. The waggonways provided a suitable alternative, but only after considerable feats of engineering (including construction of the famous Causey Arch) and a judicious picking of routeways that minimized ill-feeling and misfortune among the coal-owners. "Could we have its history writ by a good pen wd outdo Don Quixote", wrote the coal-owner, James Clavering, of the peculiarly ill-fated Bucksnook waggonway. "The new Waggonway...crosses so many Mountains & Vales...that I can compare it to nothing more properly than ye Via Appia", boasted William Bowes of his Western Way in 1721. Waggonway development reached its greatest extent in the uplands of North-west Durham with some routes stretching eight or ten miles from the riverside staithes. Nowhere can the network have had a more telling impact than on Tanfield Fell, which in 1761 (Fig 28:2) preserved the essential features that had been forged a generation earlier. Several hundreds of people worked on the unenclosed fell in the course of the eighteenth century and some appear to have founded new settlements at Tantobie, Clow Dean, Whiteley Head and Ewhurst Head.

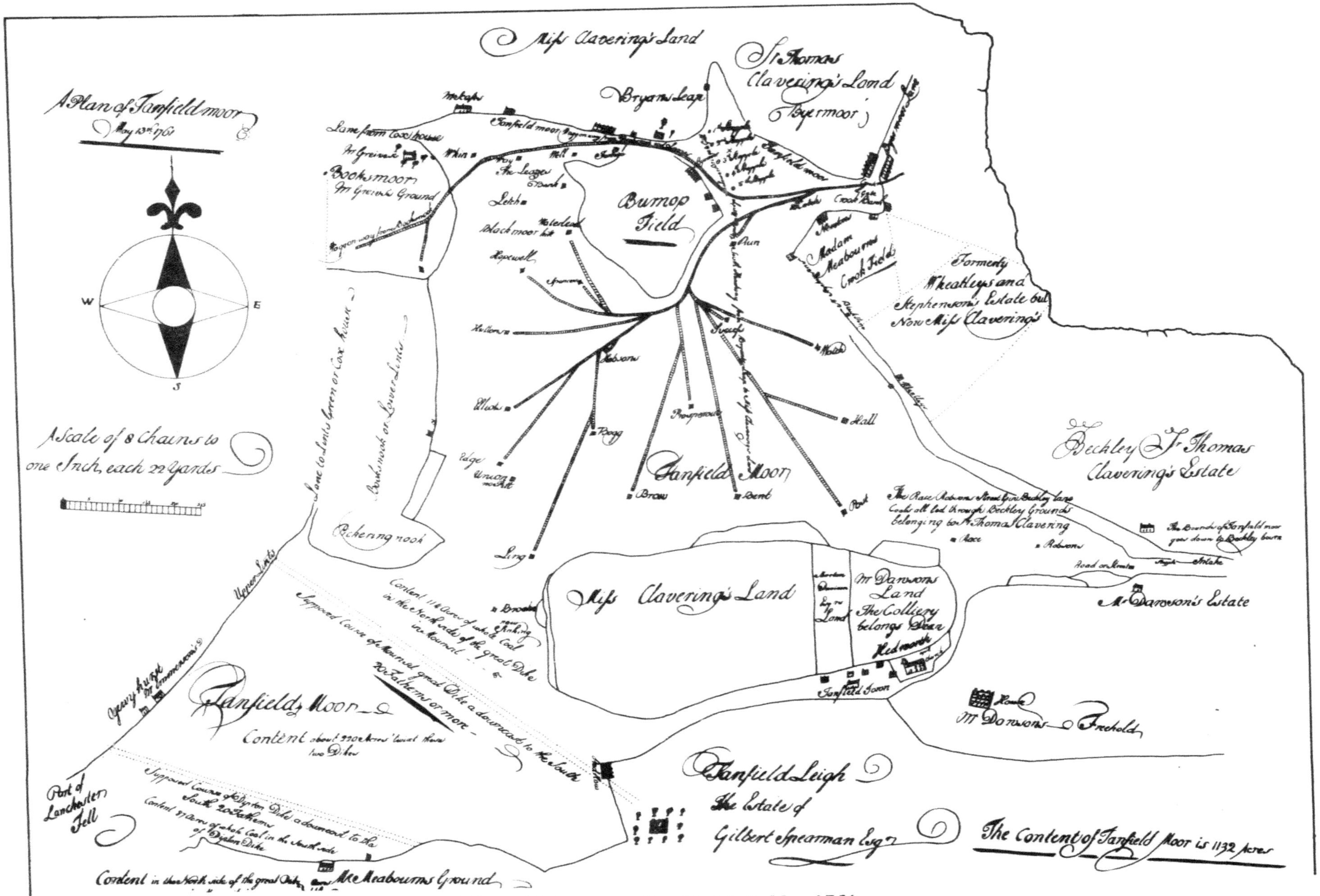

Fig. 28:2 Tanfield Fell: 1761

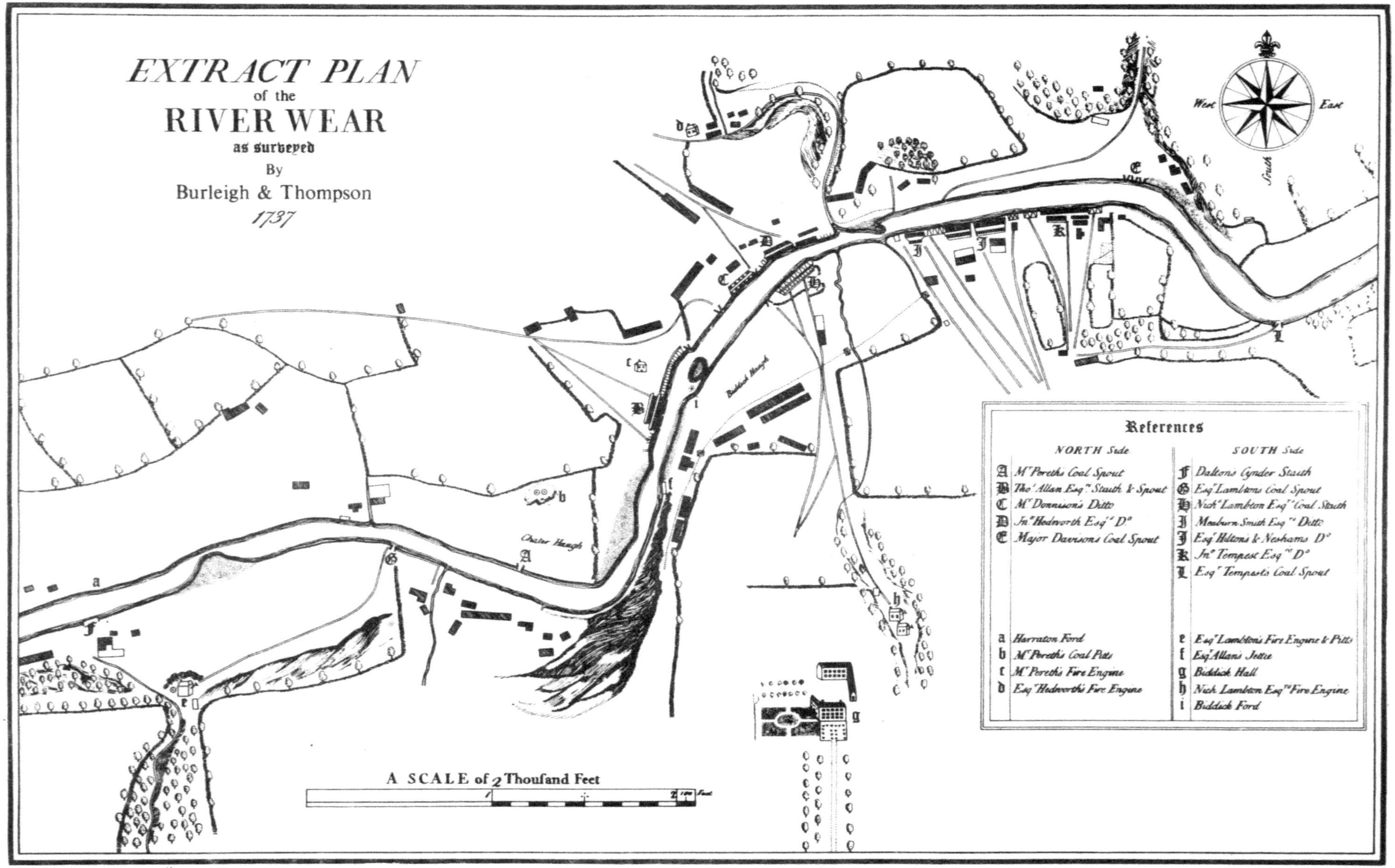

Figure 28:3 Extract plan of the River Wear as surveyed by Burleigh and Thompson, 1737.

Like waggonway development, deeper sinking was a difficult and expensive undertaking. It required the removal of vast quantities of water from the mines by the application of steam power. Thomas Newcomen's atmospheric engine, or 'fire' engine as it was commonly called, may have been employed as early as 1715 in the North Durham coalfield and by 1737 several were evidently in use at the Wearside collieries. (Fig. 28:3). Their curiously laborious antics must have fascinated a contemporary observer, as would the intense activities along the waggonways and at the staithes. Unlike the waggonways, which affected substantially the colliery location pattern prior to 1750, steam pumping only became widespread after that date. Consequently, in the course of the eighteenth century, the major focus of mining activity shifted from the western part of the North Durham coalfield to the deeper seams in the eastern part. There was probably a movement of people, too, from parishes where mines were closing to those where mines were about to open and new settlements sprang up as at New York (1761) and New Winning (1782) in Washington parish.

Consideration of the patterns of industrial and demographic change outlined above, leads to an inescapable conclusion: that extra people meant extra mouths to feed. Moreover, the numbers earning a living in the non-agricultural sector of the economy were growing at a faster rate than the numbers engaged solely in farming. How was this population to be fed? An increase in local supplies of animal products might have been expected to come with relative ease, in view of the traditional pastoral bias of North-east agriculture, but what of grains? On first impression, the Port Books and other shipping records, suggest the answer: a pattern of exchange in which the region exported vast quantities of coal and imported cereals and other foodstuffs. A simple enough arrangement but one which could cause a crisis at times of low coal prices and high grain prices, as, for instance, in the late eighteenth century, and earlier, in 1709:

> "The prices of working and drawing (coals) are considerably advanced on account of the price of corn which is double the price at least of former years and it cannot be supposed that a man's family can be mentained by less than formerly, the Price being double, he accordingly sets a value upon his Laboure and all this goes out of the Coalowner's Pocket."

In other years, too, there must have been pressure on local agriculture to be more productive and as far as can be ascertained, it made appropriate response. Some elementary calculations suggest that even in years of high grain imports, the majority of the population was still fed on locally grown corn, and it is clear that at times in the eighteenth century the North-east managed to produce a surplus for export.

Apparently this was achieved by the introduction of some revolutionary changes in agriculture which involved the abandonment of permanent tillage and permanent pasture for the cropping practices of convertible husbandry: practices that were flexible enough to accommodate the somewhat inhospitable physical environment of much of the region; practices that had been encouraged by the lowland enclosures of townfields in the late sixteenth and seventeenth centuries. It remains to be discovered where in the two counties such agricultural advances were most in evidence but it is not difficult to imagine that the greatest improvements were made in the lowlands and along the coastal plains, away from the then exploited coalfield. But what improvements could be made where agriculture had to compete increasingly, for land and capital, with the coal industry? From what has been said above, with regard to industrial activities along the Tyne valley and upon Tanfield Fell, and from the comments of Defoe in the eighteenth century and Bell in the nineteenth, it would seem that in questions of land-use priorities agriculture was inevitably the loser. One further, striking example will reinforce the point.

In 1620 a case was brought before the Court of Common Pleas in Durham. The copyholders of the township of Whickham, the most intensively mined area of the North Durham coalfield in the seventeenth century, sought redress from the coalmining adventurers whose colliery workings had laid to waste more than a third of their lands. It was claimed that corn and meadow lands were ruined by coaldust and 'cankered water', that the church and dwellings were in a state of collapse because of subsidence, and that so many pit shafts remained open and unfenced that "divers of the King's subjects have casually in the night time fallen into them and there perished". Passing judgement in the case, Sir Richard Hutton, Chancellor of the County Palatine of Durham, expressed both sympathy and general agreement with the copyholders; but his ultimate concern was that the coal trade should not suffer on account of its national importance: "The uttering and vending of coles from thence (Whickham) is become a matter of great necessitie and much concerninge the generall good of the kingdome". And so it was, that 30 years later (Fig 28.4) coal workings were still impeding agricultural practices in the townfields.

A complete contrast is afforded by events in the neighbouring parish of Gateshead in the early eighteenth century. They demonstrate how the successful marriage of farming and mining interests could be achieved through the sheer initiative and business skill of an enlightened landlord. William Cotesworth began life as a yeoman's son in Teesdale c.1668 and became a tallow chandler when he moved to Gateshead. In a matter of 15 or 20 years he developed into a leading politician of the Whig cause and an entrepreneur of the first rank. He had an annual turnover of £30,000, largely

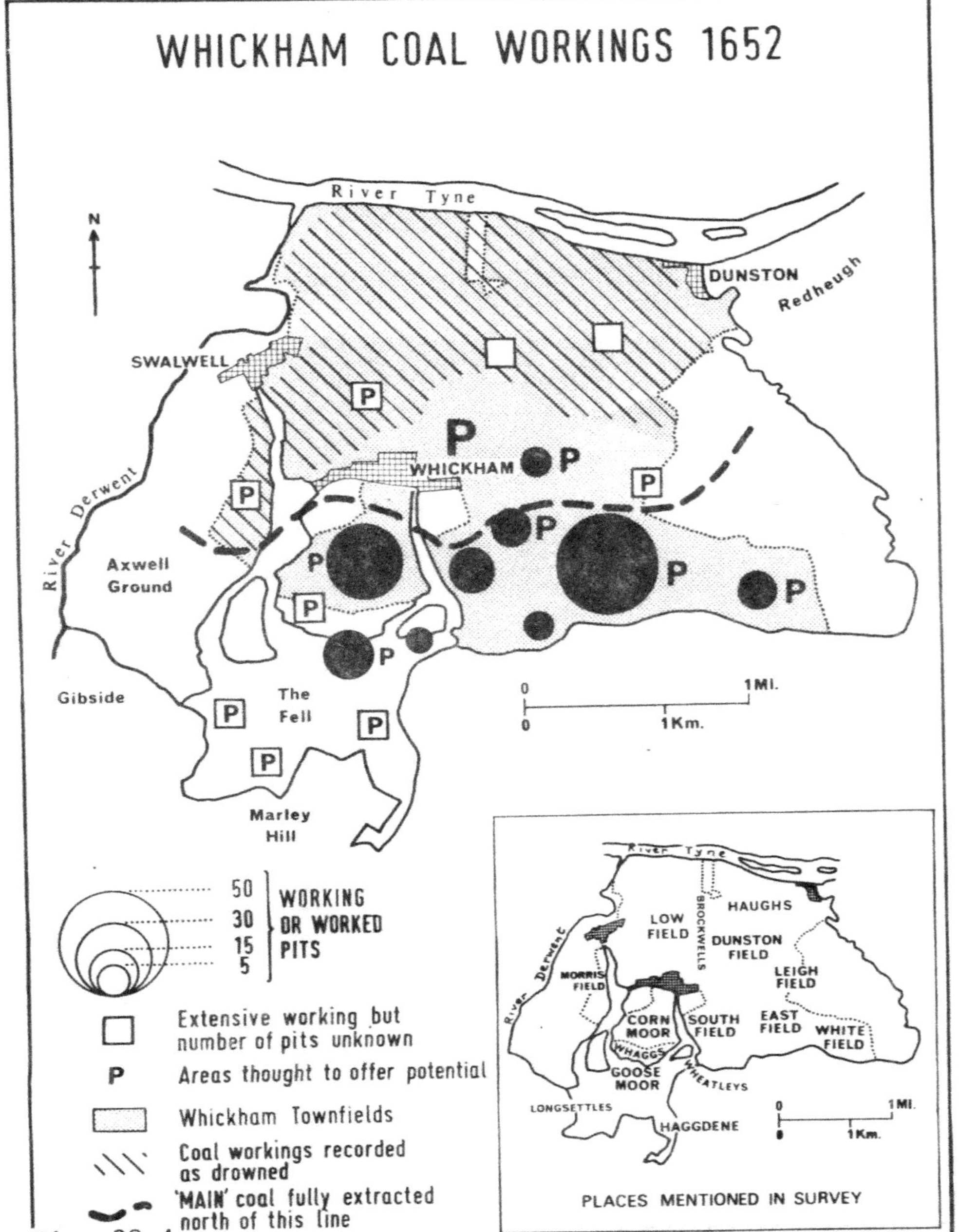

Fig. 28:4

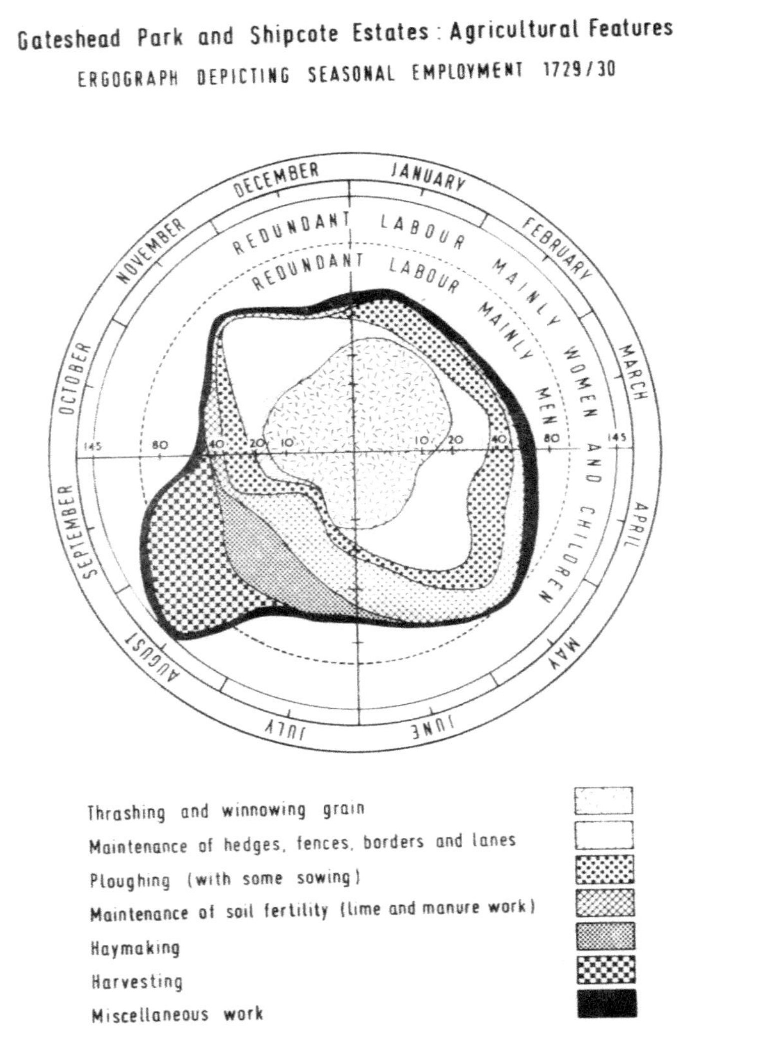

Fig. 28:5

on account of his enormously varied and time consuming dealings in the coal and salt trades, and had contacts with European businessmen in Bordeaux, Rotterdam, Amsterdam, Bremen, Hamburg, Dantzig, Konigsberg, Narva, Stockholm and Gothenburg. However, his ultimate gratification was the acquisition of landed property at home: the lordships of Hartleyburn and Bellister in Northumberland, estates at Stockton, Gateshead and Whickham in County Durham, and half shares in lands at Stella and Winlaton (also in County Durham). Significantly, at the time he acquired the ancient demesne lands at Gateshead Park (1716) and Shipcote (1719) coal mining in the parish of Gateshead was seriously impeded by drowned workings, and opportunities were thought to lie in another direction, as the Bishop of Durham's clerk was to recall some 30 years or so later:

> "When Mr Cotesworth first came possessed of this manor he found a great part of the Inclosed lands in a much worse condition than a waste, being almost covered with pit heaps, so that neither grass nor corn could well grow. He carried them off, raised new fences and improved the land to its present condition at so great an expense that he might have bought a Freehold Estate of equal yearly value."

This was indeed a remarkable achievement: in January 1720 there were no fewer than 156 pits and their spoil heaps in Shipcote alone and the estimated cost of removal was £781: by 1723 (Figs.28:6a-6d) they had apparently all gone. Once cleared, the lands were hedged with thousands of quicks; lime, manure and a short rotation of grain crops were used to 'improve' the soils and then the greater portion was put down to grasses and clover. The sequence of events from 1722 to 1724 can be reconstructed in detail (Figs.28 6a-6d). Once improved, the lands were either farmed directly or let to tenants at from £2 to £3 per acre. It was the lucrative nature of market gardening and the milk trade in this quasi-urban district which enabled tenants to afford such sums. Cotesworth's comment, that "the coal fitters must not take all the cream and leave only the blue milk", represented much more than the emotional response of a yeoman's son who hated the devastations of coal mining.

The idea of Gateshead in the forefront of the agricultural revolution of the eighteenth century in North-east England would no doubt come as something of a shock to its present-day inhabitants. Yet, in its own way, that was precisely where it stood. As far as is known, Cotesworth was the first person to grow clover in the region (by 1722) and the husbandry practices he pioneered for Gateshead soon spread to other estates in the Tyne valley where coal mining continued, though with less intensity than in former times. Thus in 1770, the novelist, Smollett, could stand on Gateshead Fell and, by ignoring the scene of dereliction immediately before him on the fell itself, note how the country on both sides of the Tyne yielded "a delightful prospect of agriculture and plantations". As Cotesworth had told his steward, Thomas Sissons, in 1723, "Anyone can follow a good method but those that go out of one are surely to blame".

If we contemplate the wisdom of Cotesworth's policy on the one hand, and the decision of Sir Richard Hutton regarding Whickham on the other, we invoke, at once, a thoroughly pragmatic approach to the study of links between agriculture and industry. We place ourselves in an historical context, posing questions and making what we hope are wise decisions. The most critical question of all might well have been - What course of action affords best advantages to me? And there is ample evidence to suggest that nearly all levels of society found agricultural/industrial links advantageous. It was, however, as intimated above, the major landowners who had most to gain from wise choice. This was especially the case in England, as opposed to some European states, because base minerals usually belonged to the owner of the soil under which they lay and so the task of determining land-use priorities could be left, in theory at least, in the hands of one person. In practice, some critical decisions were often left to estate officials and lessees, as with the extensive ecclesiastical possessions in County Durham. It was an arrangement open to abuse, but that is not to say that it was inevitably inimical to good practice.

Throughout the period c.1550 to 1810, but especially after c.1700, the estates of the Bishop of Durham and his Dean and Chapter provided their ecclesiastical overlords with a substantial and expanding source of income from both mining and agricultural interests. Between the first and last decades of the eighteenth century, the average annual income from fixed rents, lease renewals and overworkings at Dean and Chapter collieries showed a startling rise from less than £100 to more than £4,000. Bishopric income from the like source showed a more modest four-fold increase, but this was, after all, a century in which the output of coal in the region as a whole scarcely more than doubled. The record of agricultural income is similarly impressive, and argues well for the inclusion of those clauses in mining leases which were designed to protect good husbandry. While annual rents on ecclesiastical holdings hardly altered in the eighteenth century, leasehold renewal fines (levied every seven years or on the death of a named person) showed a dramatic increase, with rises never less than three-fold and in some cases as great as seven-fold. Since these fines were based upon the current valuation of the property, they reflect, in monetary terms at least, a sharp increase in agricultural productivity. As previously intimated, such an increase seems to have been based not so much upon a wide range of husbandry techniques but rather, upon efficient management of a limited range of high quality seeds and livestock and an acute awareness of

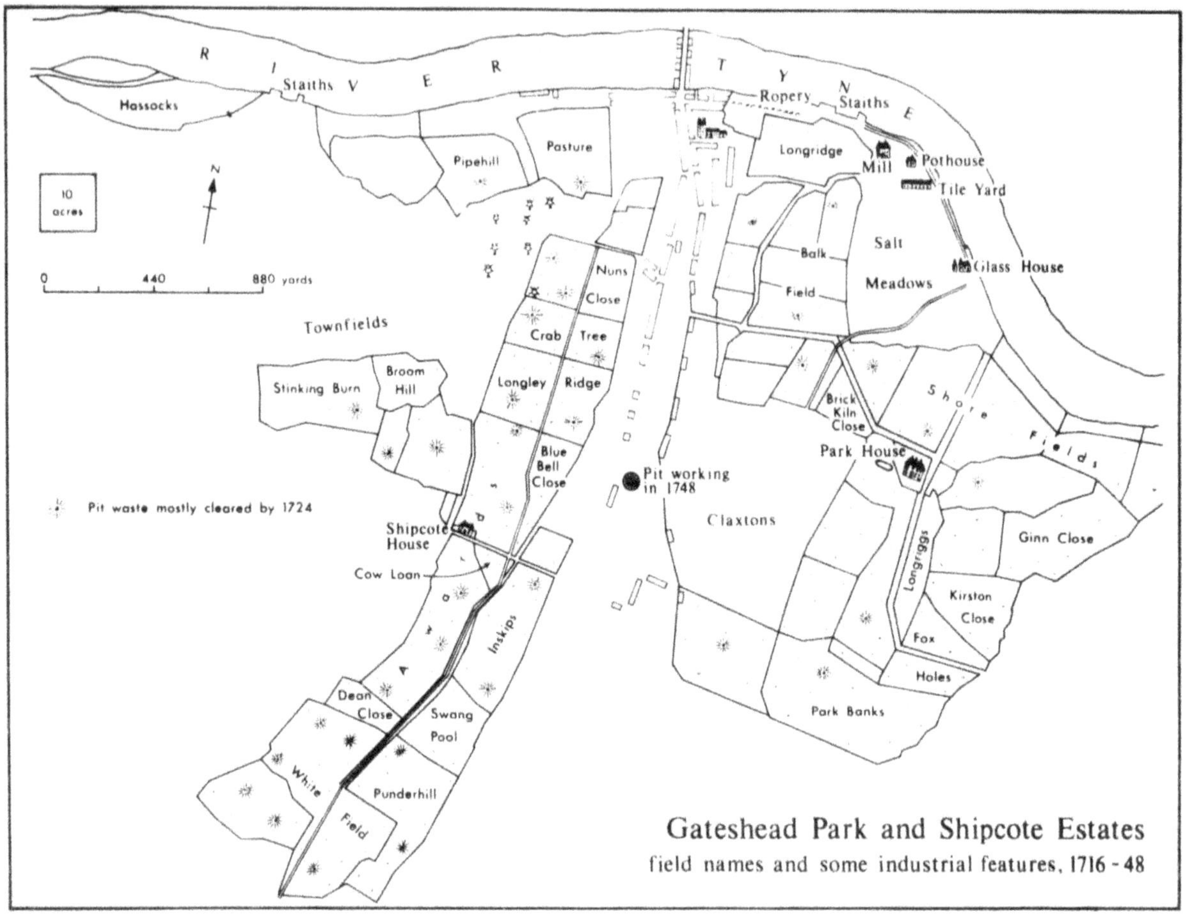

Fig. 28:6a

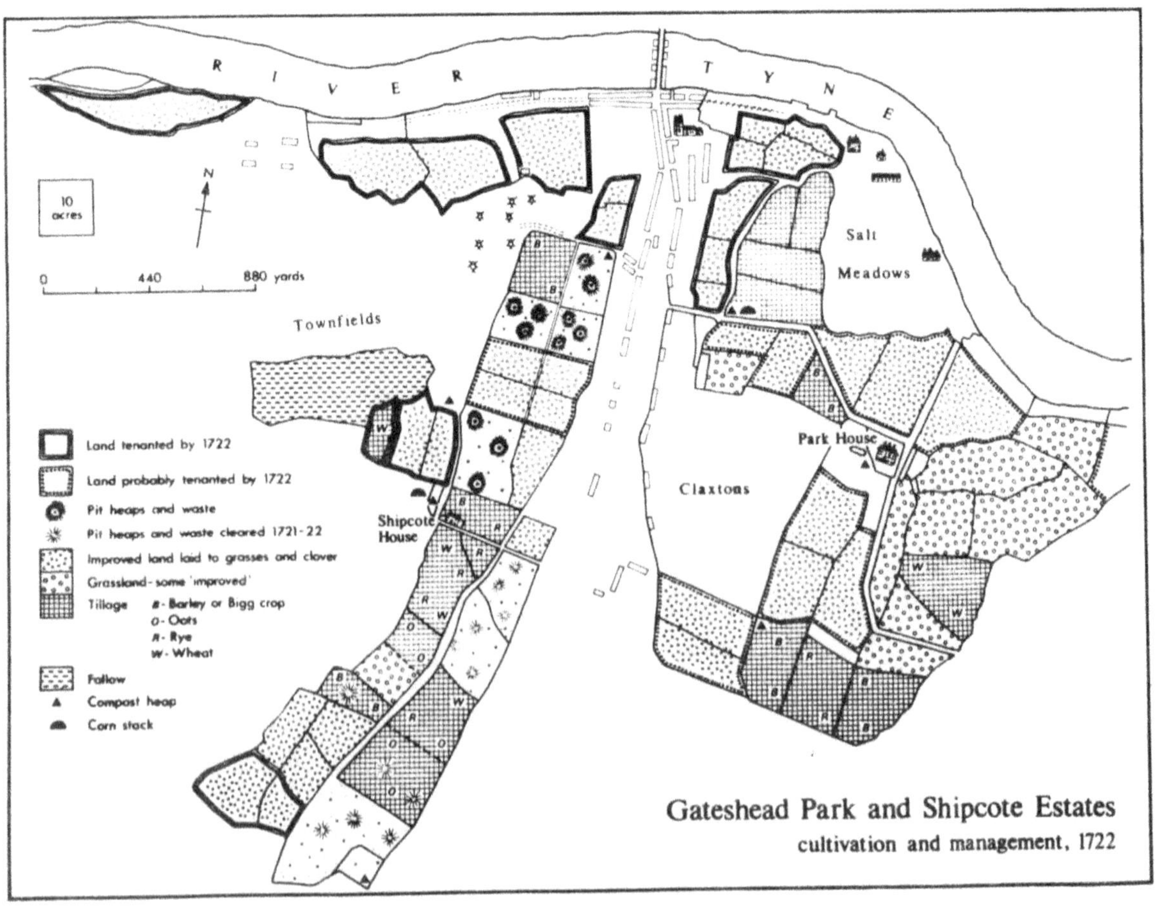

Fig. 28:6b

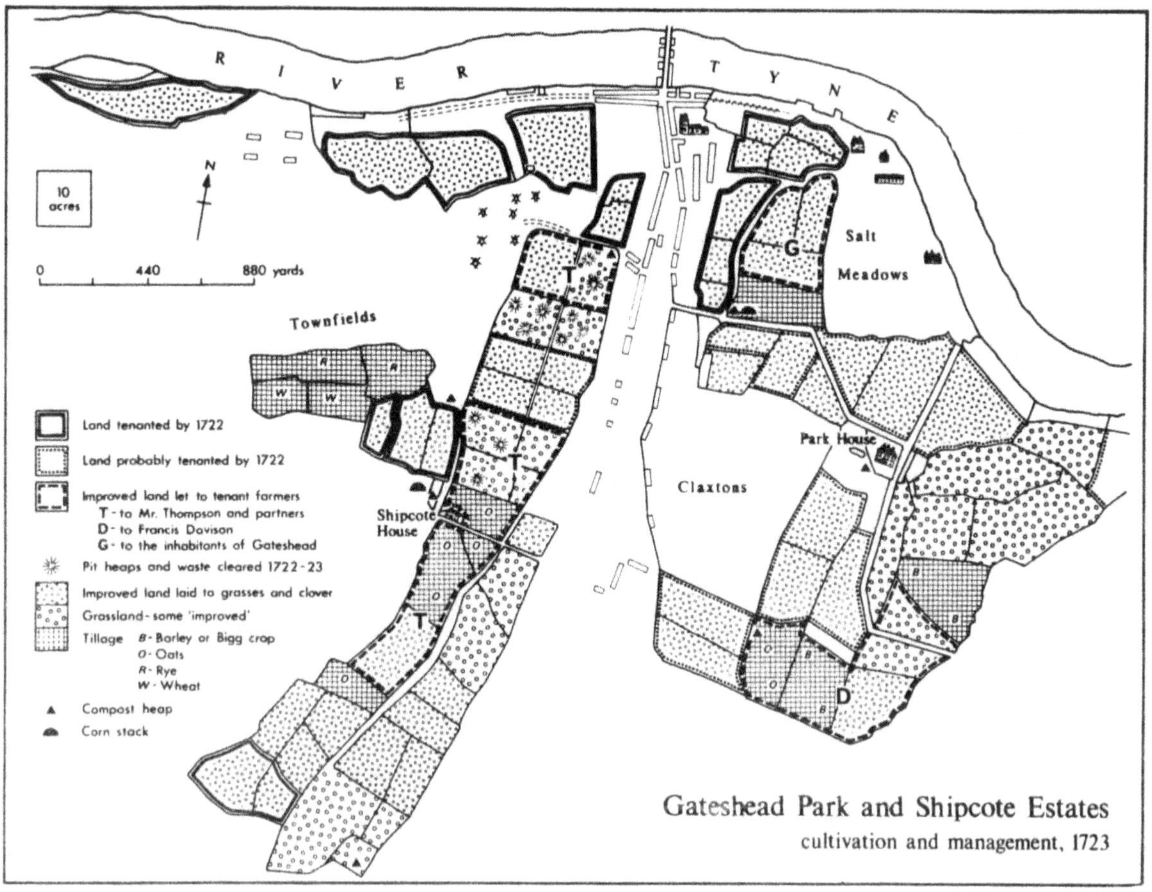

Fig. 28:6c

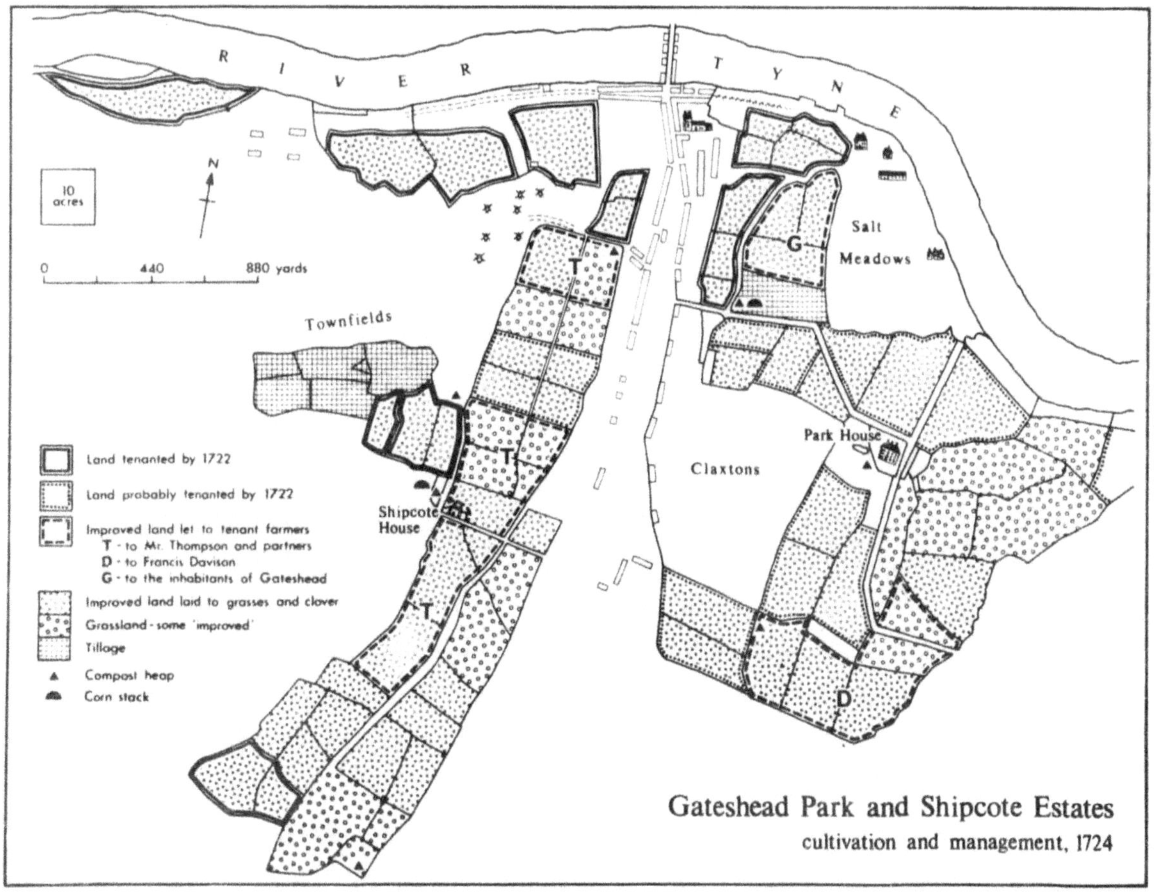

Fig. 28:6d

land capability and market opportunity.

Almost comparable with renewal fine increases on ecclesiastical estates were some rent increases on other Durham estates in the eighteenth century. In the case of the Bute and Clavering possessions in the western part of the North Durham coalfield such sums of money were probably channelled into the financing of colliery ventures at Crookbank and Andrew's House. Mark Hughes has shown that Northumberland provides a similar, and even more convincing picture of the flow of agricultural capital into mining adventures. On the Blackett, Allgood and Greenwich Hospital estates, farm rents quadrupled between 1700 and 1790, and from 1756 large deposits were made in the Tyne Bank. It was from this bank that the coal-owning cartel, the 'Grand Allies', borrowed, when their own agricultural rents were insufficient to finance the sinking of Killingworth colliery. There were times, too, when capital flowed in the opposite direction. Between 1787 and 1810, William Russell bought in County Durham estates which cost a little less than £750,000, largely on the profits of one colliery at Wallsend. In most years it yielded a net surplus of between £30,000 and £60,000.

The acquisition of landed estates as a means to respectability, political influence and a steady source of income, had been a recurrent theme in North-east England long before Russell's time. The sequestrations and confiscations of the Rising of the North (1569), the Civil War and Commonwealth (1640-60) and the Jacobite Rebellion (1715), afforded ample opportunity for the rise of a new landed interest whose fortunes had been made in commerce and whose sons and daughters maintained the connection by marrying into the merchant classes or becoming apprenticed to them. The vicissitudes of great northern families such as the Liddells, Carrs, Ellisons and Ridleys were cast in this mould. The mercantile origins of some landowners are traceable in probate records such as the inventory of John Watson of Newcastle who died in 1680. His possessions included two estates in the North Durham parishes of Whickham and Jarrow, coals resting at Stella and Team staithes on the Tyne, eight keels or coalboats, and a part share in five ships. The rise of another prominent figure, William Cotesworth, has been noted above. When and where the chance arose, the new landed interest had an obvious incentive to exploit the mineral wealth which lay beneath their property, yet a wise man also looked to the safer if comparatively meagre returns of agricultural production on the surface. For some such as William Cotesworth, who began life as a yeoman's son, this may have been a natural instinct. For others it was an awareness of the risk in channelling too much capital into mining adventures. Fortunes had been made in the coal industry, but for every success there were innumerable failures. And eighteenth century markets and meeting places still resounded with the stories of 'Master Beaumont', a southern gentleman and mining adventurer at the turn of the sixteenth century who "within a few years consumed all his money and rode home upon his light horse".

Among the lower echelon of society, and from which many of the new landed proprietors had indirectly sprung, the indivisibility of agricultural and mining interests was well represented. Most farming leases in the coalmining districts obliged tenants to provide coal cartage services. And as early as 1459 we find copyholders at Whickham, Wm. Moore and Jo. Erle, carrying 67 chaldrons of coal from mine to shipping staithe on 15 days between 1st April and 1st August. The probate inventories of yeomen in mining districts frequently list coal wains and other equipment used in the industry. In 1670 the possessions of Francis Bucke of Streetgate included "2 cole waines and 5 cole waine horses", while John Gowland of Low Pitt Houses, Houghton-le-Spring, had "5 carts, 2 cole waines and a horse waine with all other implements of husbandrie". Without earnings in the 'land carriage' it is difficult to imagine Gateshead tenants in the early eighteenth century affording £2-£3 per acre for land, even if subsequent milk sales proved profitable. The obligation to provide cartage services formed such an attractive and necessary supplement to agricultural income that for some farmers it became a cherished prerogative. When such a prerogative was abused and the Whickham coalowners introduced labourers from neighbouring parishes to perform such services, the indignant copyholders obliged these 'foreigners' to pay them 'catch' money or risk molestation.

As in most historical studies the problem of measurement remains, the problem of guaging the extent to which documentary finds indicate a trend or simply an eccentricity. While ample evidence exists of the close links between agriculture and coal mining or other industry, and of the beneficial effects these might have on different levels of society, there is equally, evidence of the poverty and misery which such links, or the breakdown of such links, might have fostered. The happy coal-carrying husbandmen of one generation could become the landless proletariate of the next. By the late seventeenth and early eighteenth centuries, signs of the long-term social consequences of industrialization were perhaps already to be seen in attempts to infill village greens with tiny cottages (as at Ryton, Whickham and Winlaton) or to settle upon the open fells (as at Gateshead, Heworth and Eighton) where Durham's banished plague victims had once ended their days. In 1700 there were 83 squatters recorded on Gateshead Fell; by 1809, when the fell was about to be enclosed, there were almost 200. John Bell, one of the enclosure commissioners responsible for their eviction, was moved to write:

> "The pulling down of these cottages was one of the
> most unpleasant acts, the Commissioners had to perform,

the greater number having to be taken down by force, whilst occupied by their distressed Owners, and the abuse and threatening language the Commissioners received in consequence was considerable".

One unfortunate aspect of life affecting the majority of the labouring population was the seasonal nature of work in both coal mining and agriculture. A unique set of early accounts for Gateshead Park and Shipcote estates allows the extent of the problem to be judged (Fig.28:5), but for agriculture only. In the absence of comparable data for other occupational groups or areas there seems little chance of gaining a reliable overall picture of seasonal employment. What is clear is that a large, if indeterminate, number of workers was laid off in the winter months. Some of these, like the keelmen at Dunston, might have returned briefly to their native Perthshire or the family home in Weardale. For others the outcome was hardship. Ideally, one would wish to study the fortunes of a large number of families, study their levels of fertility and mortality and assess their self-sufficiency as economic units, follow the progress and dispersion of their members into diverse occupations, trace the progress of the sons from yeomen to merchants to landed proprietors or view the process in reverse. That is a task constrained by research time, and more seriously, by the incomplete or unreliable nature of some surviving sources.

NOTE In the preparation of this paper I have drawn upon a range of primary and secondary sources. I must apologise for the omission of detailed references at this time. Some of the points raised are explored more fully in my following referenced publications: 'Agricultural improvement and changing regional economies in the eighteenth century' in A.R.H.BAKER & J.B HARLEY (eds) Man made the land (1973); Demographic trends in County Durham, 1560-1801: Data sources and preliminary findings with particular reference to North Durham (University of Manchester, School of Geography, Research Paper no. 5, 1978); 'The progress of enclosure in County Durham, 1550-1870' in H.S.A. FOX & R.A. BUTLIN (eds) Change in the countryside (IBG Special publication, no. 10, 1979).

Some further secondary sources include: T.G.BELL, 'Report on the agriculture of the County of Durham' Jl. Roy. Agric. Soc 17 (1856); D. DEFOE, A tour through the whole island of Great Britain (1724-6); E. HUGHES, North Country life in the eighteenth century: the North East, 1700-1750 (1952); J. LANGTON, 'Landowners and the development of coal mining in South-west Lancashire, 1590-1799' in H.S.A. FOX & R.A.BUTLIN op. cit.; I. LEISTER, The sea coal mine and the Durham miner (University of Durham, Department of Geography, Occasional Pub. (new series) no. 5, (1975); M.J.T. LEWIS, Early Wooden

Railways (1970); A.E. SMAILES, North England (1960); H. UHLIG, Die Kulturlandschaft, Methoden der Forschung und das Beispiel. Nordostenengland (1956): J.U. NEF The Rise of the British Coal Industry (1932)

ACKNOWLEDGEMENT I should like to thank Mr. N. Scarle and other members of staff of the Drawing Office, University of Manchester, for their help in the preparation of Figures 1-6. Also the staff of several record repositories, especially in Durham and Northumberland, for their unfailing assistance in tracing sources.

Figures 28:2 and 28:3 are based upon manuscript maps in the Institute of Mining and Mechanical Engineers, Newcastle upon Tyne, to whom grateful acknowledgement is made.

www.ingramcontent.com/pod-product-compliance
Ingram Content Group UK Ltd.
Pitfield, Milton Keynes, MK11 3LW, UK
UKHW061843041026
15591UKWH00015B/891

9 780860 542384